Blueprint Reading: Construction Drawings for the Building Trades

Blueprint Reading: Construction Drawings for the Building Trades

Sam A. A. Kubba, Ph.D.

New York Chicago San Francisco Lisbon London Madrid
Mexico City Milan New Delhi San Juan Seoul
Singapore Sydney Toronto

McGraw-Hill books are available at special quantity discounts to use as premiums and sales promotions, or for use in corporate training programs. To contact a special sales representative, please visit the Contact Us page at www.mhprofessional.com.

7 8 9 0 DOC/DOC 0 1 4 3

ISBN 978-0-07-154986-8
MHID 0-07-154986-2

Sponsoring Editor
Joy Bramble Oehlkers
Production Supervisor
Pamela A. Pelton
Editing Supervisor
Stephen M. Smith
Acquisitions Coordinator
Rebecca Behrens
Project Manager
Jacquie Wallace

Copy Editor
Wendy Lochner
Proofreader
Leona Woodson
Indexer
Leona Woodson
Art Director, Cover
Jeff Weeks
Composition
Lone Wolf Enterprises, Ltd.

This book is printed on acid-free paper.

This book is dedicated to
My mother and father,
Who bestowed on me the gift of life . . .
And to my wife and four children,
Whose love and affection inspired me on . . .

ABOUT THE AUTHOR

Sam A. A. Kubba, Ph.D., is an award-winning architect whose practice includes projects in the United States, the United Kingdom, and the Middle East. He has more than 30 years of experience in all aspects of design, construction, and property condition assessments. A member of the American Institute of Architects, the American Society of Interior Designers, and the Royal Institute of British Architects, he has lectured widely on architecture, interior design, and construction. Dr. Kubba is the principal partner of The Consultants' Collaborative, a firm noted for its work in architecture, interior design, and project management. He is also the author of several books, including *Mesopotamian Furniture*, *Space Planning for Commercial and Residential Interiors, Property Condition Assessments,* and *Architectural Forensics*.

CONTENTS

CHAPTER 10: INTERPRETING SPECIFICATIONS   239

CHAPTER 11: BUILDING CODES AND BARRIER-FREE DESIGN   257

FOREWORD

The ancient Egyptians and Babylonians drew them on papyrus and clay tablets; the Greeks drew them on linen cloths, carved them on wood, marble tablets, and stone. Michelangelo drew them on parchment paper with exceptional detail for his masterpieces. Important since ancient times, this tool is now called a blueprint. The blueprint as we know it today, which is essentially the cyanotype process, also known as the old monochrome photographic printing process that produces a cyan-blue print, was developed by British astronomer and photographer Sir John Herschel in 1842.

Combined with new technologies, the blueprint has evolved through the years and is now produced in electronic format and printed on white paper. Whatever it is called, or printed on, or however it is processed, the blueprint basically has one function and message: make it as shown.

Now the question is, does it contain enough correct information to make it as shown, and do the people who will implement it have sufficient qualifications to interpret the blueprint and to make it as shown? Therefore, the burden is on both the producers and the interpreters of the blueprint.

In his book, *Blueprint Reading: Construction Drawings for the Building Trades*, Sam Kubba addresses a century-old problem—producing and understanding the message that the blueprint contains. Blueprint producers are primarily the people with professional education and experience and are certified in their related fields in almost every industry. Implementers are the people who transform the information into reality. As much as the qualifications of the producer are essential, the executer is also an indispensable component—as one designs, the other applies, and the two become one creation. As Kubba explains in some detail—mainly for the building trades and the construction industry—there is ongoing concern with improving this process from inception to completion of the creation. It is to the advantage of everyone involved in this process, including generations of people who will benefit from the end product, that ideas are understood, transformed, and applied properly. Otherwise, substandard transformation and application will produce dissatisfied users of the building.

Education, training, a systematic approach, communication, coordination, understanding, and trade knowledge are the key elements to convert ideas into reality. This book addresses these issues and guides the reader on a successful path of implementation.

Guy Collette, Founder, and Judi Collette, President
Collette Contracting Inc.
A Design/Build Company

ACKNOWLEDGMENTS

A book of this scope would have been extremely difficult without the assistance and support of numerous individuals—friends, colleagues, architects/engineers, and contractors—who contributed greatly to the formation and crystallization of my thoughts and insights on many of the topics and issues discussed herein. I am also indebted to innumerable people and organizations that have contributed ideas, comments, photographs, illustrations, and other items that have helped make this book a reality instead of a pipe dream.

I must also unequivocally mention that without the unfailing fervor, encouragement, and wisdom of Mr. Roger Woodson, president of Lone Wolf Enterprises, Ltd., and Ms. Joy Bramble Oehlkers, senior editor with McGraw-Hill, this book might still be on the drawing board. It is always a great pleasure working with them. I must likewise acknowledge the wonderful work of Project Manager Jacquie Wallace for her unwavering commitment and support, and for always being there for me when I needed assistance or advice. I also wish to thank Ms. Wendy Lochner for copy editing the first drafts, and Ms. Leona Woodson for proofreading the final drafts and indexing the book. I would certainly be amiss if I did not mention my gratitude to Mr. Jeff Weeks for his excellent cover design.

Mechanical engineer Stephen Christian deserves a special thanks for reviewing Chapter 7 and for his many useful comments and suggestions, and professional land surveyor Donald Jernigan for reviewing parts of Chapter 6. I must not forget to thank my wife, Ibtesam, for her loving companionship and support and for helping me prepare many of the CAD drawings and line illustrations. Last but not least, I wish to record my gratitude to all those who came to my rescue during the final stretch of this work—the many nameless colleagues—architects, engineers, and contractors who kept me motivated with their ardent enthusiasm, support, and technical expertise. I relied on them in so many ways, and while no words can reflect the depth of my gratitude to all of the above for their assistance and advice, in the final analysis, I alone must bear responsibility for any mistakes, omissions, or errors that may have found their way into the text.

INTRODUCTION

In the construction industry the term *blueprint* generally refers to a composite of several plans, such as the foundation plan, the floor plan, elevations, sections, mechanical plans and details, etc., that are assembled into an organized set of drawings to transmit as much information about a project as can be placed on paper in one- or two-dimensional views. The completed set of drawings represents a pictorial description of a construction project prepared by the architect/designer and/or engineering consultant.

Blueprint reading is therefore basically finding and interpreting the information placed on prints. The information is displayed in the form of lines, notes, symbols, and schedules. At first glance, there is a welter of information that can appear intimidating. This innovative textbook clearly explains how blueprints and construction drawings are used to implement the construction process. It offers a comprehensive overview of construction drawing basics and covers standard construction sequence, including site work, foundations, structural systems, and interior work and finishes. A typical set of blueprints for a building project usually includes a number of drawing types in order to see the project to completion. Users of blueprints must be able to interpret the information on the drawings and must also be able to communicate that information to others.

This manual covers and explains the use of lines, dimensions, schedules, specifications, symbols, code requirements, construction drawing types, and methods of drawing organization, including CADD. Comprehensive in its coverage, this book provides updated information to reflect the most recent developments in the construction industry, enabling readers to further improve their communication skills when dealing with the technical information found in blueprint documents. This book introduces concepts essential to a basic, introductory understanding of residential and light construction, while providing hands-on experience in reading architectural working drawings. It is intended to serve as a valuable textbook and reference manual for building trade professionals as well as students with a career or interest in architectural drawing/design, residential design, construction, and contracting, and also those who are required to read and interpret information found in blueprint documents.

Updated to the latest ANSI, ISO, and ASME standards, this handbook helps individuals develop skills in reading and interpreting architectural and industrial drawings and in preparing simple field sketches. It is written to be a consumable, interactive text/workbook that provides basic principles, concepts, ANSI and SI metric drafting symbols and standards, terminology, manufacturing process notes, and other related technical information contained on a mechanical or CAD drawing.

Nothing is more essential in the construction industry than the ability to read blueprints, and this reference handbook teaches you just that. Covering a multitude of trade industries, this book provides plumbers, carpenters, roofers, electricians, and others access to all the necessary information. Readers will learn how to read and understand blueprints, sections, elevations, schedules, site plans, architectural plans, structural plans, plumbing plans, HVAC and mechanical plans, electrical plans, and more. A comprehensive glossary of terms and abbreviations is also included.

The handbook is written in an easy-to-understand, informal style with no prerequisites presumed. It is designed to appeal as a basal text to the hundreds of architecture- and construction-related programs

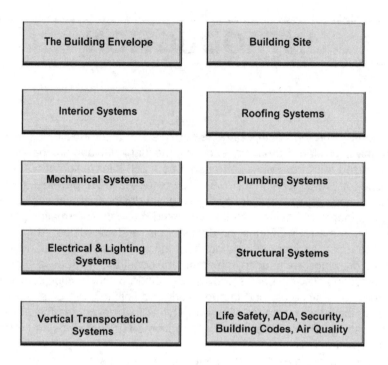

Figure I.1   Some of the many building systems in which blueprints are used to implement a building project.

across the nation and is structured to become the definitive text on the subject. *Blueprint Reading* is a unique book dealing with virtually all of the topics needed to understand construction drawings and enter the building profession.

The author is a licensed architect and general contractor with more than 30 years of experience in all aspects of design and construction. While no single handbook can provide everything you need to know regarding blueprint reading and interpretation, I feel that this book provides the best introduction and information source on the subject available at this time. It expounds a standard procedure and methodology developed over years of field experience that has proven effective.

Sam A. A. Kubba, Ph.D.

Blueprint Reading: Construction Drawings for the Building Trades

1

Blueprint Standards

1.1 GENERAL OVERVIEW

A blueprint is a type of paper-based reproduction usually of a technical drawing documenting an object, an architecture or engineering design. The term is now generally used to refer to any detailed plan of a building or object. Blueprints have for thousands of years provided a universal language by which design and construction information is transmitted to the builder, engineer, craftsperson, designer, and others. Blueprint reading therefore refers to the process of interpreting a drawing. An accurate mental picture of the object upon completion can be formulated from the information provided.

While originally blueprints did have a blue background with white lines (as a result of the process used to produce them), subsequent improvements to the copying process involved ammonia and coated papers that react to light. Additional improvements eliminated the ammonia process, leading up to today's prints. Blueprints are today usually white pages with black or blue lines. Nonetheless, the term has remained with us and probably will remain in use for a long time to come. You may also hear blueprints referred to as drawings, prints, or plans.

A blueprint is a representation of what is to be constructed. It is a drawing of what is to be built. Blueprints, however, are very precise drawings that are exact representations of what is to be built. Obviously, they are drawn much, much smaller than the proposed structure, but they are exact and detailed. Every line on a construction drawing is carefully placed. The relation of a line to another line shows distance.

Blueprints are critical forms of communication that provide a great deal of information. If the prints are not clear enough to read and use, wait until you have better ones or make your own. Otherwise you may encounter serious problems in the course of the project. Use your judgment.

Drawings have little value if they cannot be satisfactorily reproduced. Sets of prints must be provided to building departments (for approvals), as well as to contract bidders, estimators, subcontractors, and others who are concerned with the construction of the proposed building. Original drawings are typically retained by the consultant. There are several methods of making prints; these are discussed below.

1.2   PRINT BASICS

As previously mentioned, a blueprint was originally a print comprising a dark blue background with white lines (Figure 1.1). Traditional blueprints have largely been replaced by more modern, less expensive printing methods and digital displays. Indeed, today over 95 percent of businesses are producing prints generated with computer-aided drafting (CAD) software and printed on large printers. The most common methods used to reproduce drawings are outlined below:

Diazo-Print Process

In the early 1940s, cyanotype blueprints began to be supplanted by diazo prints or whiteprints, which have blue lines on a white background. These drawings are also known as Ozalid or blue-line prints. The diazo process is an inexpensive method of making prints from any type of translucent material. It is the result of a printing process that utilizes ultraviolet light passing through a translucent original drawing to expose a chemically coated paper or print material underneath. As the light does not pass through the lines of the drawing, the chemical coating beneath the lines remains unexposed. The print material is then exposed to ammonia vapor, which activates the remaining chemical coating to produce the blue lines. Diazo prints can also be produced with black or sepia lines.

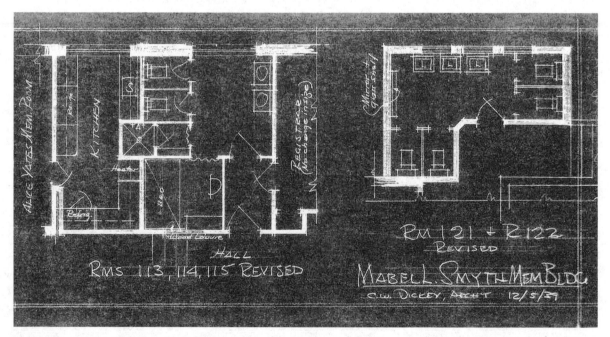

Figure 1.1   Illustration of traditional blueprint.

While modern advances in replication technology have allowed other techniques to replace the traditional blueprint, some businesses continue to maintain this process. In most cases the diazo process (which is an environmentally friendly process) has been replaced by xerographic print processes similar to standard copy-machine technology using toner on bond paper.

Plain-Paper Copies

This process uses plain-paper machines; prints can be made from opaque originals, whereas diazos require translucent originals. Plain-paper copiers are mainly used in offices where changes and modifications are needed during the development stages or where the originals need to be put on vellum or film, from which volume prints are made for distribution using the less expensive diazo process. This method is especially useful in printing CAD plots done on bond paper (Figure 1.2). Plain-paper copiers are also effective in making copies from shop-worn originals onto durable polyester film or vellum media. Some copiers can also enlarge or reduce the size of the original drawing.

Photocopy Process

The photocopy process is gaining extensive use for copying engineering and architectural drawings. With this method a picture of the original drawing is used to then produce as many prints as required, similar to a Xerox or other copying machines. This type of copier also has the added advantage of making reductions or enlargements of the original drawing. These copy machines in fact very much resemble the copiers found in the traditional office environment. The main difference is the capacity to copy large drawings. The increasing popularity of these machines is due mainly to the use of ordinary paper that does not require coating and avoids the possible hazards of ammonia.

Photographic Reproduction

Various types of photographic reproduction are available that provide greater accuracy and detail, making them easier to read, correct, and reprint. Photostats can be made from an original drawing by using a large, specially designed camera that produces enlargements or reductions from the original work. This direct print process delivers a negative with white lines and a dark background. A number of high-quality reproduction methods such as microfilming are available that use a film negative made from the original drawing. Projection prints from a photographic negative can be reproduced on matte paper, glossy paper, vellum, and Mylar (Cronoflex™). These excellent-quality prints can be enlarged or reduced in scale with accuracy and are very durable.

To microfilm a drawing, one must first convert the original drawing to a microfilm frame using a special microfilm-processing camera mounted on a frame over a platform. The camera reduces the drawing in order to fit on the microfilm; the reduction can be varied by changing the height of the camera above the drawing. After the film is developed, the negative is typically mounted on a standard-size aperture or data card and is systematically filed for future retrieval. A microfilm-enlarger reader-printer with a display screen is normally used to review the images as well as reproduce print copies of various sizes from the negative. Microfilming is also an excellent means of storing drawings, thus eliminating the need to retain cumbersome original tracings (Figure 1.3).

Figure 1.2   HP Designjet T610 blueprint plotter for printing CAD and GIS projects *(courtesy, Hewlett-Packard Development Company, L.P.)*.

Figure 1.3   Storage of blueprint paper drawings can be very cumbersome and space-consuming.

Multicolor Offset Prints

Sometimes prints are produced in several colors (usually blue and red) via offset printing. Typically the original drawings are made in smaller scales and sizes for further economy. The various colors are used to highlight new work in relation to existing construction or to display complex mechanical or electrical systems in new projects. The main advantage to the use of colors is in facilitating the reading of the drawings, resulting in fewer mistakes and requiring less time in interpreting them.

More recently, designs created using computer-aided-design techniques may be transferred as a digital file directly to a computer printer or plotter; in some applications paper is avoided altogether and work and analysis are done directly from digital displays. Even as print and display technology advances, the traditional term "blueprint" continues to be used informally to refer to each type of image.

Computer-Aided Design and Drafting

Computer-aided design (CAD) has brought a new innovation to the field of engineering design and is discussed in greater detail in Chapter 2. CAD is used to design, develop, and optimize the process to make a clean, clear, computer-generated construction drawing, accurately drafted as per specified dimensions. CAD has turned out to be an invaluable technology with great benefits, such as lower product-development costs, faster processing, and a greatly shortened design cycle. It is a technology that permits the consultant to focus on the business of design instead of wasting time searching for and redrawing old documents.

CAD drawings can be generated from most formats including hand-drawn documents, tiff files, or any other image files that can be converted to a .dwg file using AutoCAD or other CAD software.

1.3 BASIC DRAFTING STANDARDS AND STANDARDS-SETTING ORGANIZATIONS

In the United States, the American National Standards Institute (ANSI) and the International Organization for Standardization (ISO) have adopted drafting standards that are voluntarily accepted and widely used throughout the world. These standards incorporate and complement other architectural/engineering standards developed and accepted by professional organizations such as the American Institute of Architects (AIA), the American Society of Mechanical Engineers (ASME) and others. Some large firms have adopted their own standards to suit their individual needs.

ANSI Standards for Blueprint Sheets

Most architectural-drafting offices in the United States use ANSI standard sheet sizes as shown in Table 1.1. ANSI Y14.2, Y14.3, and Y14.5 are the standards that are commonly used in the U.S.:

1. In the U.S. letter-size paper is an architectural (first series) A-size sheet. It is 8.5 × 11 inches. An engineer's (second series) A-size sheet is 9 × 12 inches.
2. B-size sheets are double the size of A sheets (11 × 17—also called tabloid—or 12 × 18 inches).
3. C-size sheets are double the size of B sheets (17 × 22 or 18 × 24 inches).
4. D-size sheets are double the size of C sheets (22 × 34 or 24 × 36 inches), and so on.

Table 1.1 ANSI (American National Standards Institute) standard recommended sheet sizes.

U.S. ENGINEERING	DIMENSIONS (inches)	DIMENSIONS (mm)
ANSI A	8.5 × 11	215.9 × 279.4
ANSI B	11 × 17	279.4 × 431.8
ANSI C	17 × 22	431.8 × 558.8
ANSI D	22 × 34	558.8 × 863.6
ANSI E	34 × 44	863.6 × 1117.6
U.S. ARCHITECTURAL	**DIMENSIONS (inches)**	**DIMENSIONS (mm)**
ARCH A	9 × 12	228.6 × 304.8
ARCH B	12 × 18	304.8 × 457.2
ARCH C	18 × 24	457.2 × 609.6
ARCH D	24 × 36	609.6 × 914.4
ARCH E	36 × 48	914.4 × 1219.2
MISCELLANEOUS	**DIMENSIONS (inches)**	**DIMENSIONS (mm)**
LETTER	8.5 × 11	215.9 × 279.4
LEGAL	8.5 × 14	215.9 × 355.6
U.S. GOVERNMENT	8 × 11	203.2 × 279.4
STATEMENT	5 × 8.5	139.7 × 215.9

ISO Standards

The metric drawing sizes correspond to international paper sizes, in Europe and many other countries ISO (International Standards Organization) metric measurements are used (Table 1.1). The ISO defines a set of standard metric line widths for drafting. For example, the ISO A series of sheet sizes is based on a constant width to length ratio of 1: $\sqrt{2}$. Thus, each smaller sheet size is exactly half the area of the previous size, so that if we cut an A0 sheet in half, we get two A1 sheets; if we cut an A1 sheet in half, we get two A2 sheets; and so on. In Figure 1.2 the A0 size is defined as having an area of 1 square meter. This lets us express paper weights in grams per square meter.

The relationship of 1: $\sqrt{2}$ is particularly important for reduction onto microfilm or reduction and enlargement on photocopiers. Metric equipment including microfilm cameras, microfilm printers, photocopiers, as well as drawing pen sizes are designed around this ratio. This simplifies the process of archiving drawings, resizing, and modifying drawings.

Like the ISO A-series sheet sizes, the pen sizes increase by a factor of two, which allows additions and corrections to be made on enlargements or reductions of drawings. Each width is assigned a color code. The color code corresponds to that for the matching lettering stencil. Line widths and color codes are standardized across all manufacturers.

Other Standard-Setting Organizations

As previously mentioned, ANSI and ISO are the two main drafting-standard-setting bodies. However, there are other organizations that have developed engineering standards that complement these. Like

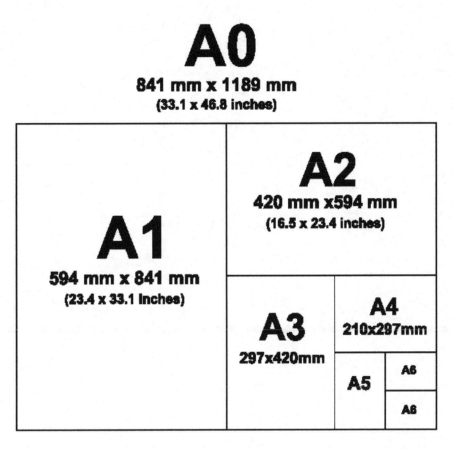

Figure 1.4 Diagram showing ISO (International Standards Organization) metric sheet sizes.

the ANSI, the American Society of Mechanical Engineers (ASME) has standardized drawing-sheet sizes that are designated by letter and published in MIL-STD-100A. ASME also published the ASME Y14.5M-1994 Dimensioning and Tolerancing standard. Other standard-setting organizations include the American Institute of Architects (AIA), the Canadian Standards Association (CSA), the American Welding Society (AWS), and others. Moreover, some large firms have adopted their own standards to suit their individual needs. In the final analysis, the drawing-sheet sizes you use are determined by the needs of your employer or client and your need to economically lay out required information.

2

Blueprints and Construction Drawings: A Universal Language

2.1   INTRODUCTION

Not too long ago, it was predicted by some that by the beginning of the 21st century, blueprints would become obsolete and no longer be used on construction sites. It was further suggested that construction information would essentially be read off computer screens rather than blueprint drawings. This would not only be more efficient, but it would also save a tremendous amount of paper. These predictions did not materialize. Although construction plans are read regularly on computer screens and are being sent via computer to job sites, paper blueprint drawings remain the preferred medium on building sites. In many parts of the world, manual drafting and blueprint drawings are still the norm.

Blueprint reading consists essentially of finding information on prints. The information may be displayed on a drawing in the form of lines, notes, symbols, and schedules. The items are typically located either in the title block or in the field of the drawing (i.e., anywhere within the border lines outside the title block). You should also keep in mind that blueprints typically come in sets. A set of prints for a single-family residence may contain no more than a few sheets, whereas on a large project a complete drawing set may contain scores of sheets for different disciplines (e.g., architectural, structural, electrical, mechanical, plumbing, etc.).

The general process and sequence for reading blueprints can be summarized as follows:

1. Verify that the set of drawings and specifications is complete. Likewise, verify that the documents in hand are the most current.

2. Start by reviewing the site or plot plan to better comprehend the setting of the building and the general topography.

3. Visually scan the architectural drawings to get a better overall understanding of the project. Look at the title block to extract any general information pertaining to the project that may be needed (consultant's name, client's name, project title, drawing number, etc.). Check for unusual or complicated features that may impact how the building is constructed. In particular review the elevations and sections and the materials used.

4. Review the foundation plan and read the general notes to get a better understanding of the construction specifications and other information relevant to the drawing. Also look at the relevant building details.

5. Review the structure's wall construction and the material and methods used. Also study the details showing how the wall is to sit on the designed foundations and which walls if any are load-bearing and which are not.

6. Review the plumbing, mechanical, and electrical drawings.

7. Check all notes on these plans to see if there have been any revisions. Check to see if the building codes have been taken into account. Ensure that the notes on the drawings are clear and that there is no ambiguity.

8. Review the specifications and compare them to the drawings. (Specifications normally have priority over drawings.) If there are discrepancies, the consultant should be notified.

2.2   TECHNICAL DRAWING

Also known as drafting, technical drawing is the practice of creating accurate representations of objects for architectural and engineering needs. A practitioner of the discipline is known as a drafter. Today the mechanics of the drafting task have considerably changed through the use of CADD computer systems, but regardless of whether a drawing is drawn manually or with computer assistance, it must be reproducible.

Manual Drafting

Basic drafting procedure consists of placing a piece of paper (or other material) on a smooth surface with right-angle corners and straight sides--typically a drafting table. A sliding straightedge commonly known as a t-square is then placed on one of the sides, allowing it to slide across the side of the table and over the surface of the paper (Figure 2.1). Parallel lines can be drawn by simply moving the t-square and running a pencil or technical pen along the edge. The t-square is also used as a means to hold other tools such as set squares or triangles. To do this, the drafter places one or more triangles of known angles on the t-square (which is itself at right angles to the edge of the table) and then draws lines to the angles chosen on the sheet. Modern drafting tables (which in the United States are rapidly being replaced by CAD stations) come equipped with a parallel rule that is supported on both sides of the table to slide over the tracing paper. Since it is secured on both sides of the drafting table, lines drawn along the edge are parallel.

The drafter also has other tools at his/her disposal that are used to draw curves and circles, including compasses (used for drawing simple arcs and circles) and French curves (which typically consist of a piece of plastic with complex curves on it). Another tool used is the spline, which is a piece of rubber-coated articulated metal that can be manually bent to most curves. Figure 2.2 is an illustration of some of the templates and instruments used in manual drafting.

Drafting templates allow the drafter to consistently recreate recurring objects in a drawing without continuously having to reproduce them from scratch. This is particularly useful when using common symbols: for example, in the context of stagecraft, a lighting designer will typically draw from the United States Institute for Theatre Technology (USITT) standard library of lighting-fixture symbols to indicate the position of a common fixture across multiple positions. Templates can usually be purchased from various vendors, usually customized to a specific task, but it is also not uncommon for a drafter or designer to create customized templates.

This basic drafting system requires an accurate table, and careful attention should be given to the positioning of the tools. A common error is to allow the triangles to push the top of the t-square down slightly, thereby throwing off all the angles. Even tasks as simple as drawing two angled lines meeting at a point require a number of moves of the t-square and triangles. In general, drafting often proves to be a time-consuming process.

A solution to these problems was the introduction of the so-called mechanical drafting machine, an application of the pantograph that allowed the drafter to make an accurate right angle at any point on the page quite quickly. These machines often included the ability to change the angle, thereby removing the need for the use of triangles as well.

In addition to the complete mastery of the mechanics of drawing lines, circles, and text onto paper (with respect to the detailing of physical objects), the drafting effort requires a proficient understanding of geometry, trigonometry, spatial comprehension, and above all a high standard of precision and accuracy as well as close attention to detail.

Figure 2.1 A drafter using a typical drawing board, t-square, and other instruments *(after Muller, Edward J., et al.: Architectural Drawing and Light Construction, 5th ed., Englewood Cliffs: Prentice Hall, 1999).*

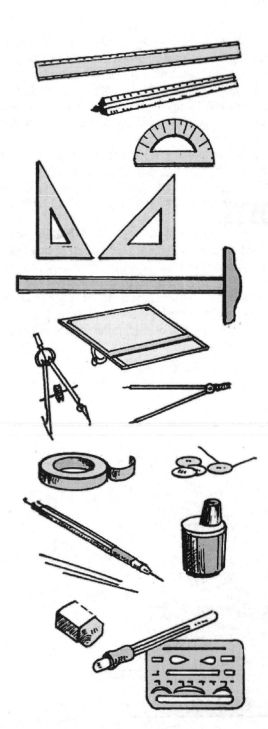

Ruler 18" Clear

Triangular Scale

Protractor

Triangles 30/60 & 45 degrees

T square

Drafting Table with Paralliner

Compass and Dividers

Drafting Tape and Drafting Dots

Drawing Pencils with Leads and
Lead Pointer

Erasers

Eraser Shield

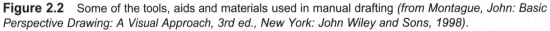

Figure 2.2   Some of the tools, aids and materials used in manual drafting *(from Montague, John: Basic Perspective Drawing: A Visual Approach, 3rd ed., New York: John Wiley and Sons, 1998).*

Lettering

Lettering is used on construction drawings as a means to provide written information. A construction sketch or drawing without lettering rarely communicates an adequate description of the object. It is almost always necessary to provide additional labels, notes, and dimensions to clarify the size, type of materials, and location of the component. Drawings are therefore a means of communication of information to others, and text generally is one of the main mediums to transmit information in the form of notes, titles, dimensions, etc. Lettering should enhance a drawing by making it easy to interpret and pleasant to look at; it should not detract from the drawing or be illegible or unsightly to look at. Legibility and consistency are the key ingredients to good lettering. Architectural lettering is often done using all uppercase letters, and abbreviations are commonly used to save space and drafting time (refer to the abbreviation list in the Appendix). When necessary, specific notes can be placed close to the item being identified or connected to it with leader lines.

Although most of the lettering done today on construction drawings is computer-generated, skill in freehand lettering adds style and individuality to a designer's work. And in any style of lettering, uniformity is important. This applies to height, proportion, strength of lines, spacing of letters, and spacing of words. Letters should be spaced by visually equalizing the background areas between the letterforms and not by mechanically measuring the distance between the extremities of each letter (Figure 2.3). The use of light horizontal guidelines (using a hard lead such as a 4H) should be practiced to control the height of letters, while light vertical or inclined guidelines are required to keep the letters uniform. Words should be spaced well apart, while letters should be spaced closely within words.

A cursory examination of the various alphabets and typefaces in use today clearly shows that the vast majority fit into one of four basic classifications as outlined below (Figure 2.4):

- Roman: perfected by the Greeks and Romans and later modernized in the 18th century, the Roman alphabet displays enormous grace and dignity and is considered by many to be the most elegant typeface family.

- Gothic: this alphabet is the base from which our single-stroke technical lettering has evolved and is the primary style used by the majority of today's designers. It is an easily read and simply executed style that has been in use for many years as a commercial, block-type letter. Its main characteristic is the uniformity in width of all of the strokes. Modifications of this letter include inclined, squared, rounded, boldface, lightface, and serif.

- Script: script alphabets are cursive in nature and resemble handwriting. The lowercase letters are interconnected when used within words or sentence beginnings. Their characteristic free-flowing strokes impart a sense of delicacy and personal temperament and are not considered appropriate for general use in technical drawing.

- Text (Old English): originally used by European monks for recording religious manuscripts, this alphabet is characterized by the use of strokes of different width, due to the original employment of a flat quill pen. This alphabet is rarely used in modern work and is not considered suitable for technical drawing because it is difficult to read and draw.

All of the above typefaces can be produced in italic (which has inclined, lightface, and curved characteristics). The character of the typeface used should always be appropriate to the design being presented. Today there is a large body of well-designed typefaces available in the form of pressure-sensitive dry-transfer sheets in addition to computerized typography.

TYpogRaphY

abcdefghijklmnopqrstuvwxyz

0123456789

abcdefghijklmnopqrstuvwxyz
0123456789

ABCDEFGHIJK

lmnopqrstuvwxyz

0123456789

SPACING   SPAC

Correct spacing of equal areas   Incorrect spacing of letter forms

SERIFS Serifs

Figure 2.3 Examples of some of the more popular typefaces used today.

ROMAN
GOTHIC
Script
Text

Figure 2.4 Most alphabets currently in use can be classified into four basic categories: Roman, gothic, script and text.

2.3   COMPUTER-AIDED DRAFTING (CAD) AND COMPUTER-AIDED DESIGN AND DRAFTING (CADD)

Moving from manual drafting to computer-aided design and drafting was an important step in terms of the potential of today's technology. CAD originally meant computer-aided drafting because of its original use as a replacement for traditional drafting. Now it usually refers to computer-aided design to reflect the fact that modern CAD tools do more than just drafting (Figures 2.5A, B, and C). Related acronyms are CADD, which stands for computer-aided design and drafting; CAID, for computer-aided industrial design; and CAAD, for computer-aided architectural design. All of these terms are essentially synonymous, but there are a few subtle differences in meaning and application. CAM (computer-aided manufacturing) is also often used in a similar way or as a combination (CAD/CAM).

When we use computer-aided design and drafting (CADD), certain questions arise that we never think of when working on the drawing board. Although with CAD or CADD we do not use the typical drawing-board tools, we are still required to design or make a drawing.

CADD is an electronic tool that enables you to rapidly create accurate drawings with the use of a computer. In fact, an experienced computer drafter can normally produce a construction drawing in less time than it would take if it was done manually. Moreover, unlike the traditional methods of making drawings on a drawing board, with CADD systems you can create professional drawings just by using a mouse and clicking buttons on the keyboard. Furthermore, drawings created with CADD have many advantages over traditionally produced drawings. In addition to the fact that they are neat, accurate, and highly presentable, they can be easily modified and converted to a variety of formats. In addition, CADD-generated drawings can be saved on the computer, a flash, a CD, or an external hard drive in lieu of vellum or Mylar sheets that require the use of large storage cabinets for filing.

A decade ago CADD was largely used for specific engineering applications that required high precision, and because of CADD's high production costs, not many professionals could afford it. In recent years, however, computer prices have dropped significantly, and professionals are increasingly taking advantage of this by adopting CADD programs.

There is a wealth of CADD programs available on the market today. Some are intended for general drawing work, while others are designed for specific engineering applications. There are programs that enable you to do 2D drawings, 3D drawings, renderings, shadings, punch lists, space planning, structural design, piping layouts, HVAC, plant design, project management, and other applications. Today we can find a CADD program for almost every engineering discipline that comes to mind.

CADD Presentations

Although CADD is primarily intended for single-line drafting and has very limited capabilities to create artistic impressions, CADD's 3D and rendering features are quite impressive. A 3D model of an object can be created and viewed from various angles and, with correct shading and rendering, can be made to look very realistic. With CADD you can create fine drawings with hundreds of colors, line types, hatch patterns, presentation symbols, text styles, and other features. Even if you don't like something about your presentation after you have finished it, you can instantly change it (Figure 2.6).

Most CADD programs have a number of ready-made presentation symbols and hatch patterns available that can be used to enhance the look of drawings. When drawing a site plan, for example, a draftsperson can instantly add tree symbols, shrubs, pathways, human figures, and other landscape elements to create a professional-looking site plan. Similarly, an architect can use ready-made symbols

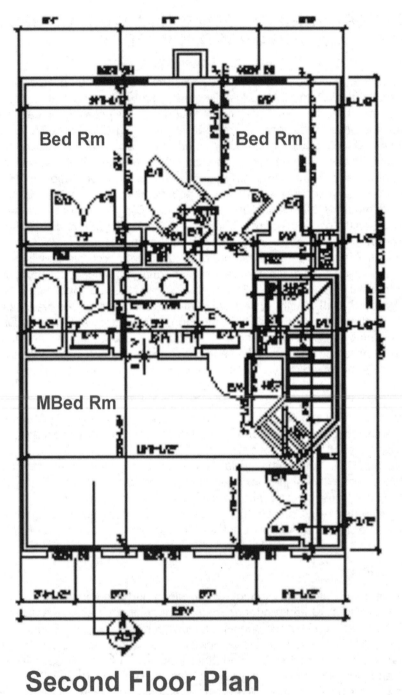

Second Floor Plan

Scale: 1/4" = 1' 0"

Figure 2.5A Example of a type of CADD-generated drawing.

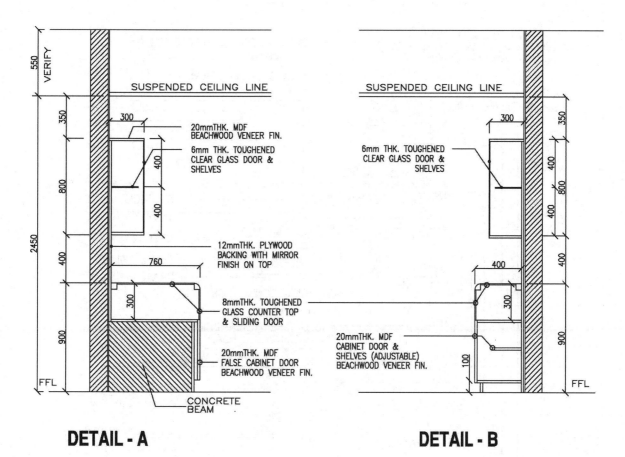

DETAIL - A   **DETAIL - B**

Figure 2.5B Example of a type of CADD-generated drawing.

of doors, windows, and furniture to make a presentation. Architects and designers also sometimes design their own symbols when working with CADD.

In addition to preparing impressive presentations on paper, CADD can be used to make on-screen presentations. Ideas can be presented on-screen by merely plugging the computer to a projector. Advanced CADD programs also allow you to created animated images. You can show how a building would appear while walking through it or how a machine assembly will operate as different machine parts move.

Editing Flexibility

One of the main advantages of CADD is that it allows quick alterations to drawings. Modifications can be made with pinpoint accuracy. It takes only seconds to do a job that would otherwise take hours on a drawing board to produce. In many cases you may not even have to erase a section to make the change.

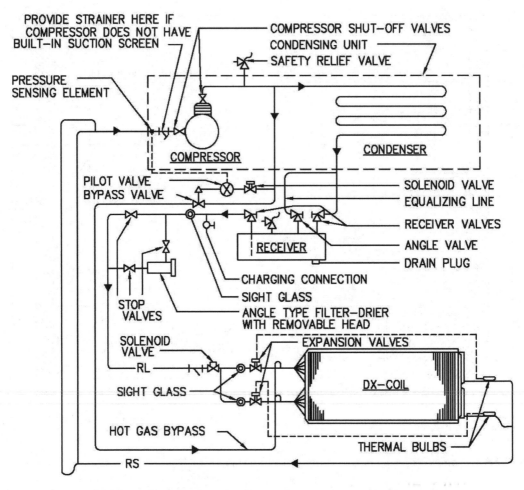

REFRIGERANT PIPING DETAIL (A/C)

Figure 2.5C   Example of a type of CADD-generated drawing.

You can often rearrange the existing components of the drawing to fit the new shape. This enables the designer to compare various options with minimal effort.

The following are some of the main editing capabilities of most CADD systems:

- Move, copy, mirror, or rotate drawing elements with ease
- Enlarge or reduce elements of a drawing
- Make multiple copies of a drawing element
- Add one or more drawings to another drawing
- Change font style and size

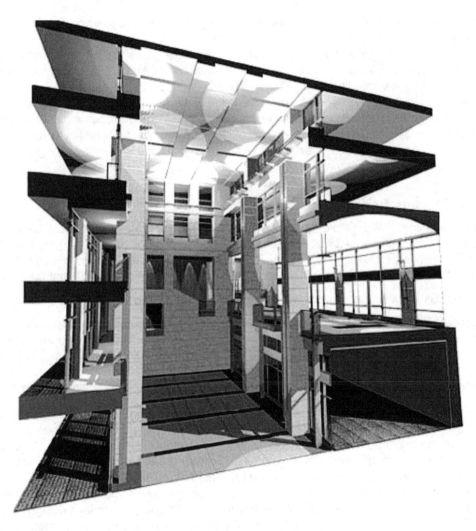

Figure 2.6   Example of a presentation perspective using ArchiCAD software.

- Change units of measure of dimensions
- Stretch drawings to fit new dimensions
- Convert CADD drawings to other formats

Units and Accuracy Levels

CADD allows you to work with greater accuracy. If you need to create highly accurate geometrical shapes, while avoiding time-consuming mathematical calculations, CADD is the answer. Computer-software programs like CADD allow the designer to work with different units of measure, such as architectural units, engineering units, scientific units, and surveyor's units. These units can be represented in various formats commonly used by professionals.

When working with engineering units, the designer or drafter can specify whether all the dimensions should be represented in inches, feet-inches, centimeters, or meters. Angular units of measurement such as decimal degrees, minutes, seconds, or radians can also be chosen.

In general, when there is a need to work on a large-scale drawing such as a plan of a township, a high degree of accuracy may not be warranted, and it may be decided to set a lesser degree of accuracy—say ,1 foot, 0 inches. The computer will then round off all the measurements to the next foot, which avoids the use of any fractions less than a foot. Where minute detail is required, a higher degree of accuracy such as 1/8 or 1/64 inch may be set.

Drawings Storage and Access

As previously mentioned, it is simple, quick, and convenient to generate or organize a CADD drawing on a computer. Most computer hard drives have the capacity to store thousands of drawings (and this storage capacity is continuously increasing with advancing technology), and they can be opened within seconds.

Some of the advantages of a computer's electronic filing system over traditional filing include:

- It enables and encourages the creation of a highly organized and efficient environment.
- It contributes to large reductions in general working space.
- An electronic drawing does not age or become faded. Whenever a new drawing is required, it can be printed from disks. (It is important that the program used to store the CADD files be regularly updated to avoid becoming obsolete. With continuously advancing technology, some storage methods have already become outdated, such as zip drives and some types of disks.)

Through networking electronic drawings can be shared by several users, allowing them to coordinate their tasks and work as a team. Different professionals such as architects, engineers, and construction managers can use the same electronic drawings to coordinate building services. When a modification is made to a drawing, this information becomes instantly available to all the team members. It has thus become far easier to share information. Professionals located in different cities or countries can now instantly transmit electronic drawings to one another. They can also publish these drawings on the Internet for anyone to access. Most CADD programs incorporate special functions designed to allow you to export drawings in a format (such as .gif or .jpeg) that can be viewed on the Internet.

Project Reporting

The computer is an ideal instrument for generating project reports, cost estimates, and other business documents. CADD's database capabilities include the ability to link specific nongraphic information (such as text or financials) with the graphic elements of the drawing. The nongraphic information is stored in a database and can be used to prepare reports.

An architect, for example, can attach text attributes associated with the door and window symbols in a drawing. These attributes include the door's size, material, hardware, cost, and so forth. Equipped with this information, the computer can then automatically generate a door schedule listing all the doors and windows in the drawing!

Nongraphic information is directly linked with the graphics on the screen so that, when a change is made to the drawing, the values in the reports are automatically updated. This provides a useful means to manage large projects from design through project completion.

2.4   TYPES OF BLUEPRINTS

Types of Construction Drawings

Providing accurate construction drawings up front helps ensure that construction projects will proceed in an orderly manner, reducing costly and time-consuming rework by contractors and subcontractors down the line. Construction drawings are generally categorized according to their intended purpose. Types commonly used in construction may be divided into five main categories based on the function they intend to serve:

1. Preliminary drawings
2. Presentation drawings
3. Working drawings
4. Shop/assembly drawings
5. Specialized and miscellaneous drawings

Preliminary-Design and Concept Drawings

At the initial promotional stages of a project, the architect or designer often prepares preliminary sketches, which are essentially schematic design/concept-development drawings. These provide a convenient and practical basis for communication between the designer and the owner in the idea formulation stage. During the design phase, these drawings go through many alterations, helping the client to determine the most aesthetically attractive and functional design. These drawings are not meant for construction but rather for exploratory purposes, providing an overall concept that reflects the client's needs, as well as functional studies, materials to be used, preliminary cost estimates and budget, preliminary construction approvals, etc. Preliminary drawings are also typically used to explore with other consultants concepts relating to the mechanical, plumbing, and electrical systems to be provided. These are followed by formal design-development drawings prior to the working-drawing or construction-document stage.

Presentation Drawings

The purpose of presentation drawings is to present the proposed building or facility in an attractive setting at the proposed site for promotional purposes. They usually consist of perspective views complete with colors and shading, although they may also contain nicely drawn elevation views with shadows and landscaping (Figures 2.7A and B). Presentation drawings are therefore essentially selling tools, a means to sell the building or project before it reaches the working-drawing stage, and are used in brochures and other outlets. This phase is also where the schematic design is developed, finalized, and approved by the client.

Working Drawings

Also called project and constructions drawings, working drawings include all the drawings required by the various trades to complete a project. These drawings are technical and are intended to furnish all the necessary information required by a contractor to erect a structure. Working drawings show the size, quantity, location, and relationship of the building components. They are typically prepared in considerable detail by the architect or engineer, and the amount of time and effort expended on them comprises a major portion of the consultant's design services.

Two proposals for Marriott Hotel in Abu Dhabi

Figure 2.7A Presentations can play a pivotal role in convincing a client to approve a particular design. In this illustration, two design proposals by the author for a hotel, reflecting different architectural treatments, were submitted to both the client and to Marriott for approval.

Cabinet of Ministers design proposal - Reception Area

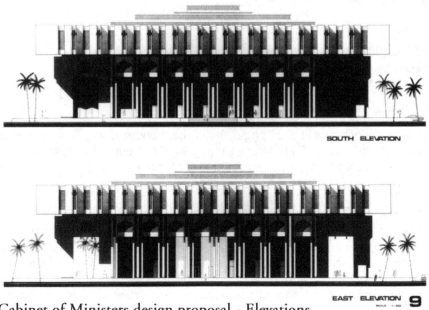

SOUTH ELEVATION

EAST ELEVATION **9**
SCALE 1:100

Cabinet of Ministers design proposal - Elevations

Figure 2.7B International competition submission for a cabinet of ministers complex in Abu Dhabi by the author that won first prize.

Sometimes the printing may be difficult to read or important information may be missing from the drawings. Occasionally entire pages may be missing, or the contractor may have received only a set of plans or specifications. If the prints are incomplete or of poor quality, the consultant should be immediately notified and asked to address the problem.

Working drawings serve many functions:

1. They are the means for receiving a building permit. Before construction begins, the local building authority has to review the working drawings to ensure that they meet required building codes. A building permit will be issued after approval of the drawings.

2. They are used for competitive bidding. They allow contractors to study the documents and make bids based on their review of the drawings and other documents, thus providing the owner with the most economical cost for construction.

3. They provide instructions for construction. Working drawings should contain all the necessary information to build the structure.

4. They are used for material take-offs. Labor, material, and other estimates are made from working drawings prior to commencement of construction.

5. They provide a permanent record for future use (such as remodeling and dispute resolution).

6. They can be used as a basis for leasing purposes.

7. After the project is awarded, the drawings form the basis of the contract between the contractor, subcontractor, and client.

The pages in a set of blueprints are usually carefully lettered and numbered. The letters shown here are the ones most commonly used in the industry:

- A: Architectural pages
- S: Structural pages
- P: Plumbing pages
- M: Mechanical pages
- E: Electrical pages

Thus if a set of blueprints consist of 30 pages, it may be numbered as follows: A1 through A8 (eight architectural pages); S1 through S10 (10 structural pages); P1 through P3 (three plumbing pages);M1 through M4 (four mechanical pages); E1 through E5 (five electrical pages).

Shop and Assembly Drawings

Shop and assembly drawings are technical drawings prepared by various contractors, subcontractors, and suppliers participating in the construction process to show how their product is to be made. Since many products contain more than one component, shop and assembly drawings (also called fabrication drawings) identify each component and show how they all fit together. These drawings should contain all the necessary information on the size, shape, material, and provisions for connections and attachments for each member, including details, schedules, diagrams, and other related data to illustrate a material, product, or system for some portion of the work prepared by the construction contractor, subcontractor, manufacturer, distributor, or supplier. Product data includes items such as brochures, illustrations, performance charts, and other information by which the work can be evaluated. The information provided must be in sufficient detail to permit ordering the material for the product concerned and its fabrication in the shop or yard. In practice the consultant often has to rely on these specialists to furnish precise information about their components.

In most projects, whether large or small, contractors and subcontractors are frequently required to draft shop drawings even for minor shop and field projects such as doors, cabinets, and the like. Thus, for example, if complex cabinetwork is required, it must be built to exact size and specifications. A shop drawing becomes necessary to ensure that the cabinetwork will fit into the structure and that the structure will accommodate it. In Figure 2.8A we see shop-drawing details for a restaurant waiter station and Figure 2.8B shows how a cabinet is to be assembled. Approval of the shop drawings usually precedes the actual fabrication of the component. Shop drawings also help the consultant check the quality of other components that subcontractors propose to furnish.

Details

Detail drawings provide information about specific parts of the construction and are on a larger scale than general drawings. They show features that do not appear at all or are on too small a scale in general drawings. The wall section and elevator details in Figures 2.9A and B are typical examples and are drawn to a considerably larger scale than the plans and elevations.

Framing details at doors, windows, and cornices, which are the most common types of details, are nearly always shown in sections. Details are included whenever the information given in the plans, elevations, and wall sections is not sufficiently "detailed" to guide the craftsmen on the job. Figure 2.10 shows some typical door and eave details.

A detail contains both graphic and written information. An area of construction is drawn at a larger scale in order to clearly show the materials, dimensions, method of building, desired joint or attachment, and so on.

Details are often drawn as sections. It is as if a slice is made through a specific area and the inner components are visible. In Figures 2.11A and B we see an example of a typical bay window detail.

There are many types of details, all of which are drawn as needed to clarify specific aspects of a design. A drawing sheet will often show several details. The complexity of the project will determine which areas need to be shown at a larger scale.

Details are always drawn to scale. A typical scale for a detail is 3 inches to 1 foot (scale: 3 inches = 1 foot, 0 inches). The scale for each detail will vary depending on how much information is required to make the construction clear to the builder. Each detail will have the scale noted below.

Specialized and Miscellaneous Drawing Types

There are numerous other types of drawings used by architects and engineers in the construction industry.

Freehand sketches are drawings made without the aid of any type of drawing instruments. Sketches can be an extremely valuable tool for architects, designers, builders, and contractors. It is often the quickest and most economical method to communicate ideas (Figures 2.12A, B, and C), construction methods, and concepts or to record field instructions. It is an ideal method to sell an idea to a client and get preliminary approval for a design. Likewise, when installing mechanical or electrical systems and circuits, you may sometimes have to exchange information about your job with others. A freehand sketch can be an accurate and appropriate method to communicate this information. This type of drawing is informal in character, may or may not be drawn to scale, and need not follow any particular format. A

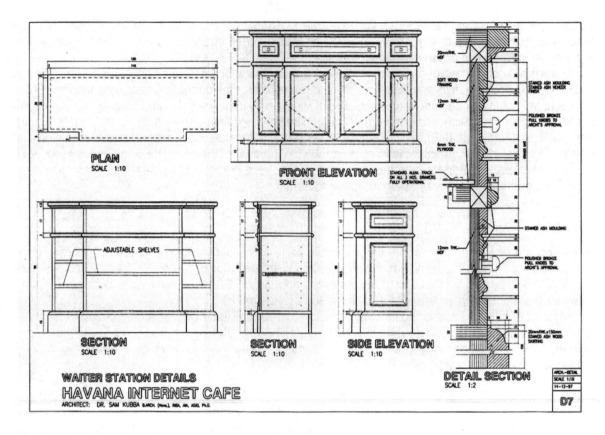

Figure 2.8A   Drawing showing the details for a restaurant waiter station.

sketch can be used in many ways. Another example of where to use a sketch is to show a field change that must be made. No matter how well a project is planned, field changes may sometimes need to be made, and a sketch will often go a long way to helping the builder visualize the designer's intention or the construction techniques to be used. Sketches may include dimensions, symbols, and other information needed to convey your idea of the required change to someone else (such as the project supervisor or project chief).

Erection drawings, or erection diagrams, indicate the location and position of the various members in the finished structure. Erection drawings are especially useful to builders performing the erection in the field. The information erection drawings show includes supplying the approximate weight of heavy pieces, the number of pieces, and other helpful data.

Framing drawings are necessary to show the layouts and provide other relevant information about the various framing components. These include floor joists, trusses, beam locations, and other structural elements. Framing layouts are drawn to scale but don't normally get into the details of each stud location in the walls, because framing contractors are required to follow certain rules and regulations to assure that the structure meets the required building-code specifications.

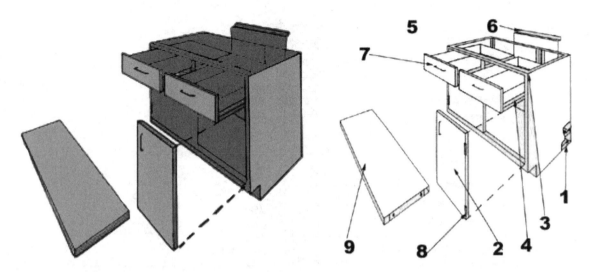

Figure 2.8B   Illustrations showing the components of a cabinet.

Falsework drawings show temporary supports of timber or steel that are required sometimes in the erection of difficult or important structures. When falsework is required on an elaborate scale, drawings similar to the general and detail drawings already described may be provided to guide construction. For simple falsework field sketches may be all that is needed.

Master-plan drawings are commonly used in the architectural, topographical, and construction fields. They show sufficient features to be used as guides in long-range area development and usually contain a considerable amount of information including section boundary lines, contour lines, acreage, existing utilities, rights-of-way and appurtenances, horizontal and vertical control data, locations and descriptions of existing and proposed structures, existing and proposed surfaced and unsurfaced roads and sidewalks, streams, and north-point indicator (arrow).

2.5   TITLE BLOCKS

It is standard practice to include a title block on each page of a set of blueprints. It is typically located in the bottom right-hand corner of the drawing frame. Many firms, however, are using customized sheets that extend the title block from the lower right to the upper right-hand side of the sheet (Figure 2.13). The title block should show the name of the project and the drawing and sheet numbers as well as the drawing title (e.g., "site plan"). The drawing number is especially important, both for purposes of filing the blueprint and for locating the correct drawing when it is specified on another blueprint. The title block also typically shows the name of the consultant, architect, engineer, or designer and the signature of the approving authority. The title block should normally show the date the drawing was made and the initials of who made it. This information is important because using an outdated set of drawings can cause serious problems. Any revisions should be noted within the title block.

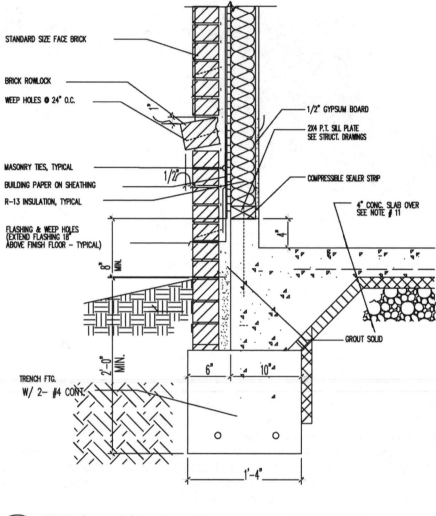

STANDARD SIZE FACE BRICK

BRICK ROWLOCK

WEEP HOLES @ 24" O.C.

1/2" GYPSUM BOARD

2X4 P.T. SILL PLATE
SEE STRUCT. DRAWINGS

MASONRY TIES, TYPICAL

BUILDING PAPER ON SHEATHING

R-13 INSULATION, TYPICAL

COMPRESSIBLE SEALER STRIP

4" CONC. SLAB OVER
SEE NOTE # 11

FLASHING & WEEP HOLES
(EXTEND FLASHING 18"
ABOVE FINISH FLOOR – TYPICAL)

GROUT SOLID

TRENCH FTG.
W/ 2- #4 CONT.

2 / DT TYPICAL FOUNDATION DETAIL W/ BRICK
Scale: 1 1/2" = 1' 0"

Figure 2.9A   AutoCAD drawing showing a typical foundation/wall section detail for a new residence.

To summarize, the title block should generally contain the following information:

- Name of the consultant, company, or organization with address and phone number.
- Title of the drawing. This is an identification of the project by client name, company, or project name.
- Drawing number. This can be a specific job or file number for the drawing.

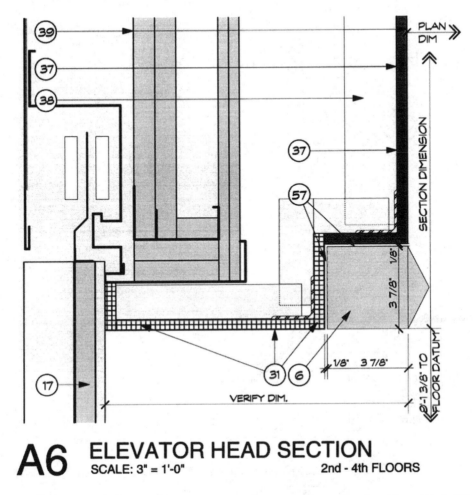

A6 ELEVATOR HEAD SECTION
SCALE: 3" = 1'-0"   2nd - 4th FLOORS

Figure 2.9B Example of an elevator detail for a commercial building produced on VectorWorks/MiniCAD using a Macintosh computer *(courtesy Herring and Trowbridge, Architects).*

- Scale. Some consultant firms provide a location for the general scale of the drawing in the title block, although most firms now omit the scale from the title block and place it on the sheet directly below the title of each individual plan, view, or detail. Where more than one scale is used, as is found on detail sheets, the space for indication of scale should read "as noted" or "as shown."

- The signature or initials of the drafter, checker, approving officer, and issuing officer, with the respective dates, should be shown.

- Drawing or sheet identification. Each sheet should be numbered in relation to the entire set of drawings. Thus, if the set consists of ten sheets, each consecutive sheet is numbered 1 of 10, 2 of 10, and so on.

- Other information as required.

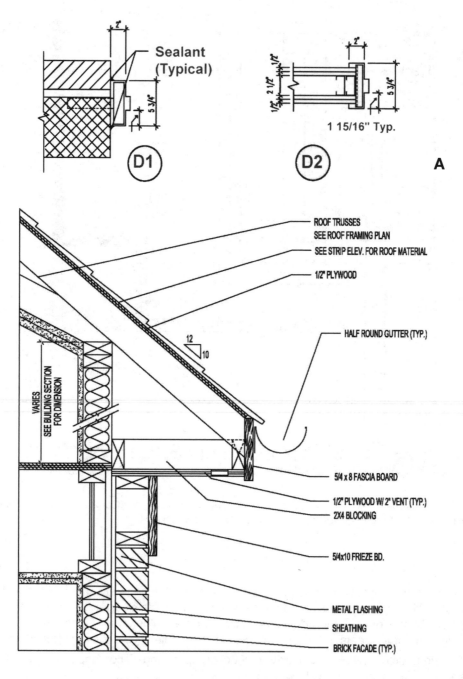

Sealant
(Typical)

D1

D2

1 15/16" Typ.

A

ROOF TRUSSES
SEE ROOF FRAMING PLAN

SEE STRIP ELEV. FOR ROOF MATERIAL

1/2" PLYWOOD

HALF ROUND GUTTER (TYP.)

12
10

VARIES
SEE BUILDING SECTION
FOR DIMENSION

5/4 x 8 FASCIA BOARD

1/2" PLYWOOD W/ 2" VENT (TYP.)

2X4 BLOCKING

5/4x10 FRIEZE BD.

METAL FLASHING

SHEATHING

BRICK FACADE (TYP.)

EAVE DETAIL (WITH BRICK FACADE)
Scale: 1 1/2" = 1' 0"

B

Figure 2.10   Typical AutoCAD-generated door (A) and eave (B) details.

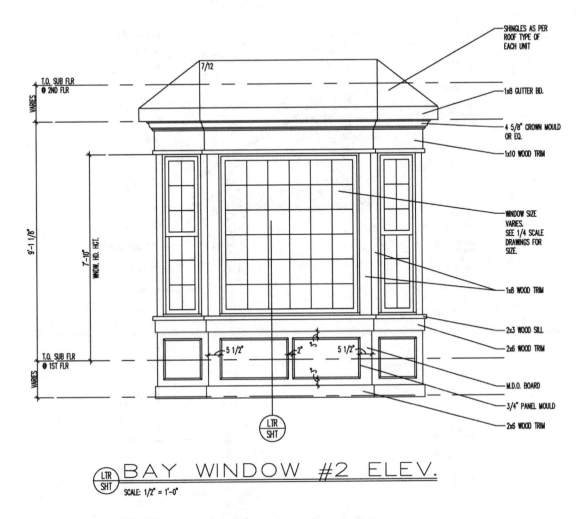

Figure 2.11A   An AutoCAD drawing of a residential bay window showing an elevation.

The following information is also normally required in title blocks: the name and location of the activity, the specifications and contract numbers (if any), the preparing activity including the architect-engineer (A-E) firm if applicable, and the surnames of the personnel concerned in the preparation of the drawings.

Revision Block

Many firms provide a revision column in which drawing changes are identified and recorded. The revision block is usually located in the right-hand corner of the blueprint. All revisions are noted in this block and dated and identified by a letter and an optional brief description of the revision with the initials of the individual making the change (Figure 2.14). If changes are made on the face of the drawing after it has been released for construction, a circle with a revision number or letter should accompany the change.

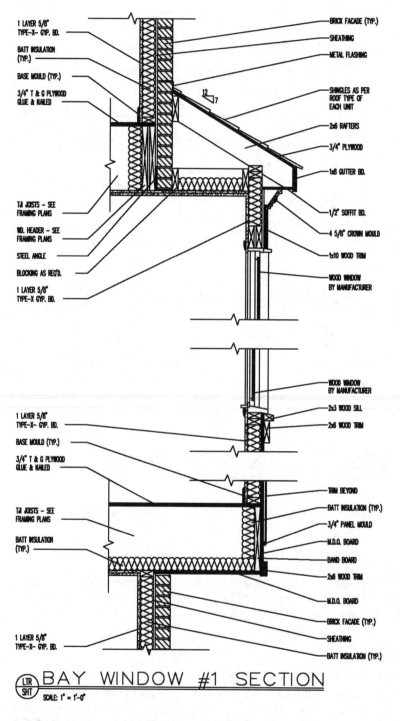

1 LAYER 5/8"
TYPE-X- GYP. BD.

BATT INSULATION
(TYP.)

BASE MOULD (TYP.)

3/4" T & G PLYWOOD
GLUE & NAILED

BRICK FACADE (TYP.)

SHEATHING

METAL FLASHING

SHINGLES AS PER
ROOF TYPE OF
EACH UNIT

2x6 RAFTERS

3/4" PLYWOOD

1x8 GUTTER BD.

TJI JOISTS – SEE
FRAMING PLANS

WD. HEADER – SEE
FRAMING PLANS

STEEL ANGLE

BLOCKING AS REQ'D.

1 LAYER 5/8"
TYPE-X GYP. BD.

1/2" SOFFIT BD.

4 5/8" CROWN MOULD

1x10 WOOD TRIM

WOOD WINDOW
BY MANUFACTURER

WOOD WINDOW
BY MANUFACTURER

2x3 WOOD SILL

2x6 WOOD TRIM

1 LAYER 5/8"
TYPE-X- GYP. BD.

BASE MOULD (TYP.)

3/4" T & G PLYWOOD
GLUE & NAILED

TJI JOISTS – SEE
FRAMING PLANS

BATT INSULATION
(TYP.)

TRIM BEYOND

BATT INSULATION (TYP.)

3/4" PANEL MOULD

M.D.O. BOARD

BAND BOARD

2x6 WOOD TRIM

M.D.O. BOARD

BRICK FACADE (TYP.)

SHEATHING

BATT INSULATION (TYP.)

1 LAYER 5/8"
TYPE-X- GYP. BD.

BAY WINDOW #1 SECTION

LTR
SHT

SCALE: 1" = 1'-0"

Figure 2.11B   An AutoCAD drawing of a residential bay window showing detail. The information shown on the detail allows the builder to construct the bay window as intended.

Figure 2.12A   Sketch of a president's suite bedroom—Crowne Plaza, Abu Dhabi.

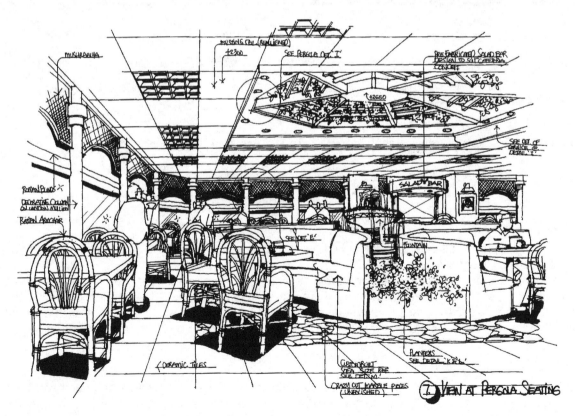

Figure 2.12B   Sketch of Havana Internet Cafe, Abu Dhabi.

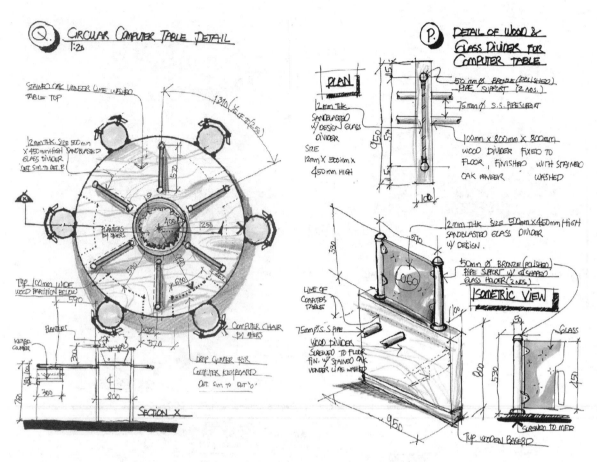

Figure 2.12C Sketch of computer desk details, Havana Internet Cafe, Abu Dhabi.

Scale

The graphic representation of the project is drawn to some proportion of the actual size of the project; usually 1/8 inch is equal to a foot.

Although the original drawing may be scaled accurately, the print will be a copy of that original and is not likely to be the same size as the original drawing. The copy may have been reduced slightly. Likewise, the paper size is often affected by temperature and humidity and may therefore stretch or shrink. Due to these and other factors, avoid relying on measurements taken by laying a rule on the drawing.

Zoning

For large projects, a drawing may sometimes be divided into a grid using letters and numbers. When zoning is used, it is typically located inside the drawing frame. Zoning allows easy reference to various parts of the drawing by referencing a coordinate such as B4.

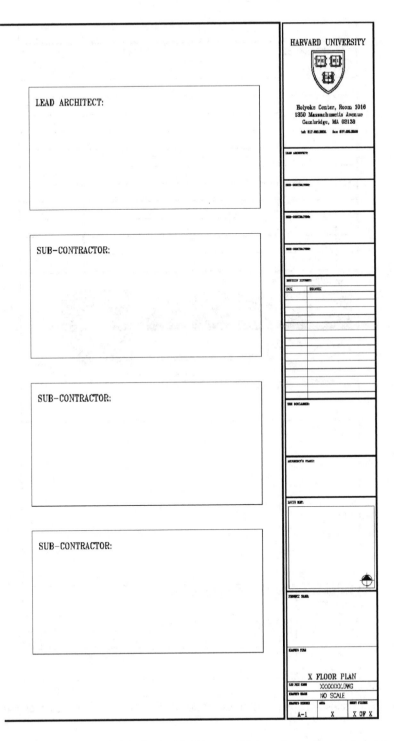

Figure 2.13 Cover sheet including title block used by Harvard University on its working drawing documents.

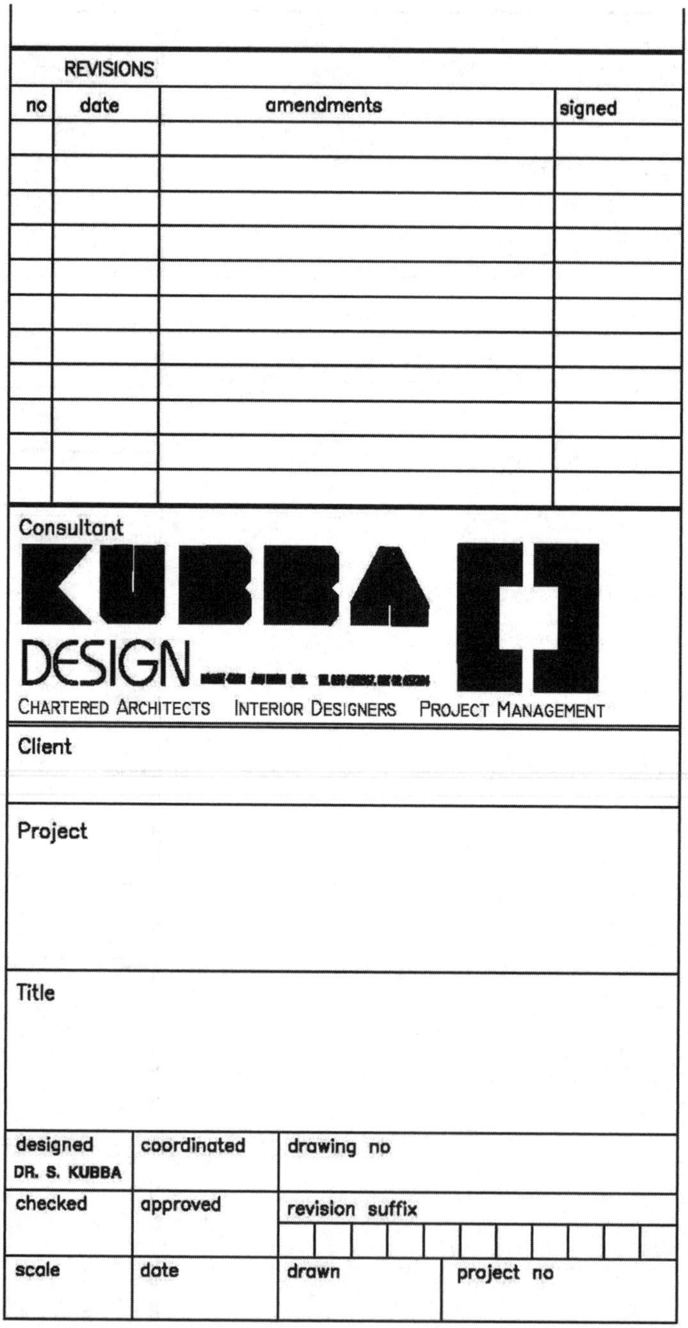

REVISIONS				
no	date	amendments		signed

Consultant

KUBBA

DESIGN

CHARTERED ARCHITECTS   INTERIOR DESIGNERS   PROJECT MANAGEMENT

Client

Project

Title

designed	coordinated	drawing no	
DR. S. KUBBA			
checked	approved	revision suffix	
scale	date	drawn	project no

Figure 2.14   Title block showing position of revision block on a consultant's typical working drawings.

3

Understanding Line Types

3.1 INTRODUCTION—THE ALPHABET OF LINES

The alphabet of lines is the universal language of the technician, architect, and engineer. In reality, lines are the basis of all construction drawings. To read and understand blueprints, you need to understand the use of lines. By combining lines of different thicknesses, types, and lengths, it is possible to graphically describe objects in sufficient detail to allow a person with a basic understanding of blueprint reading to accurately visualize their size and shape. As will be explained, line characteristics such as width, breaks, and zigzags all have meaning, and each line has a specific design and thickness that distinguishes it from other lines.

Drafting is an international graphic language that uses lines, symbols, and notes to describe a structure to be built, and lines themselves are expressive tools on well-executed drawings. Certain lines are drawn thick so they stand out clearly from other information on the drawing, whereas other lines are drawn thin. Thin lines are not necessarily less important than thick lines, but they are subordinate for identification purposes. Drawings with only lines of the same intensity are often difficult to interpret and usually monotonous to read.

As mentioned in Chapter 1, there are a number of organizations such as the American National Standards Institute (ANSI) and the International Organization for Standardization (ISO) that have voluntarily adopted certain drafting standards that are generally accepted and widely used around the world.

3.2 LINE WEIGHTS AND TYPES

Line weights are generally subject to the following conventions:

1. For manual pencil drafting using drawing boards, finished work includes bold object lines (2H to B pencils), light dimension lines, center lines, section lines, and so on (3H to 5H pencils). Temporary construction lines, guidelines for lettering, and other types should be kept very light (7H or 8H). Border lines for the drawing sheet and title block should be made bold (3B to 6B).

2. For inked or computer-plotted drawings different pen widths are used to achieve similar effects (see Figure 3.1). When plotting (printing) using a laser or inkjet printer from a computer drawing in

AutoCAD, these line-boldness conventions are replicated by configuring each line color as certain line widths.

3. Border lines are roughly twice as bold as object lines, which in turn are roughly twice as bold as dimension lines. In addition, AutoCAD drafting uses colors to emulate each thickness (black or white for objects, green for dimensions, blue for borders, etc.). In AutoCAD's print-dialogue box you are allowed to designate different line widths for each different color.

Line Types: The major line types and line thicknesses using pens or plotter machines are outlined below. It becomes obvious from the table below that larger sheets require the use of thicker lines than smaller sheets.

Object Lines

Object lines, also known as visible lines, are solid lines used mainly to define the shape and size of a structure or object. They are continuous prominent lines representing the edges of surfaces or the intersection of two surfaces, as shown in Figure 3.2. An object/visible line is typically drawn thick (dark) and solid so that the outline or shape of the object (e.g., wall, floor, elevation, detail, or section) clearly stands out on the drawing with a definite contrast between these lines and secondary lines on the drawing. They are heavier than hidden lines, dimension lines, center lines, and broken lines. As we shall see later in the chapter, blueprint drawings often contain different solid line types that are not object lines.

Dashed Lines

Dashed or hidden lines serve more than one purpose in construction drawings. They are comprised of medium- or light-weight, uniformly sized broken lines consisting of evenly spaced short dashes and are generally intended to represent hidden surfaces or intersections of an object. On floor plans they may be used to represent features that lie above the plane of the drawing, such as high wall cabinets in a kitchen. You may vary the lengths of the dashes slightly in relation to the size of the drawing.

On remodeling-job drawings, they are also used to indicate the position of preexisting construction. In some cases they are used for relationship clarification or to show alternative positions of a movable component. To be complete, a drawing must include lines that represent all the edges and intersections of the surfaces in the object. Many of these lines are invisible to the observer because they are covered by other portions of the object. For example, in Figure 3.3A the dashed lines indicate the location of blocking hidden behind the wall.

In architectural drafting dashed lines may be applied in different weights to reflect their purpose (e.g., to reflect importance or distance from the main view) while showing drawing features that are not visible in relationship to the view or plan. These dashed features can be subordinated to the main emphasis of the drawing. Hidden lines should typically begin and end with a dash, in contrast with the visible lines from which they start, except when a dash would form a continuation of a visible line. Dashes should be joined at corners; likewise, arcs should begin with dashes at tangent points. Hidden lines should be omitted when not required for the clarity of the drawing. Although features located behind transparent materials may be visible, they should be treated preferably as concealed features and shown with hidden lines. Examples of dashed-line representations include beams and headers, upper kitchen cabinets, undercounter appliances (e.g., dishwasher or refrigerator), or electrical circuit runs, as shown in Figure 3.3B. Figure 3.3C is another example of hidden-line use.

Letter	Type of Line	*Recommended Line Thickness in mm. (inch equivalent in brackets)			Application
		A0	A1	A2 A3 A4	
A	Continuous - thick	0.7 (0.028)	0.5 (0.020)	0.35 (0.014)	• Object Visible) outline lines • Border line
B	Continuous - thin	0.35 (0.014)	0.25 (0.010)	0.18 (0.007)	• Dimension lines • Leaders • Imaginary intersection of surfaces • Extension lines, and intersection lines • Section lines & Hatching • Adjacent parts and tooling • Short center lines • Contour lines
C	Continuous - thin, freehand or ruled with zig-zag	0.35 (0.014)	0.25 (0.010)	0.18 (0.007)	• Indication of repeated detail • Break lines (other than on an axis)
D	Dashed - medium Space (S) = 1 mm minimum A = 2S to 4S	0.50 (0.020)	0.35 (0.014)	0.25 (0.010)	• Hidden surfaces • Surfaces above plane
E	Chain - thin Space (S) = 1mm minimum A = 2S to 3S B = 2A to 10A	0.35 (0.014)	0.25 (0.010)	0.18 (0.007)	• Centerlines • Pitch lines • Alternative position of moving part • Path lines for indicating movement • Features in front of a cutting plane • Developed views • Material to be removed
F	Chain - thick	0.7 (0.028) 0.35 (0.014)	0.5 (0.020) 0.25 (0.010)	0.35 (0.014) 0.18 (0.007)	• Cutting planes

* A0 = 1189mm x 841mm (46.8 in x 33.1 in);   A1 = 841 mm x 594 mm (33.1 in x 23.4 in);
A2 = 594 mm x 420 mm (23.4 in x 16.5 in);   A3 = 420 mm x 297 mm (16.5 in x 11.7 in)
A4 = 297 mm x 210 mm (11.7 in x 8.3 in);   A5 = 210 mm x 149 mm (8.3 in x 5.9 in).

Figure 3.1 Major types of line used in construction drawings. Line weights are clearly impacted by sheet size. The larger the sheet being used, the heavier the line weight required. An object line on an AO sheet for example should be twice as thick as an object line drawn on an A2, A3, or A4 sheet.

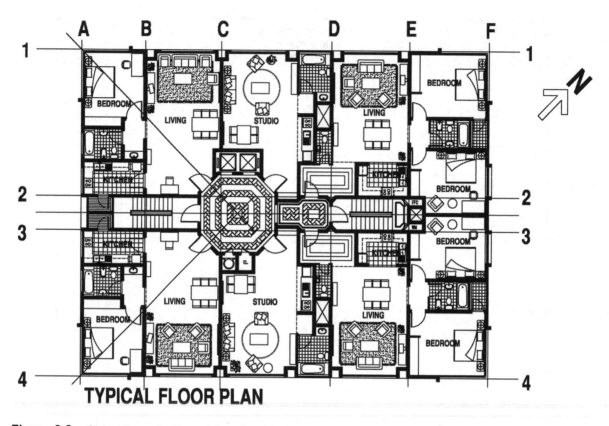

Figure 3.2 Object lines identify and describe the shape of an object. For example, in this drawing object lines are used to portray the furniture as well as the various architectural elements in the design (e.g., stairs, elevators, walls, etc).

Section Lines

Section lines (also called cross-hatch lines) indicate the cut surface in a section view. They usually consist of fine linework (thinner than object lines), typically angular (45 degrees), giving a tone to sectioned surfaces (Figures 3.4A and B. Section lines are used to emphasize the cutting-plane surface and make the view easier to visualize.

Center Lines

Center lines consist of thin (light), broken lines of alternating long and short dashes. They are used to identify the centers of symmetrical objects such as a column, wall, or window. Center lines are also used to indicate the center of a whole circle or part of a circle and to show that an object is symmetrical about a line (Figures 3.5A and B). The symbol is typically drawn as such: (C/L). Center lines are symmetrical on the axis of circular features and are also used to locate centers of windows and door symbols on floor plans.

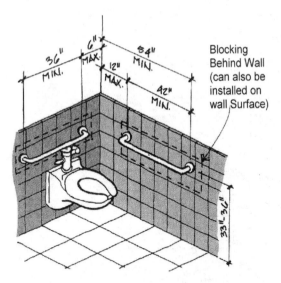

Figure 3.3A Hidden lines represent invisible edges and surfaces—in this case, blocking hidden behind a wall.

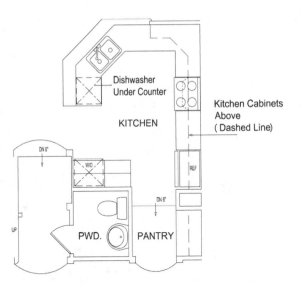

Figure 3.3B Hidden lines here represent an undercounter appliance and hung kitchen wall cabinets.

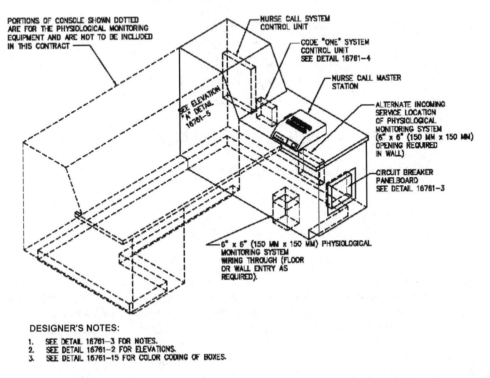

Figure 3.3C An illustration of a nurses' console showing extensive use of hidden lines.

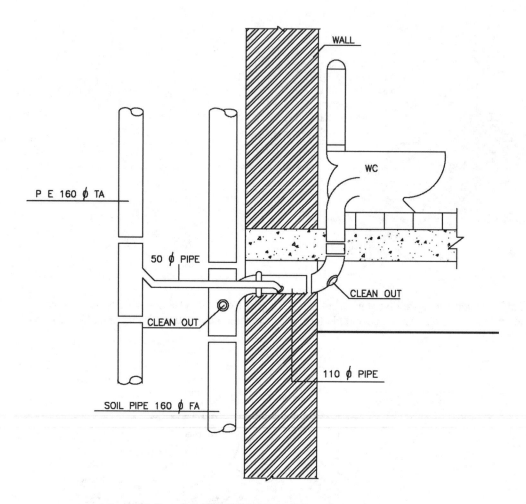

W.C. CONNECTION DETAILS

Figure 3.4A   Sectional lines add emphasis to the walls in this illustration.

The long dashes of center lines may vary in length, depending upon the size of the drawing. Center lines should begin and end with long dashes, and they should not intersect at the spaces between the dashes. They should extend uniformly and distinctly a short distance beyond the object or feature of the drawing unless a longer extension line is required for dimensioning or for some other purpose. Very short center lines may be unbroken if there is no confusion with other lines.

Symmetry lines are center lines used as axes of symmetry for partial views. To identify the line of symmetry, two thick, short parallel lines are drawn at right angles to the center line. Symmetry lines are

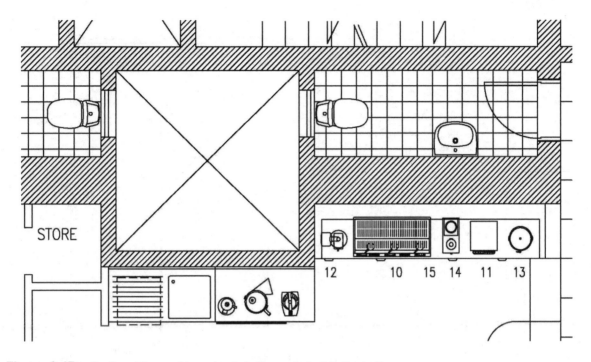

Figure 3.4B   Sectional lines add emphasis to the walls in this illustration.

used to represent partially drawn views and partial sections of symmetrical parts. Symmetrical-view visible and hidden lines may be extended past the symmetrical line if this improves clarity.

Extension Lines

Extension lines are used in dimensioning to show the size of an object. Extension lines are thin, dark, solid lines that extend from an object at the exact locations between which dimensions are to be placed. A space of approximately 1/16 inch is usually allowed between the object and the beginning of the extension line (Figure 3.6). Plan to avoid crossing other extension lines and/or dimension lines.

Extension lines are fine lines that relate the dimension lines to their features. They do not touch the features; instead they start about 1/6 inch (4.23 mm) from the feature and extend about 1/8 inch (3.175 mm) beyond the arrowheads of the dimension line.

Extension lines show the extent of a dimension, and dimension lines show the length of the dimension and terminate at the related extension lines with slashes, arrowheads, or dots. The dimension number in feet and inches is placed above and near the center of the solid dimension line.

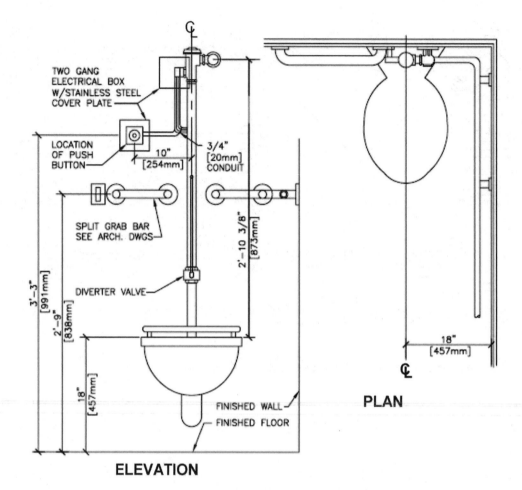

TWO GANG
ELECTRICAL BOX
W/STAINLESS STEEL
COVER PLATE

LOCATION
OF PUSH
BUTTON

10"
[254mm]

3/4"
[20mm]
CONDUIT

SPLIT GRAB BAR
SEE ARCH. DWGS

2'-10 3/8"
[873mm]

DIVERTER VALVE

3'-3"
[991mm]

2'-9"
[838mm]

18"
[457mm]

FINISHED WALL
FINISHED FLOOR

18"
[457mm]

PLAN

ELEVATION

Figure 3.5A   Center lines indicate the centers of circles, arcs, and symmetrical objects.

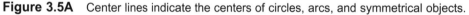

Dimension Lines

Dimension lines are solid lines similar in weight to extension lines that are used to indicate length. They are drawn from one extension to the next, representing the distance between the extension lines. Once the external shape and internal features of a part are represented by a combination of lines, further information is provided by dimensions. Fractional, decimal, and metric dimensions are used on drawings to give size descriptions. Each of these three systems of dimensioning is used throughout the text.

   Dimension lines are fine lines that are often broken at the dimension and ending with arrowheads, dots, or a small diagonal line (Figures 3.6, 3.7A and B). The tips or points of these arrowheads indicate the exact distance, referred to by a numerical dimension placed either at a break in the line or directly above the lines near their center.

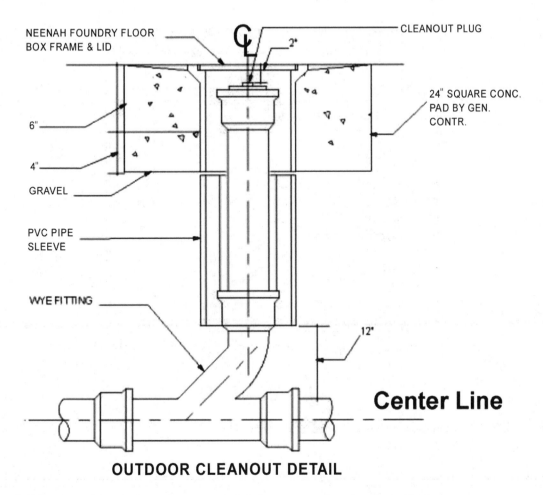

NEENAH FOUNDRY FLOOR
BOX FRAME & LID

CLEANOUT PLUG

2"

24" SQUARE CONC.
PAD BY GEN.
CONTR.

6"

4"

GRAVEL

PVC PIPE
SLEEVE

WYE FITTING

12"

Center Line

OUTDOOR CLEANOUT DETAIL

Figure 3.5B   Center lines indicate the centers of circles, arcs, and symmetrical objects.

Different types of arrowheads are used in dimensioning. The point or tip of the arrowhead touches the extension line. The size of the arrow is determined by the thickness of the dimension line and the size of the drawing. Closed and open arrowheads are the two shapes generally used. The closed arrowhead is preferred. The extension line usually projects 1/16 inch beyond a dimension line. Any additional length to the extension line is of no value in dimensioning.

When dimensions can be added together to come up with one overall dimension, they are known as chain dimensions. Chain dimensions are usually expressed in a single line whenever possible (Figure 3.7B).

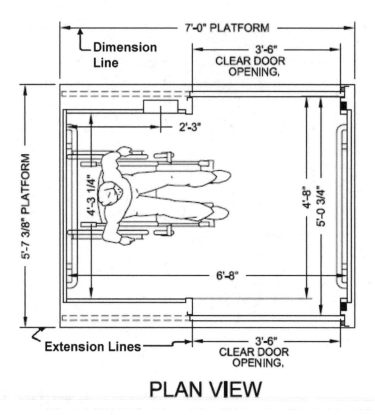

PLAN VIEW

Figure 3.6 Extension lines are typically used for dimensioning.

Phantom Lines

Phantom lines are thin, dark lines that consist of a long dash and two short dashes. They look like cutting-plane lines except that they are lighter. Phantom lines are used mainly to show alternative positions of fixtures, movable partitions, motion (e.g., alternative door swings) or future construction additions (as well as existing structures to be removed). They are also used to indicate repeated details and materials prior to machining.

Leader Lines and Arrowheads

Leader lines are fine lines terminating with an arrowhead or dot at one end to relate a note or callout to its feature. They are often drawn at an angle or straight from the principal lines on the drawing or in a free-curved manner to distinguish them easily from object lines.

Leader lines are used to label elements by connecting an object to a note or abbreviation or a dimension to the object it represents on a drawing (Figures 3.8A, B, and C).

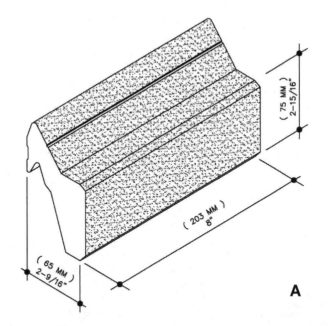

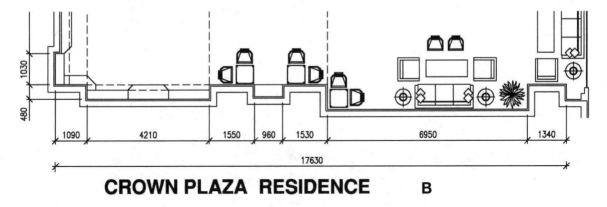

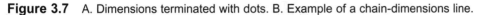

CROWN PLAZA RESIDENCE B

Figure 3.7 A. Dimensions terminated with dots. B. Example of a chain-dimensions line.

Cutting-Plane and Viewing-Plane Lines

These are very prominent broken lines (usually two dots and a dash) that are used to show the location of cutting planes for sectional views. Arrows on their ends indicate the direction in which the section is observed (Figure 3.9A). Lines or circular symbols are sometimes used at their ends to relate the cutting planes to their section views. Like property lines, these lines are normally heavier than any other lines on a drawing.

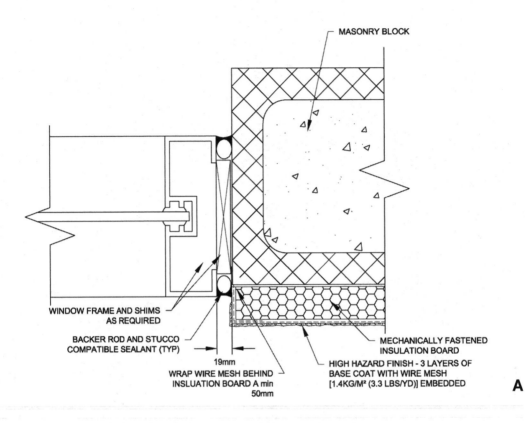

MASONRY BLOCK

WINDOW FRAME AND SHIMS
AS REQUIRED

BACKER ROD AND STUCCO
COMPATIBLE SEALANT (TYP)

19mm

WRAP WIRE MESH BEHIND
INSLUATION BOARD A min
50mm

MECHANICALLY FASTENED
INSULATION BOARD

HIGH HAZARD FINISH - 3 LAYERS OF
BASE COAT WITH WIRE MESH
[1.4KG/M² (3.3 LBS/YD)] EMBEDDED

A

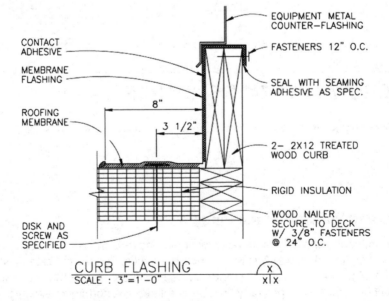

EQUIPMENT METAL
COUNTER—FLASHING

FASTENERS 12" O.C.

SEAL WITH SEAMING
ADHESIVE AS SPEC.

2- 2X12 TREATED
WOOD CURB

RIGID INSULATION

WOOD NAILER
SECURE TO DECK
W/ 3/8" FASTENERS
@ 24" O.C.

CONTACT
ADHESIVE

MEMBRANE
FLASHING

ROOFING
MEMBRANE

8"

3 1/2"

DISK AND
SCREW AS
SPECIFIED

CURB FLASHING
SCALE : 3"=1'-0"

x / x|x

B

Figure 3.8A,B   Examples of leader lines terminating with arrowheads.

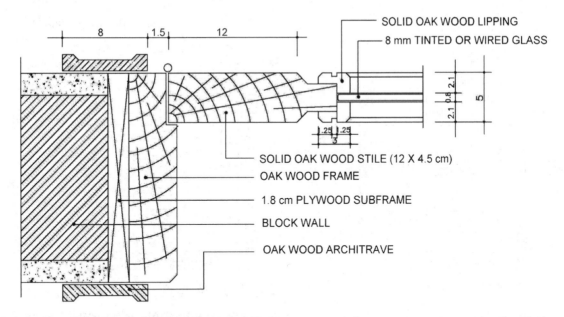

Figure 3.8C   Example of a leader line terminating with a dot.

Break Lines

These are fine straight lines with zig-zag-zig offsets to show a break or termination of a partial view or to omit portions of an object. Architects frequently use break lines to eliminate unimportant portions of details, thus allowing the important portions to be made larger. They are also used on detail and assembly drawings. Small break lines are often fine, ragged lines.

A straight, thin line made with freehand zigzags is used for long breaks, a thick freehand line for short breaks, and a jagged line for wood parts. Special breaks may be used to show cylindrical and tubular parts and to indicate that an end view is not shown; otherwise, the thick break line is typically used. The type of break line normally associated with architectural drafting is the long break line. Break lines are used to terminate features on a drawing when the extent of the feature has been clearly defined.

Contour Lines

These are fine lines that are used mainly to delineate variations in a site's elevation. If a site is fairly level, there will be few if any contour lines on the drawing, whereas if it has a significant slope, it is likely that the plot plan will show a number of contour lines. Sometimes a model is used to depict the topography of a site (Figure 3.11).

Property Lines

Property lines define the boundaries of a property. These lines are normally heavier than other lines on site or plot plans.

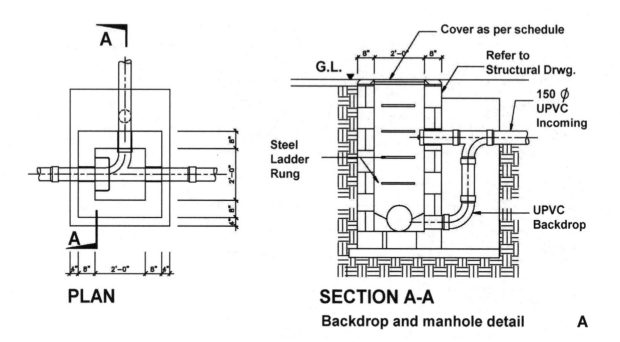

PLAN

SECTION A-A

Backdrop and manhole detail

A

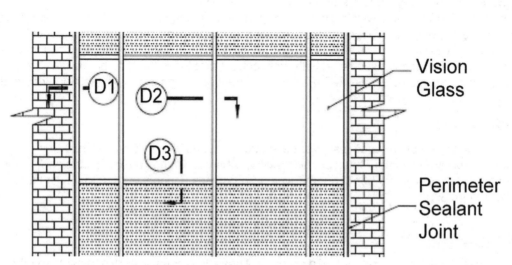

B

Figure 3.9A,B   Examples of cutting-plane lines.

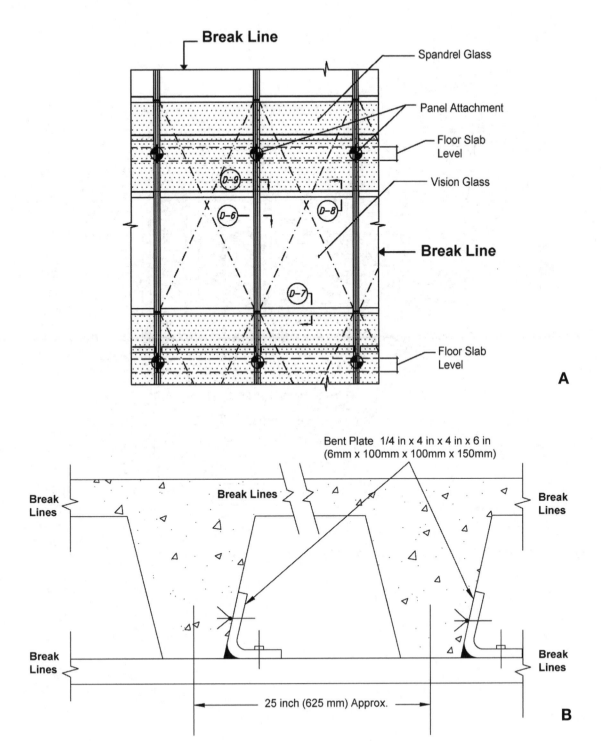

Figure 3.10A,B Examples of the use of break lines.

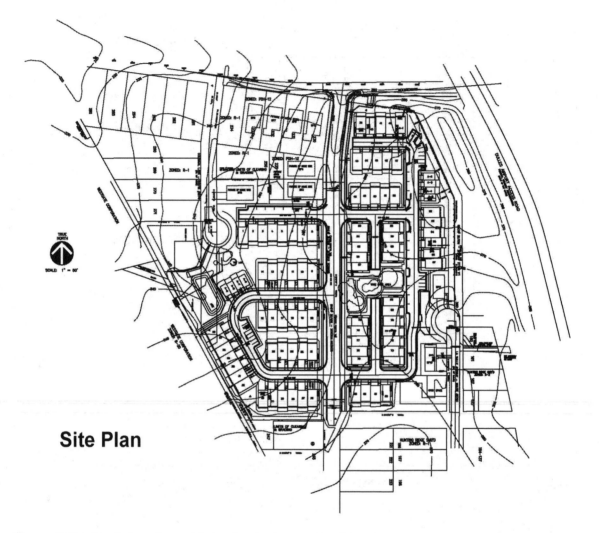

Site Plan

Figure 3.11   Illustration of a model showing the topography of the site.

4

Understanding Dimensions

4.1   TYPES OF DIMENSIONS

A dimension is a numerical value expressed in an appropriate unit of measure. It is indicated on drawings along with lines, arrows, symbols, and notes to define the size and specification of an object.

Dimensioning is thus a process of incorporating numerical values onto a drawing to enable the sizing of different elements and the location of parts of a building or object. Drawings should be fully dimensioned so that a minimum of computation is required and all the parts can be built without having to scale the drawings to determine an object's size. Duplication of dimensions should whenever possible be avoided unless it adds clarity. Figure 4.1 shows examples of different types of dimensions.

Distances may be indicated with either of two standardized forms of dimension: linear and ordinate. Linear Dimensions are used for displaying and measuring length along the X or Y axis. As the name suggests it can only be aligned along the X or Y axis. These dimensions are typically used to show absolute lengths along the X or Y axis (Figure 4.1A). With linear dimensions, there are two parallel lines, called "extension lines," that are spaced at the distance between two features and shown at each of the features (Figure 4.2A). The dimension line consists of a line perpendicular to the extension lines; it is shown between the extension lines and terminates at these lines typically with arrows, slashes, or dots. (For examples of different endpoints see Figures 4.2A, 4.2B, and 3.8C.) The distance is indicated numerically at the midpoint of the dimension line, either adjacent to it or in a gap provided for it.

Ordinate dimensions are used for measuring the length along any X or Y axis and displaying length as a text with the use of a leader. Generally, these dimensions are used to show lengths of entities using leader lines (Figure 4.1B).

Radial Dimensions are used for measuring the radius of arcs, circles, and ellipses and displaying it with a leader line. Radial dimensions often use an "R" followed by the value for the radius (Figure 4.3); Diametral dimensions generally use a circle with a forward-leaning diagonal line through it, called the diameter symbol, followed by the value for the diameter. A radially aligned line with arrowhead pointing to the circular feature, called a leader, is used in conjunction with both diametral and radial dimensions. All types of dimensions are typically composed of two parts: the nominal value, which is the "ideal" size of the feature, and the tolerance, which specifies the amount that the value may vary above and below the nominal.

The Metric System

The metric system originated in France in the 1790s as an alternative to the traditional English units of measurement. It was intended to standardize the units of measurement to assist in the expansion of trade and commerce throughout continental Europe. Today the majority of countries around the world have adopted an exclusively metric system of measurement. The United States, however, remains one of the few countries that have not adopted this system. It is expected that the construction industry will eventually convert to the metric system. Some of its many advantages include the elimination of fractions on drawings, simpler calculations, and international uniformity.

Dimension Conventions

There are a number of conventions that relate to dimensioning that the drafter and blueprint reader should be aware of:

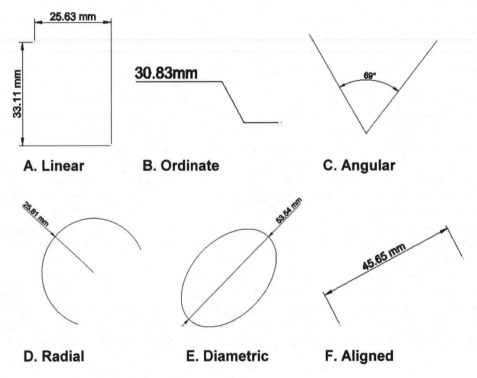

Figure 4.1   A diagram showing the different types of dimensions.

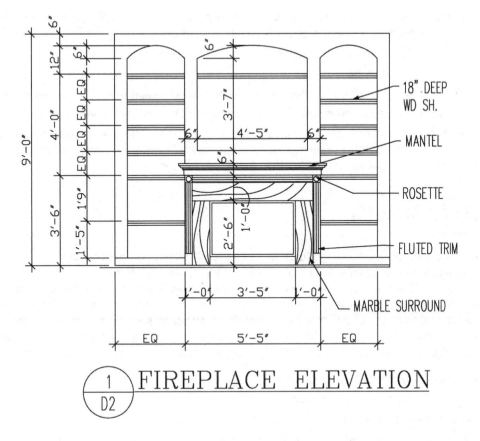

Figure 4.2A A drawing of a fireplace elevation showing various dimensions including a chain dimension that is outside the smaller dimensions.

1. Dimension lines are generally continuous with the number being centered and slightly above the line. Alternatively, the dimension line is broken (typically on engineering drawings), and the numerical dimension is positioned in the break.

2. Both feet and inches need to be shown (e.g., 10 feet, 6 inches). Even if the dimension has no inches, the zero remains as part of the designation (e.g., 10 feet, 0 inches).

3. When dimensions are small (less than 1 foot), only inches are used.

4. Overall or chain dimensions are placed outside the smaller dimensions (Figure 4.1).

5. Arrowheads, dots, or slashes are placed at the extremities of dimension lines to indicate the limits of the dimension (Figures 4.2A, 4.2B, and 3.8C).

6. Dimensions reflect actual building sizes irrespective of the scale used.

7. Where grid lines are used on a drawing (such as on a modular drawing), it is not necessary to dimension all the grids. Only one grid is normally dimensioned.

8. Similar dimensions need not be duplicated on the various views.

9. Door dimensions may be indicated in the floor-plan symbols or given in a door schedule.

10. Curved or angular leaders are often used to eliminate confusion with other dimension lines.

11. Dimensions showing location are given to centerlines of doors and windows on plan views.

12. To avoid costly mistakes, it is strongly advised to study all the information available before making a determination regarding the dimensions in question. Check that there is no conflict or discrepancy with the information shown in other views.

4.2 USING SCALES

Because building projects are too large to be drawn to actual size on a sheet of paper, everything needs to be drawn proportionately smaller to fit. The views in a set of blueprints are normally drawn at a reduced scale by the drafter. Scale notations are given with each drawing. The scale of a drawing is usually noted in the title block or just below the view when it differs in scale from that given in the title block. When drawing buildings to a specific scale, the drawings retain their relationship to the actual size of the building or object using a simple ratio. This practice of using an accepted standard ratio between full size and what is seen on the drawings is referred to as scale. The scale used on construction drawings depends on:

1. The actual size of the building or object.

2. The amount of detail required to be shown.

3. The size of sheet selected for the drawing.

4. The amount of dimensioning and notation needed.

5. Common practices that regulates certain scales (e.g., normal residential structures are generally drawn at ¼ inch = 1 foot, 0 inches).

Scales require distinct machine-divided markings coupled with sharp edges to achieve accurate measurements. The shape of the scales may be triangular, flat, or beveled, and they come in various sizes: they are generally about 12 inches in length although 6-inch scales are also available (Figure 4.4). The three commonly used drawing scales for reading construction drawings and for the development of plans are: the architectural scale, the engineering scale, and the metric scale. Many architects and engineers include a statement on their drawings stating that "dimensions shown on the plans take precedence over scaling." Some of the problems associated with scaling a drawing include size changes due to reproduction methods, last-minute forced dimension changes, and varying degrees of drafting accuracy depending on the skills of the drafter.

Architect's Scale

Architectural scales are normally flat or triangular in shape and come in different lengths, the 12-inch (30-cm) triangular shape being the most popular. All three sides of the triangle scale (except those with a 12-inch scale) contain two scales on each usable surface. Each of these scales uses the full

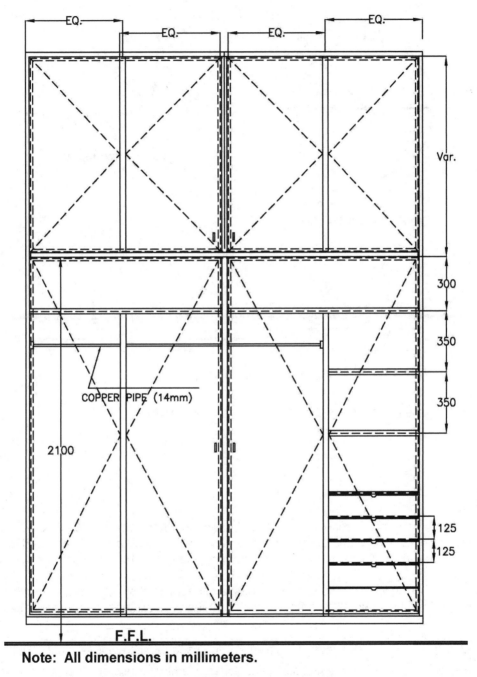

Note: All dimensions in millimeters.

WARDROBE ELEVATION

Figure 4.2B   A drawing showing dimension lines terminating at the extension lines with arrows at the endpoints. Figure 4.1 shows slashes at the endpoints, and Figure 3.8C shows dots at its end points.

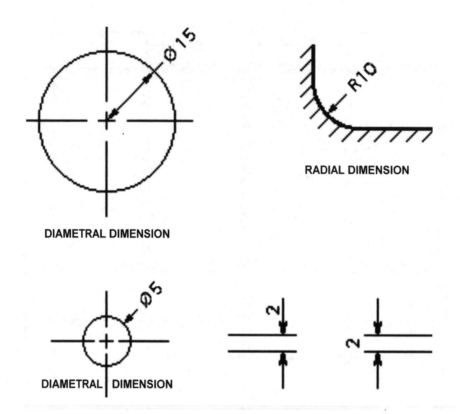

Figure 4.3   A drawing with examples of diametral and radial dimensions.

length of the instrument: one is read from left to right and the other from right to left. Likewise, a scale is usually either half or double that of the scale it is paired with. For example, if one end is a 1/8-inch scale, the opposite end is a 1/4-inch scale; the opposite end of a 1 1/2-inch scale would be a 3-inch scale.

The scales are designed to measure feet, inches, and fractions of an inch. Below are the most common scales found on the triangular architect's scale with their approximate metric equivalent in parentheses:

- 1/32 inch = 1 foot (1/400 metric equivalent—often used for site plans—actual 1/384)
- 1/16 inch = 1 foot (1/200 metric equivalent—often used for large projects and small site plans—actual 1/192)
- 1/8 inch = 1 foot (1/100 metric equivalent—actual 1/96)
- 1/4 inch = 1 foot (1/50 metric equivalent—actual 1/48)

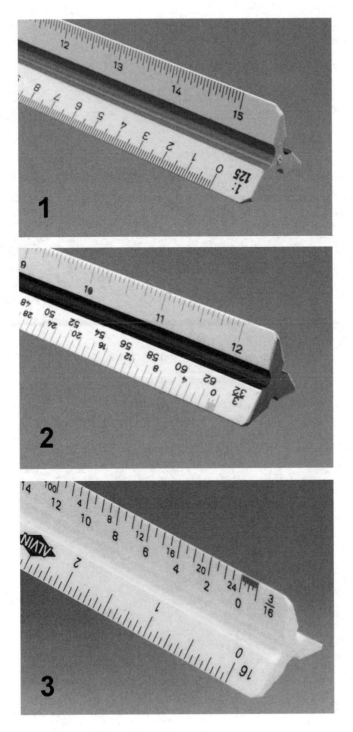

Figure 4.4A An Illustration showing different triangular scales.

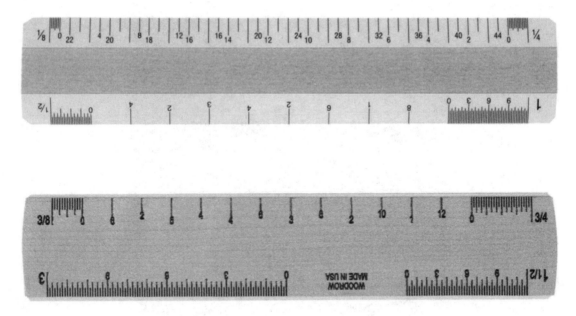

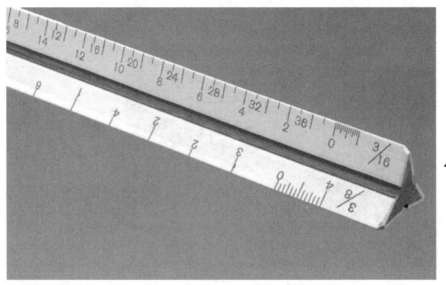

Figure 4.4B   An Illustration of 6-inch scales.

END VIEW

Figure 4.4C   A typical architect's scale showing end view.

- 3/8 inch = 1 foot (no precise metric equivalent—actual 1/32)
- 1/2 inch = 1 foot (1/20 or 1/25 metric equivalent—actual 1/24)
- 3/4 inch = 1 foot (no precise equivalent—actual 1/16)
- 1 inch = 1 foot (one-twelfth full size—approximate equivalent 1/10)
- 1 1/2 inches = 1 foot (one-eighth full size)
- 3 inches = 1 foot (one-quarter full size)

As an example, when used on a typical floor plan that is 1/8-inch scale, each 1/8 inch on the drawing represents 1 foot of actual size. The same applies for a 1/4-inch scale in that each 1/4-inch segment on the drawing represents 1 foot of actual size. The same approach applies to other scales.

There is no strict convention for which scale is used on which drawings, although certain parts of a set of drawings are traditionally drawn to certain scales. For example, most floor plans and elevations are in 1/8- or 1/4-inch scale, depending on the size of the building and sheet. For residential structures, the 1/4-inch scale is usually used, roughly equivalent to 1:50 in metric scale (Figure 4.5). For large commercial buildings, smaller scales may be used. Exterior elevations are often drawn to 1/4-inch scale. Sections and cross-sections may be drawn to 1/4-, 1/2-, or 3/4-inch scale if the section is complex. Depending on the amount of information presented, construction details can vary from 1/2- to 3-inch scale and even full-size scale for certain millwork details (Figure 4.6).

Civil Engineer's Scale

The engineer's scale is typically used to measure distance on site and land-related plans such as construction site plans, among other uses. Land measurement on site and plot plans differs slightly from measurement on building structures. It is conventional to show land measurement in feet and decimal parts of a foot, carried out to three places (55.478 feet) without the use of inches.

Most engineering scales are physically very similar and are based on the same principles as the architectural scales, except that measurements are divided into tenths, twentieths, etc., rather than halves, quarters, and eighths. The engineer's scale has six scales: 10, 20, 30, 40, 50, and 60. Other specialty scales are divided into even small increments such as 100. The 10 scale refers to 10 feet per inch; the 20 scale is 20 feet per inch, and so on.

Sometimes the architect or engineer may include a detail strictly for visual clarification. These details are labeled "NTS" (meaning "not to scale"). This basically means that the detail is for illustration purposes only and not for extracting quantities and measurements.

Metric Scales

According to the American National Standards Institute (ANSI), the International System of Units (SI) linear unit commonly used on drawings is the millimeter. The American National Metric Council, in its publication American Metric Construction Handbook, recommends the following with reference to metric drawings:

1. Architectural working drawings are to be dimensioned in millimeters (mm) and meters (m).
2. Plot plans and site plans are to be dimensioned in meters (m) or possibly kilometers (km), depending upon the scale, with accuracy to only three decimal places.

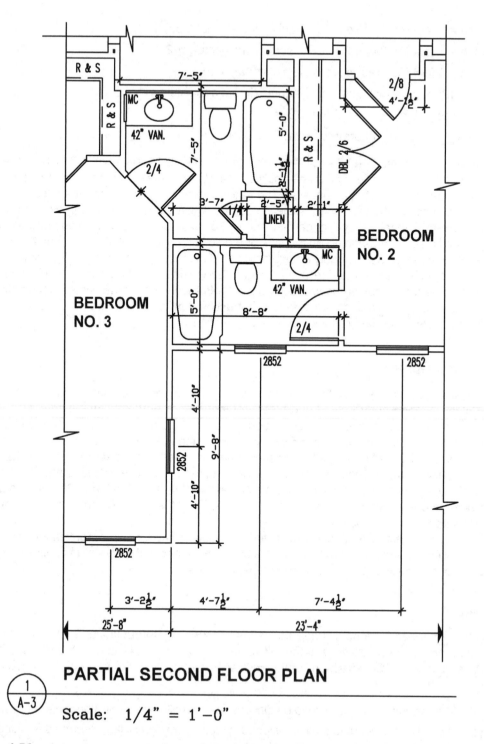

PARTIAL SECOND FLOOR PLAN

1
A-3

Scale: 1/4" = 1'-0"

Figure 4.5A   A drawing showing scales in imperial (1/4" = 1'-0").

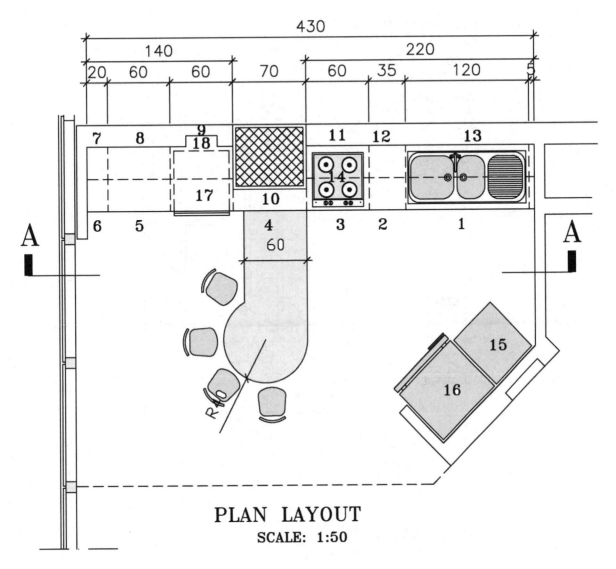

PLAN LAYOUT
SCALE: 1:50

Figure 4.5B A drawing showing scales in metric (1:50).

3. No periods are used after the unit symbols.

4. Scale on drawings is to be shown by a ratio (1:1, 1:10, 1:50, etc.).

The preferred method of metric dimensioning is called a soft conversion, in which common metric modules are used. Thus, 2-x-4 lumber is 40 x 90 mm using soft conversion. This method is far more convenient when drawing plans and measuring in construction. When measuring plywood thickness in met-

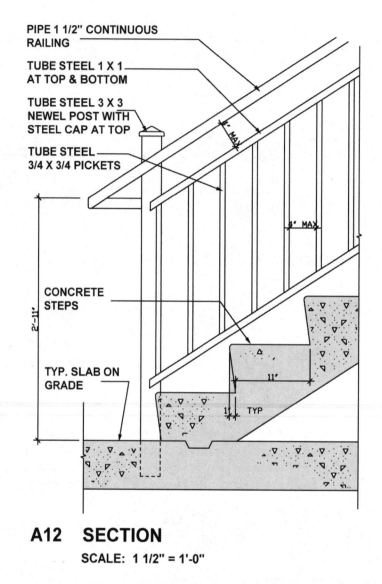

PIPE 1 1/2" CONTINUOUS RAILING

TUBE STEEL 1 X 1 AT TOP & BOTTOM

TUBE STEEL 3 X 3 NEWEL POST WITH STEEL CAP AT TOP

TUBE STEEL 3/4 X 3/4 PICKETS

CONCRETE STEPS

TYP. SLAB ON GRADE

A12   SECTION

SCALE: 1 1/2" = 1'-0"

Figure 4.6A   A drawing of a stair detail drawn to a scale of 1 1/2" = 1'-0".

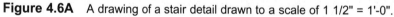

ric units, 5/8 inch thick equals 17 mm and 3/4 inch thick equals 20 mm. Using the same method, the length and width of sheet of plywood change from 48 x 96 inches to 1200 x 2400 mm. In countries that have adopted the metric system, the dimensioning module is 100 mm, whereas in the United States architectural design and construction modules normally used are 12 or 16 inches. Thus, construction members that are spaced 24 inches on center (O.C.) in the United States translate into 600 mm O.C. spacing in Canada or the United Kingdom.

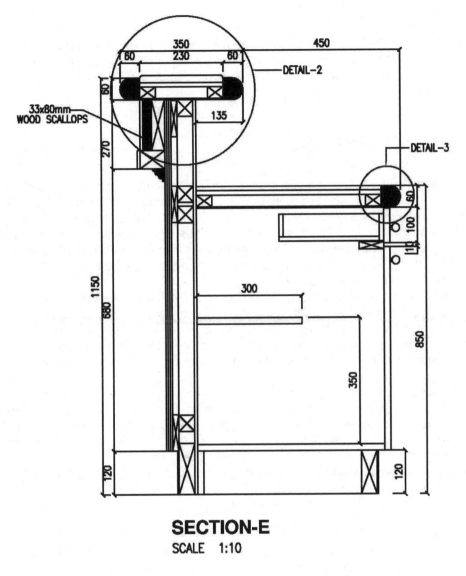

SECTION-E
SCALE 1:10

Figure 4.6B A drawing of a counter detail drawn to a metric scale of 1:10.

When reading metric dimensions on a drawing, all dimensions within dimension lines are normally in millimeters, and the millimeter symbol (mm) is omitted unless more than one dimension is quoted, in which case the symbol (mm) is included after the last dimension. Drawings produced in metric, such as floor plans, elevations, and sections, are normally drawn to a scale of 1:50 or 1:100, as opposed to the 1/4 inch = 1 foot, 0 inches or 1/8 inch = 1 foot, 0 inches scale used in the imperial system. Figure 4.5 shows examples of drawings drawn to 1:50 and 1/4 inch = 1 foot, 0 inches scales.

Indication of Scale

The scale should normally be noted in the title block of a drawing. When more than one scale is used, they should be shown close to the views to which they refer, and the title block should read "scales as shown." If a drawing uses predominantly one scale, it should be noted in the title block together with the wording "or as shown."

4.3   LINEAR DIMENSIONS

Linear dimensions, or linear units, are used to measure the distance between two points. Since two points define a line, the units of distance are sometimes called "linear" units or dimensions. In the metric system linear dimensions are generally in millimeters. To avoid having to specify "mm" after every dimension, a label such as "all dimensions in mm" or "unless otherwise stated all dimensions are in mm" is usually contained in the title block. If the dimension is less than one, a leading zero should be used before the decimal point: e.g., 0.5. Linear units can also be in centimeters and inches, meters and feet, or kilometers and miles, to name a few.

In architectural and engineering drawing, the most important dimensions determine subsequent dimensions, providing a dimension standard. Thus, if a wall is dimensioned to its center first, all following dimensions using this wall as a reference point should be dimensioned at the center.

Projection lines are used to indicate the extremities of a dimension. Dimension lines are used to label a particular dimension. Thin lines are used for both projection lines and dimension lines (Figure 4.7).

Area dimensions are two-dimensional and measure area. They are often but not always expressed as squares of linear dimensions: square inches or inches squared (in2), square feet or feet squared (ft2), and square meters or meters squared (m^2). A rectangle that is 8 feet long by 4 feet wide, for example, has an area of 32 square feet (8 linear feet times 4 linear feet). Examples of units of area that are not a square of a linear unit are the acre and hectare. There are others.

Volume dimensions are three-dimensional and are expressed as the cube of linear units. A cube that measures 2 feet on each edge has a volume of 2 x 2 x 2 = 8 cubic feet.

Two methods of dimensioning are in common use:

- Unidirectional: The dimensions are written horizontally.
- Aligned: The dimensions are written parallel to their dimension lines. Aligned dimensions should always be readable from the bottom or the right of the drawing.

When several dimensions make up an overall length, the overall dimension may be shown outside the component dimensions. When specifying an overall dimension, one or more noncritical component dimensions must be omitted (Figure 4.8A).

When all of the component dimensions must be specified, an overall length may still be specified as an auxiliary dimension. Auxiliary dimensions are never toleranced and are shown in brackets (Figure 4.8B).

Dimensions that are not drawn to scale are underlined (Figure 4.9A). When a dimension line cannot be completely drawn to its normal termination point, the free end is terminated in a double arrowhead (Figure 4.9B).

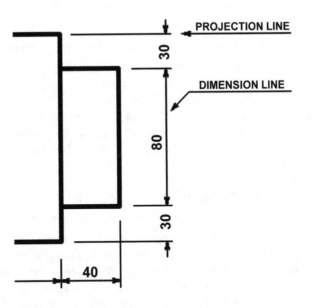

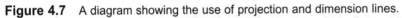

Figure 4.7   A diagram showing the use of projection and dimension lines.

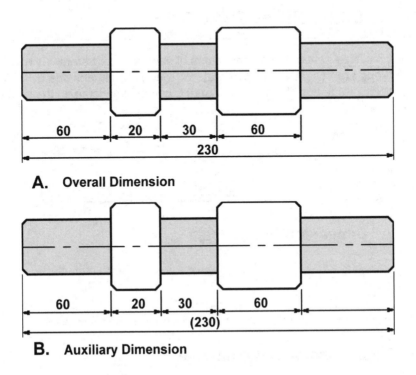

A.   Overall Dimension

B.   Auxiliary Dimension

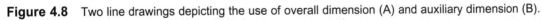

Figure 4.8   Two line drawings depicting the use of overall dimension (A) and auxiliary dimension (B).

4.4   ANGULAR DIMENSIONS

Angular dimensions are used for measuring and displaying inside and outside angles. These angles are measured in degrees (Figure 4.1C). Where greater accuracy is required, the degree is divided into 60 minutes and the minute into 60 seconds. For reference, a circle has 360 degrees. Inch and millimeter measurements both measure angles in degrees. Normally, angular dimensions are specified in decimal degrees, degrees and minutes, or degrees, minutes, and seconds (For example, 24.5°, 24° 30', 24° 30' 16"). When the angle is less than 1 degree, a leading zero should be used (for example, 0.5°, 0° 30').

4.5   REFERENCE DIMENSIONS

A reference dimension is used only for information purposes. The indication "REF" should be noted immediately under or beside the dimension. Normally it is either a duplication of a dimension or the accumulated value of other dimensions. Reference dimensions are typically located within parentheses. Reference dimensions do not have tolerances and should not be used to manufacture or inspect parts. The purpose of reference dimensions is to provide additional information. Some prints may not show reference dimensions within parentheses and instead show "REF" or "REFERENCE" after the dimension (Figure 4.10).

4.6   NOMINAL SIZE AND ACTUAL SIZE

In architecture, "nominal size" refers to the dimensions of sawn lumber before it is dried or surfaced.
Nominal dimensions are essentially approximate or rough-cut dimensions by which a material is generally called or sold in trade but which differs from the actual dimension. In the lumber trade, for ex-

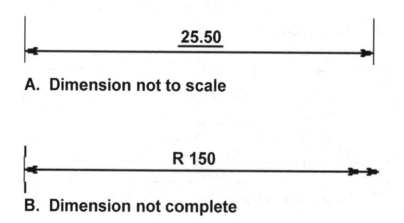

A. Dimension not to scale

B. Dimension not complete

Figure 4.9   A diagram showing "not to scale" dimension (A) and "not complete" dimension (B).

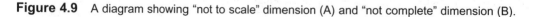

ample, a finished (dressed) 2 x 4 piece is less than 2 inches thick and less than 4 inches wide. The nominal size is usually greater than the actual dimension.

When sawn lumber (nominal size) is seasoned, dressed, or planed, the size becomes smaller; this is the "actual size." For example, a piece of 2 x 4 lumber (100 x 50) may become approx 11/2 x 31/2 inches (90 x 45) when it has been dressed or planed. The actual size therefore refers to the minimum acceptable size after it has been dressed and seasoned. A nominal 2 x 4 can have a minimum actual size of 1.5 x 3.5 inches. When referring to a specific piece of lumber, the nominal size is used. Figure 4.11 shows examples of nominal and actual sizes of lumber and boards.

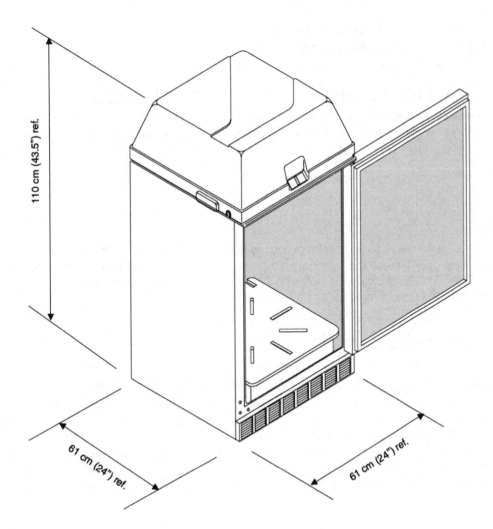

Figure 4.10 A drawing showing the use of reference dimensions.

NOMINAL SIZE	ACTUAL SIZE	NOMINAL SIZE	ACTUAL SIZE
2 X 2	1 1/2 x 1 1/2	1 X 2	3/4 X 1 1/2
2 X 3	1 1/2 x 2 1/2	1 X 3	3/4 X 2 1/2
2 X 4	1 1/2 x 3 1/2	1 X 4	3/4 X 3 1/2
2 X 6	1 1/2 x 5 1/2	1 X 5	3/4 X 4 1/2
2 X 8	1 1/2 x 7 1/4	1 X 6	3/4 X 5 1/2
2 X 10	1 1/2 x 9 1/4	1 X 8	3/4 X 7 1/4
2 X 12	1 1/2 x 11 1/4	1 X 10	3/4 X 9 1/4
4 X 4	3 1/2 x 3 1/2	1 X 12	3/4 X 11 1/4
4 X 6	3 1/2 x 5 1/2		
4 X 10	3 1/2 x 9 1/4		
6 X 6	5 1/2 x 5 1/2		

A. FRAMING LUMBER   B. BOARDS

Figure 4.11 A table showing examples of nominal and actual sizes for framing lumber and boards.

In design engineering, size is used for purposes of general identification; the actual size of a part will be approximately the same as the nominal size but need not necessarily be exactly the same. For example, a rod may be referred to as 1/4 inch, although the actual dimension on the drawing is 0.2495 inch; in this case 1/4 inch is the nominal size.

4.7 TOLERANCES

The phrase "geometric dimensioning and tolerancing" (GDT) refers to the allowable variation of a dimension. It represents the difference between the maximum and minimum acceptable limits. All dimensions except basic, reference, nominal, maximum, or minimum dimensions must have a tolerance. GDT is a method of specifying the functional geometry of an object.

Understanding the international engineering language of GDT is essential for communicating in today's highly competitive global marketplace. You are required to read and interpret GDT symbols, which provide detailed information about the function and relationship of the various part features.

5

Types of Views

5.1 INTRODUCTION

Drawings are the main vehicle of communication in the construction industry, and in learning to read and interpret blueprints, it is necessary to develop the ability to visualize the object to be built. Orthographic projection principles are basic to all fields in the construction industry. In order to be capable of interpreting a drawing, one must be able to relate the different views.

It is often not possible to read a blueprint by looking at a single view; two or three views may be needed to correctly read and visualize the form. Figure 5.1 shows three example of objects requiring only two views to read. Figure 5.2 shows a drawing using three views. Whether sketching or drawing, the objective is the same. The goal is to communicate the necessary detail to the targeted audience, whether it is the builder, manufacturer, or client.

Without the ability to communicate, architects, and engineers cannot function in a team. Competency in drawing and sketching are essential communication tools for architects, engineers, and those involved in the building trades and manufacturing industries.

In the interpretation of complex objects, even three drawings are not usually adequate to convey all the necessary information. Additional special views may be required, including pictorials, auxiliary views, sections, and exploded views. A view of an object is technically known as a projection.

Pictorials are an ancillary category within orthographic projection. Pictorials show an image of an object as viewed from a skew direction in order to reveal all three directions (axes) of space in one picture. Orthographic pictorial instrument drawings are often used to approximate graphical perspective projections, but there is attendant distortion in the approximation. Because pictorial projections innately have this distortion, great liberties are often taken for economy of effort and best effect. Pictorials are discussed later in this chapter.

5.2 ORTHOGRAPHIC (MULTIVIEW) DRAWINGS AND PROJECTIONS

For many years architects and engineers have utilized a system known as orthographic projection to accurately represent three-dimensional objects graphically on paper. In recent years the term "multiview

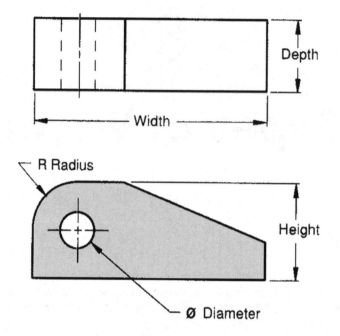

Figure 5.1A   An example of a simple drawing of objects that essentially requires only two views to read.

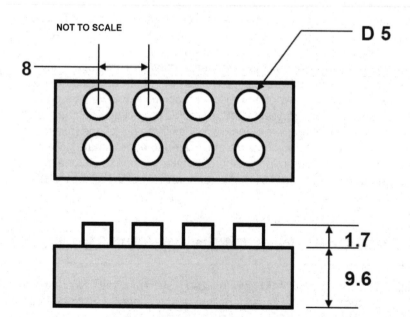

Figure 5.1B   An example of a simple drawing of objects that essentially requires only two views to read.

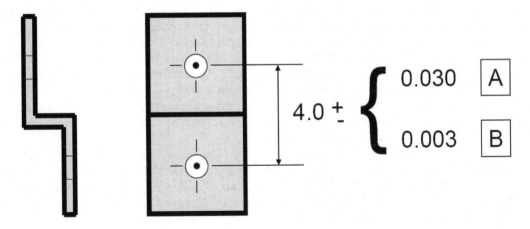

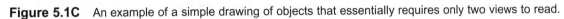

Figure 5.1C An example of a simple drawing of objects that essentially requires only two views to read.

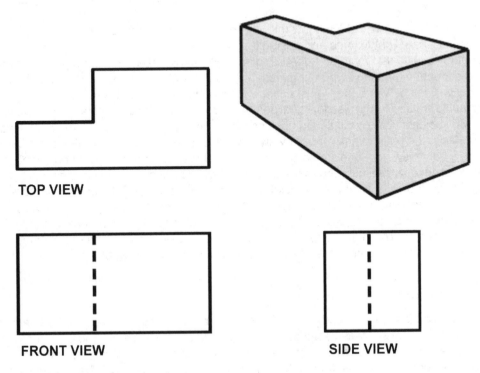

Figure 5.2 A drawing of an object requiring three views to interpret correctly.

drawing" has come into general use, indicating that more than one view is used to illustrate an object, but the terms are essentially synonymous. "Orthographic" comes from the Greek word for "straight writing (or drawing)." Orthographic projection shows the object as it looks from the front, right, left, top, bottom, or back, and different views are typically positioned relative to each other according to the rules of either first-angle or third-angle projection. Ortho views depict the exact shape of an object seen from one side at a time as you are looking perpendicularly to it without showing any depth.

A single view of an object is rarely adequate to show all necessary features. Figure 5.3 is an example of orthographic projection showing the six principal views used by architects and engineers in construction and industrial drawings.

Common types of orthographic drawings include plans, elevations, and sections. The most obvious attribute of orthographic drawing is its constant scale—that is, all parts of the drawing are represented without foreshortening or distortion, retaining their true size, shape, and proportion. Thus, in an orthographic drawing, a window shown to be 8 feet wide by 4 feet high will always be drawn at this size, no matter how far it is from our viewpoint (Figure 5.4).

Plans are really orthographic views of an object as seen directly from above. Floor plans are the most common form of plan; they delineate the layout of a building. A floor plan is represented by a horizontal section taken through the building or portion of a building just above the windowsill level. In addition to the arrangement of rooms and spaces, floor plans need to show the location of various architectural elements such as stairs, doors, and windows and details such as wall and partition thickness. Generally, the greater the scale of a drawing, the more detail that it is expected to contain (Figure 5.5). Thus, a drawing at a scale of 1/4" = 1'0" will typically contain more information and show more detail than a drawing at a 1/8" = 1'0" scale. Likewise, a scale of 1:2 is greater than that of 1/4 inch = 1 foot, 0 inches. Other types of plans used in building construction may include site plans, which typically show the layout of a site; foundation plans. which show the building structure; and reflected ceiling plans, which are normally used to locate light fixtures and design features.

Two important rules that must be adhered to in orthographic drawing are the placement and alignment of views, depending on the type of projection to be used. These rules are discussed below. In addition, projection lines between the views must be aligned horizontally and vertically.

Orthographic (multiview) projection is a generally accepted convention for representing three-dimensional (3D) objects using multiple dimensions (2D) of the front, top, bottom, back, and sides of the object. In practice, the minimum number of views possible is used to describe all the details of the object. Usually, a front view, top, and single side view are sufficient and are oriented on the paper according to accepted convention. Figure 5.6 represents a multiview projection for a simple house. The projection clearly shows that it is a form of parallel projection, and the view direction is orthogonal to the projection plane. Isometric projection attempts to represent 3D objects using a single view. Instead of the observer viewing the object perpendicular to it, the object is rotated both horizontally and vertically relative to the observer. There are rules and conventions to guide the creation of both types of projections. Additionally, either of them can be supplemented with various types of dimensions.

First-Angle Projection

First-angle projection is the ISO standard and is used mostly in Europe and Asia. If we imagine projecting a 3D object into a transparent plastic cube, the main object surfaces are projected onto the cube's walls so that the top view is placed under the front view and the right view is placed at the left of the front view, a two-dimensional representation of the object is formed by "unfolding" the box and viewing all of the interior walls as is shown in Figure 5.7A.

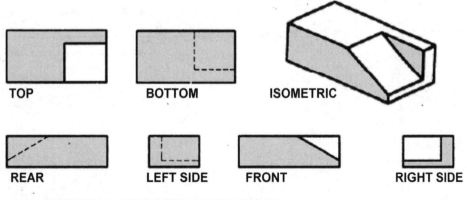

TOP   **BOTTOM**   **ISOMETRIC**

REAR   **LEFT SIDE**   **FRONT**   **RIGHT SIDE**

SIX PRINCIPAL ORTHOGRAPHIC VIEWS

Figure 5.3 An illustration of an object showing the six principal views in orthographic projection based on third-angle projection.

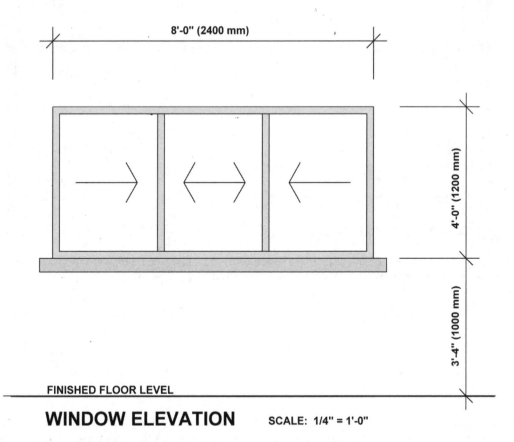

8'-0" (2400 mm)

4'-0" (1200 mm)

3'-4" (1000 mm)

FINISHED FLOOR LEVEL

WINDOW ELEVATION   SCALE: 1/4" = 1'-0"

Figure 5.4 An elevation of a window drawn in an orthographic format retaining its true size, shape, and proportion and showing no distortion or foreshortening.

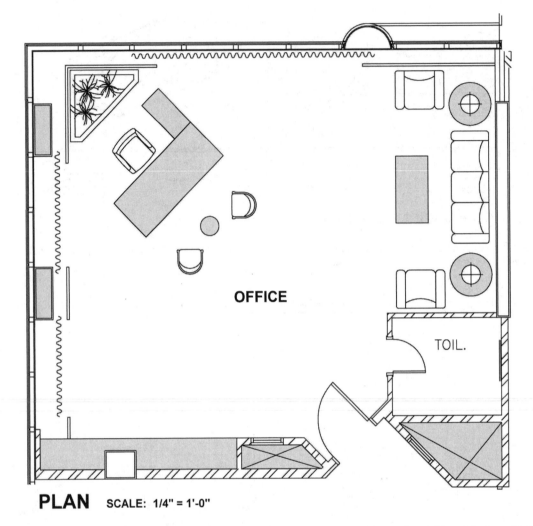

PLAN   SCALE: 1/4" = 1'-0"

Figure 5.5A   Design drawing plan of an office space drawn to a scale of 1/4 inch = 1 foot, 0 inches. Notice that there are no dimensions as would be found in a typical orthographic working drawing.

In first-angle projection, the object lies above and before the viewing planes; the planes are opaque; and each view is pushed through the object onto the plane furthest from it. Extending to the six-sided box, each view of the object is projected in the direction of sight of the object onto the interior walls of the box; that is, each view of the object is drawn on the opposite side of the box.

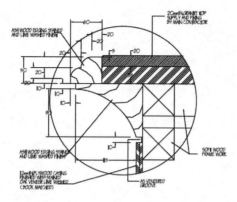

DETAIL 1 SCALE: 1:2

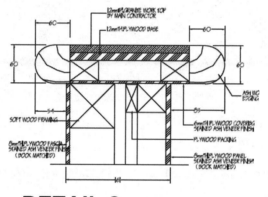

DETAIL 2 SCALE: 1:2

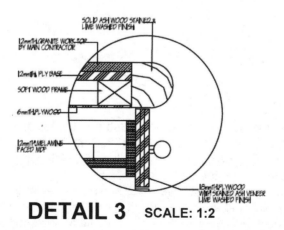

DETAIL 3 SCALE: 1:2

Figure 5.5B Counter detail drawing to a scale of 1:2. This drawing is to a greater scale and has more detail than the ¼ inches = 1 foot, 0 inches drawing.

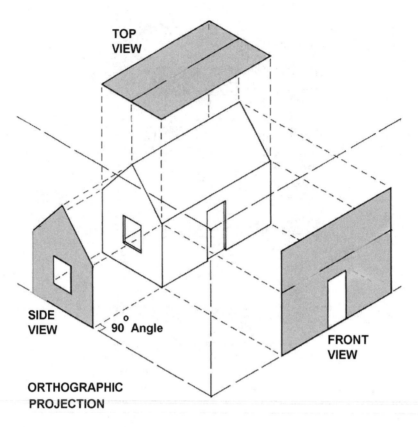

Figure 5.6 A multiview projection for a simple house. This is the format most used by architects and engineers.

Third-Angle Projection

In third-angle projection the left view is placed on the left and the top view is placed on the top (Figure 5.7 B and C). This type of projection is primarily used in the United States and Canada. It should be noted that not all views are necessarily used, and determination of which surface constitutes the front, back, top, and bottom varies depending on the projection used.

In third-angle projection, the object lies below and behind the viewing planes. The planes are transparent, and each view is pulled onto the plane closest to it. Using the six-sided viewing box, each view of the object is projected opposite to the direction of sight onto the exterior walls of the box; that is, each view of the object is drawn on the same side of the box. The box is then unfolded to view all of its exterior walls, as shown in Figure 5.7C.

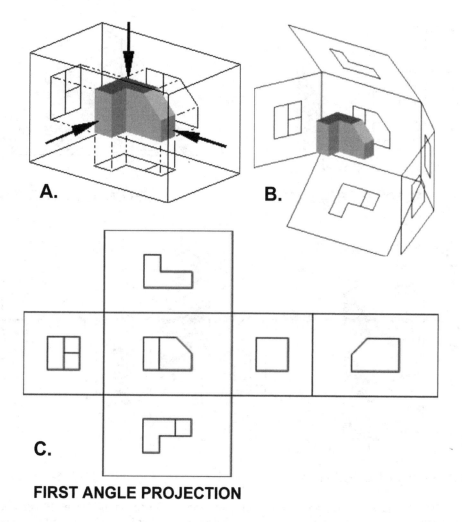

FIRST ANGLE PROJECTION

Figure 5.7A Views of an object being projected according to first-angle projection which is the ISO standard and is primarily used in Europe and Asia (source: Wikimedia Commons).

Single-View Drawings

One-view drawings are commonly used in the manufacturing industry to represent parts that are uniform in shape. These drawings are often supplemented with notes, symbols, and written information. They are normally used to describe the shape of cylindrical, cone-shaped, rectangular, and other symmetrical parts. Leaders are often used to relate a note to a particular feature, as in Figure 5.8. Thin, flat objects of uniform thickness are typically represented by one-view drawings.

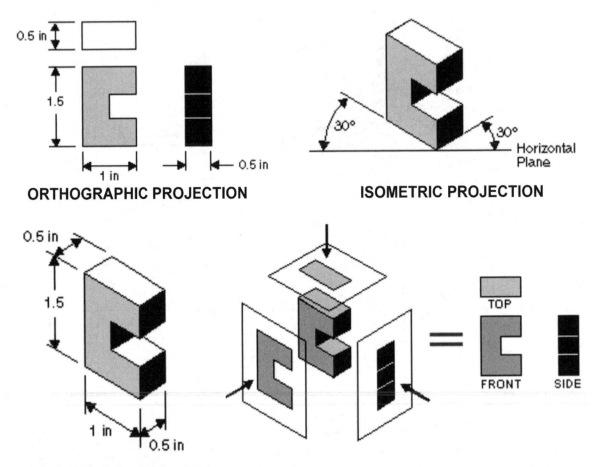

0.5 in

1.5

1 in

0.5 in

ORTHOGRAPHIC PROJECTION

30°

30°

Horizontal Plane

ISOMETRIC PROJECTION

0.5 in

1.5

1 in

0.5 in

TOP

FRONT

SIDE

ISOMETRIC PROJECTION

Figure 5.7B Examples of third angle projection that is primarily used in the United States and Canada. This type of projection produces two plan views and four side views. In third angle projection the left view is put on the left and the top view on the top.

Two-View Drawings

Simple, symmetrical flat objects and cylindrical parts, such as sleeves, shafts, rods, or studs, require only two views to show the full details of construction. The two views usually include the front view and either a right side or left side view or a top or bottom view.

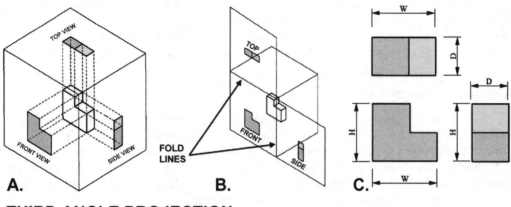

THIRD ANGLE PROJECTION

Figure 5.7C   More examples of third angle projection that is primarily used in the United States and Canada. This type of projection produces two plan views and four side views. In third angle projection the left view is put on the left and the top view on the top.

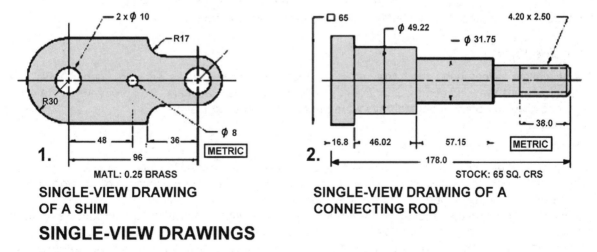

1. MATL: 0.25 BRASS

**SINGLE-VIEW DRAWING
OF A SHIM**

2. STOCK: 65 SQ. CRS

**SINGLE-VIEW DRAWING OF A
CONNECTING ROD**

SINGLE-VIEW DRAWINGS

Figure 5.8   Two examples of single-view drawings, which are commonly used in industry to represent parts that are uniform in shape.

Features on both sides of a centerline shown on a drawing are the same size and shape. These equal-length short, parallel lines are placed outside the drawing of the object on its centerline (Figure 5.9). A hidden detail may be straight, curved, or cylindrical. Whatever the shape of the detail and regardless of the number of views, it is represented by a hidden edge or invisible edge line.

Three-View Drawings

Regularly shaped flat objects that require only simple machining operations may often be adequately described with notes on a one-view drawing. However, when the shape of the object changes, portions are cut away, or complex machining or fabrication processes must be shown, the single view would normally be inadequate to describe the object accurately. The combination of front, top, and right side views represents the method most commonly used by drafters to describe simple objects (Figure 5.10). For building construction, other views would typically be needed.

The Front View

Before an object is drawn, it is examined to determine which views will best furnish the information required to construct the object. The surface shown as the observer looks at the object is called the front view. To draw this view, the drafter goes through an imaginary process of raising the object to eye level and turning it so that only one side can be seen. If an imaginary transparent plane is placed between the eye and the face of the object and parallel to the object, the image projected on the plane is the same as that formed in the eye of the observer.

The Top View

To draw a top view, the drafter goes through a process similar to that required to obtain the front view. However, in third-angle projection, instead of looking squarely at the front of the object, the view is seen from a point directly above it. When a horizontal plane on which the top view is projected is rotated so that it is in a vertical plane, the front and top views are in their proper relationship. In other words, the top view is always placed immediately above and in line with the front view.

The Side View

A side view is developed in much the same way that the other two views were obtained. That is, the drafter imagines the view of the object from the side that is to be drawn and then proceeds to draw the object as it would appear if parallel rays were projected upon a vertical plane.

Three-Dimensional Graphics

Depicting three dimensions on a flat piece of paper is a very important skill for designers, enabling them to communicate their ideas to other people. This is especially useful when showing your design to non-professionals such as managers and marketing personnel.

There are several tried and tested three-dimensional drawing systems used to produce a realistic representation of an object. Some techniques, such as isometric projection, are based on mathematical systems; others try to convey a larger degree of realism by applying perspective to the drawing. Amongst the methods covered in this tutorial are oblique, isometric, axonometric, and perspective drawing.

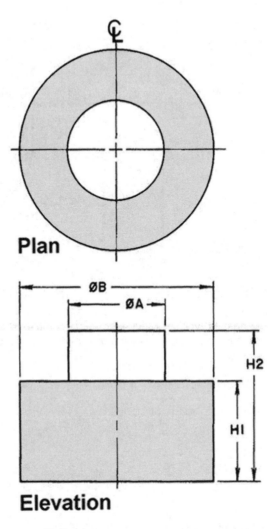

Figure 5.9 Two-view drawings are used mainly for simple, symmetrical flat objects and cylindrical parts. The views usually include a front view and either a right or left side view or a top or bottom view.

5.3 PROJECTION SYMBOLS

The International Organization for Standardization (ISO) recommends the incorporation of projection symbols on drawings that are produced in one country for use in many countries (Figure 5.11). The projection symbols are intended to promote the accurate exchange of technical information through drawings.

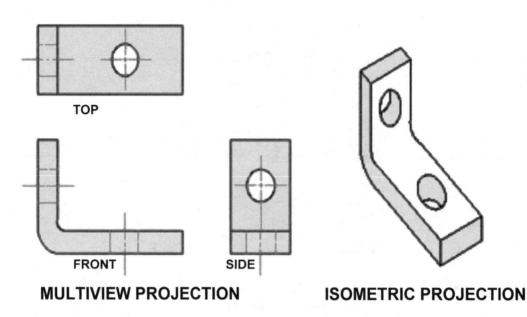

TOP

FRONT | **SIDE** |

MULTIVIEW PROJECTION   ISOMETRIC PROJECTION

Figure 5.10   Three-view drawings are typically required for more complex shapes. The combination of front, top, and right side views reflects the method normally used by drafters to describe these objects.

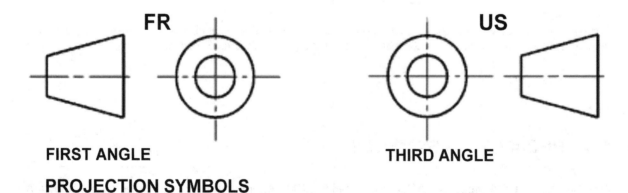

FR   **US**

FIRST ANGLE   **THIRD ANGLE**

PROJECTION SYMBOLS

Figure 5.11   The ISO projection symbols indicate whether a first-angle or third-angle projection is being used in a drawing.

As mentioned earlier, the United States and Canada use the third-angle system of projection for drawings, whereas other countries use a different system known as first-angle projection. The purpose of introducing the ISO projection symbols is to indicate that there is a continuously increasing international exchange of drawings for the production of interchangeable parts. Thus, the symbol indicates whether the drawing follows the third- or first-angle projection system.

The ISO projection symbol, the notation on tolerances, and information on whether metric and/or inch dimensions are used on the drawing should appear as notes either within the title block or adjacent to it. The designation of projection type is not always included with the symbol on a drawing. It is rarely used on architectural drawings but is usually included on engineering drawings.

When drawings are transferred from one convention to another, considerable confusion is encountered in drafting rooms and engineering departments. On engineering drawings, the projection angle is denoted by an international symbol consisting of a truncated cone, respectively, for first-angle and third-angle and whether the cone is to the right or left of the two concentric circles symbol. The 3D interpretation of the symbol can be deduced by envisioning a solid truncated cone standing upright with its large end on the floor and the small end upward. The top view is therefore two concentric circles ("doughnut"). In particular, the fact that the inner circle is drawn with a solid line instead of dashed lines designates this view as the top view, not the bottom view.

Both first-angle and third-angle projections result in the same six views; the difference between them lies in their arrangement around the box.

5.4   OBLIQUE DRAWINGS

Oblique projection is a simple form of parallel graphical projection used mainly for producing pictorial, two-dimensional images of three-dimensional objects. Oblique drawings are similar to isometrics except that the front view is shown in its true shape on the horizontal line—i.e., when drawing an object in oblique, the front view is drawn flat (Figure 5.12). Thus, it projects an image by intersecting parallel rays from the three-dimensional source object with the drawing surface. In oblique projection (as in orthographic projection), parallel lines from the source object produce parallel lines in the projected image. The projectors intersect the projection plane at an oblique angle to produce the projected image, as opposed to the perpendicular angle used in orthographic projection.

A 45-degree angle is the most commonly used to draw the receding lines from the front view, but other angles are acceptable. In an oblique sketch, circular lines that are parallel to the frontal plane of projection are drawn at their true size and shape. Hence, circular features appear as circles and not as ellipses. This is the main advantage of the oblique sketch. The three axes of the oblique sketch are drawn at the horizontal, vertical, and a receding angle that can vary from 30 to 60 degrees.

Whereas an orthographic projection is a parallel projection in which the projectors are perpendicular to the plane of projection, an oblique projection is one in which the projectors are not perpendicular to the plane of projection. With oblique projection all three dimensions of an object can be shown in a single view.

Oblique drawing is a primitive form of 3D drawing and the easiest to master. It is not a true 3D system but a two-dimensional view of an object with contrived depth. Instead of drawing the sides full size, they are only drawn at half the depth, creating a suggested depth that adds an element of realism to the object. Even with this contrived depth, oblique drawings look very unconvincing to the eye. In Figure 5.13 the side views are drawn at a 45-degree angle. In oblique projection the side views are typically fore-

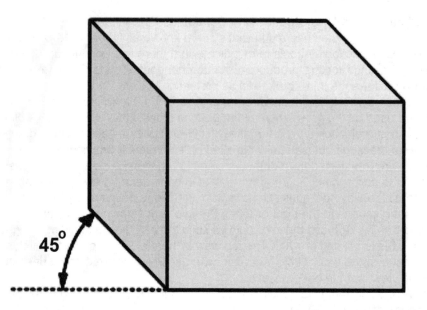

Figure 5.12 Technique for producing an oblique drawing. The front view is shown in its true form on the horizontal line and should be drawn flat.

shortened to provide a more realistic view of the object. To foreshorten the side views, the object's side measurements are normally halved. In this case, the sides are 50 mm (2 inches) long, but they have been drawn at 25 mm (1 inch). Because the oblique drawing is not realistic, it is rarely used by professional architects and engineers.

5.5 AUXILIARY VIEWS

Occasionally we find surfaces or features on drawings that are oblique to the principal planes of projection and still are shown in their true shape. Other features in modern construction are also designed at various angles to the principal planes of projection. To show these features or surfaces in true shape for accurate description, auxiliary views are used. Auxiliary views are appropriate to obtain a true size view; similar techniques to standard views unfolding about an axis are used. Auxiliary views are usually partial views and show only the inclined surface of an object. In Figure 5.14 the true size and shape of the object are shown in the auxiliary views of the angular surface. An auxiliary view is similar to an orthographic view except that it is projected to a plane parallel to the auxiliary surface and not to the customary orthographic planes. It is thus drawn at an angle to best view an object but not one of the primary or-

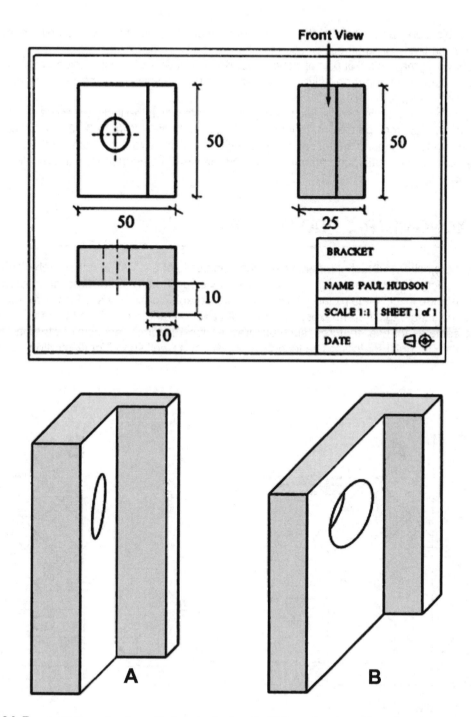

Figure 5.13A,B   A. Oblique drawing with foreshortening. B. Oblique drawing with no foreshortening. Notice how the circle looks elongated (source: Paul Hudson).

thographic projection views. The resulting view then reflects the true shape of the oblique feature while eliminating much time-consuming projection for the drafter.

Auxiliary views may be full or partial views. Rounded surfaces and circular holes, which are distorted in the regular views and appear as ellipses, appear in their true sizes and shapes in an auxiliary view, as can be seen in Figure 5.14B.

Auxiliary views are named according to the position from which the inclined face is seen. For example, the auxiliary view may be a front, top, bottom, left, or right view (Figure 5.15). In drawings of complex parts involving compound angles, one auxiliary view may be developed from another auxiliary view. The first auxiliary view is called the primary view, and the views developed from it are called secondary auxiliary views.

5.6 AXONOMETRIC PROJECTION

Axonometric projection is a technique used in orthographic pictorials. Within orthographic projection, axonometric projection shows an image of an object as viewed from a skew direction in order to reveal more than one side in the same picture, unlike other orthographic projections, which show multiple views of the same object along different axes. Because with axonometric projections the scale of distant features is the same as for near features, such pictures will look distorted, especially if the object is mostly composed of rectangular features. The technique, however, is well suited for illustration purposes.

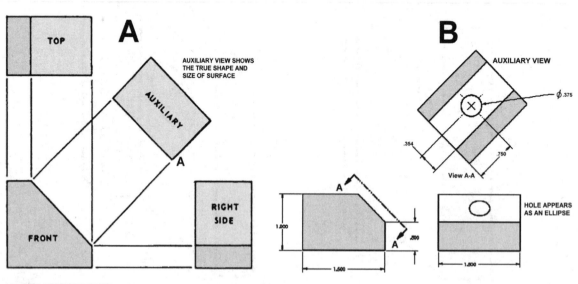

Figure 5.14A,B Two examples of auxiliary-view projections. The illustrations show that the auxiliary views are not one of the primary views of the orthographic projection.

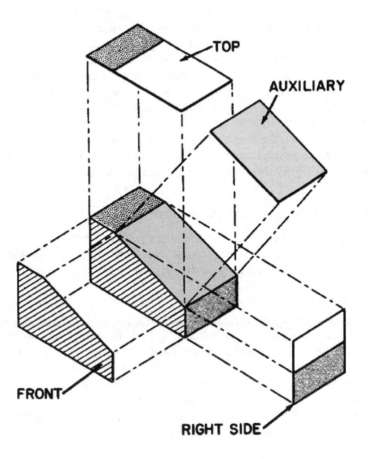

Figure 5.15   Another example of an auxiliary-view projection showing how it relates to an orthographic projection.

The distinguishing feature between projections and drawings is the unit of measurement employed. In projections a scale is constructed, which is used for measurements takeoff. The units of measurement used however are variable depending on the projection, and standard units of measurement are not used; in drawings, however, standard units of measurement (e.g. inches, feet, centimeter, etc.) are always used. The scales constructed for isometric, dimetric, and trimetric projections are always smaller than the standard units of measurements from which they are derived. This basically means that axonometric projections are always smaller than axonometric drawings. An axonometric drawing of an object, although slightly distorted, is nevertheless visually as satisfactory as an axonometric projection of it. Axonometric drawings are usually preferred to axonometric projections because no time is wasted constructing the scales needed to generate the axonometric drawings.

An axonometric drawing is one that is accurately scaled and depicts an object that has been rotated on its axes and inclined from a regular parallel position to give it a three-dimensional appearance. The principal advantage of axonometric drawing is that one can use an existing orthographic plan without any redrawing. The plan is simply tilted to the desired angle. It should be noted that in much of Europe, an axonometric drawing always has its axis at a 45-degree angle; an isometric axis is either 30/30 degrees or 30/60 degrees. The most common axonometric drawings are isometric, dimetric, and trimetric (Figure 5.16). Typically in axonometric drawing, one axis of space is shown as the vertical.

Axonometric, or planometric, drawing, as it is sometimes called, is a method of drawing a plan view with a third dimension. It is used by interior designers, architects, and landscape gardeners. A plan view is drawn at a 45-degree angle, with the depth added vertically. All lengths are drawn at their true lengths, unlike oblique drawing. This gives the impression that you are viewing the objects from above. One advantage of axonometric drawing is that circles drawn on the top faces of objects can be drawn normally.

Isometric Drawing and Isometric Projection

The term "isometric" is derived from the Greek for "equal measure," reflecting that the scale along each axis of the projection is the same, which is not true of some other forms of graphical projection. One of the advantages of isometric perspective in engineering drawings is that 60-degree angles are easy to construct using only a compass and straightedge.

The isometric drawing is most commonly used in its true form giving "equal measure" and foreshortened views of three sides of the object. An isometric drawing is one form of pictorial drawing. Hidden lines are not normally inserted. Isometric drawing is a method of visually representing three-dimensional objects in two dimensions, in which the three coordinate axes appear equally foreshortened and the angles between any two of them are 120 degrees. Isometric projection, like orthographic projection. is used in engineering drawings. An isometric drawing can be easily constructed by using a 30-60-90-degree triangle and T-square or with CAD programming. Figure 5.17A and B shows two examples of isometric drawings in an architectural context. Figure 5.17C gives an example of an architectural drawing using both orthographic projection (elevation) and isometric projection (details).

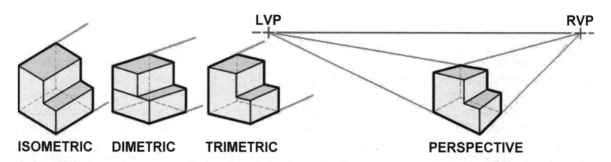

ISOMETRIC   DIMETRIC   TRIMETRIC   PERSPECTIVE

Figure 5.16 Different types of axonometric projections (isometric, dimetric, and trimetric) and perspective. In much of Europe, an axonometric uses a 45-degree angle as opposed to the 30/60-degree angles used in isometric drawing.

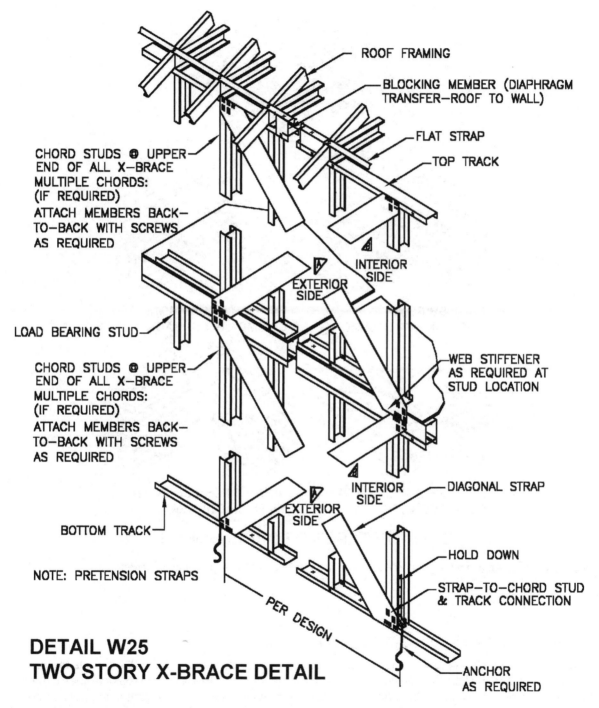

ROOF FRAMING

BLOCKING MEMBER (DIAPHRAGM TRANSFER—ROOF TO WALL)

FLAT STRAP

TOP TRACK

CHORD STUDS @ UPPER END OF ALL X—BRACE MULTIPLE CHORDS: (IF REQUIRED) ATTACH MEMBERS BACK-TO-BACK WITH SCREWS AS REQUIRED

INTERIOR SIDE

EXTERIOR SIDE

LOAD BEARING STUD

CHORD STUDS @ UPPER END OF ALL X—BRACE MULTIPLE CHORDS: (IF REQUIRED) ATTACH MEMBERS BACK-TO-BACK WITH SCREWS AS REQUIRED

WEB STIFFENER AS REQUIRED AT STUD LOCATION

INTERIOR SIDE

DIAGONAL STRAP

EXTERIOR SIDE

BOTTOM TRACK

HOLD DOWN

NOTE: PRETENSION STRAPS

STRAP-TO-CHORD STUD & TRACK CONNECTION

PER DESIGN

DETAIL W25
TWO STORY X-BRACE DETAIL

ANCHOR AS REQUIRED

Figure 5.17A   An example of the use of isometric drawings in architecture and engineering (source: North American Steel Framing Alliance).

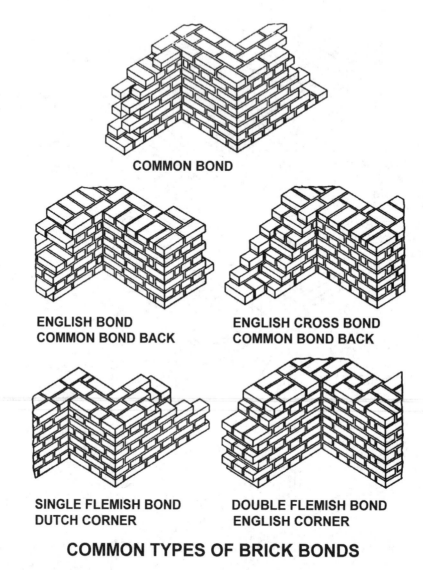

COMMON BOND

ENGLISH BOND
COMMON BOND BACK

ENGLISH CROSS BOND
COMMON BOND BACK

SINGLE FLEMISH BOND
DUTCH CORNER

DOUBLE FLEMISH BOND
ENGLISH CORNER

COMMON TYPES OF BRICK BONDS

Figure 5.17B   An example of the use of isometric drawings in architecture and engineering.

Comparison of Isometric and Orthographic Drawings

Compare the simple rectangular block shown in the orthographic representation (third-angle projection) in Figure 5.18 and the three-dimensional isometric representation. Notice that the vertical lines of the orthographic and isometric drawings (views A and B) remain vertical. The horizontal lines of the orthographic drawing are not horizontal in the isometric drawing but are projected at 30- and 60-degree angles; the length of the lines remains the same in the isometric and in the orthographic drawings.

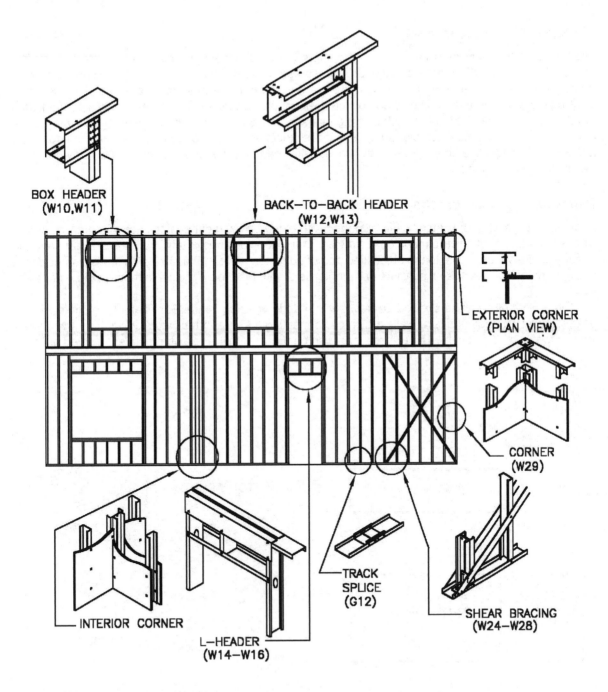

BOX HEADER
(W10,W11)

BACK—TO—BACK HEADER
(W12,W13)

EXTERIOR CORNER
(PLAN VIEW)

CORNER
(W29)

INTERIOR CORNER

L—HEADER
(W14—W16)

TRACK
SPLICE
(G12)

SHEAR BRACING
(W24—W28)

WALL FRAMING ELEVATION - DETAIL W2

Figure 5.17C   An example of the use of isometric drawings in architecture and engineering (source: North American Steel Framing Alliance).

Purpose of Isometric Drawing

The task of an isometric drawing is primarily to show a three-dimensional picture in one drawing. It is like a picture that lacks artistic details. Many utilities workers have difficulty in clearly visualizing a piping or ducting installation when they are working from a floor plan and an elevation drawing. The isometric drawing facilitates understanding by combining the floor plan and the elevation. It clearly communicates the details and clarifies the relationship of the pipes in an installation. Although isometric drawings are not normally drawn to scale on blueprints, some architects and engineers prefer drawing them to scale. Isometric drawings, like other types of drawings, follow certain rules and conventions to show three dimensions on a flat surface.

Dimensioning Isometric Drawings

An isometric drawing, or sketch, is dimensioned with extension and dimension lines in a manner somewhat similar to that of a two-dimensional drawing. The extension lines extend from the drawing, and the dimension lines are parallel to the object line and of equal length to it. Dimensioning the isometric drawing is more difficult because it consists of a single view, with less room available than on three separate views.

Circles or holes will be skewed or drawn within an isometric square. For example, a circle will appear elliptical in shape and is actually drawn by connecting a series of four arcs, drawn from the centerlines of the isometric square. The ellipses may also be drawn with the use of templates. Curved or round

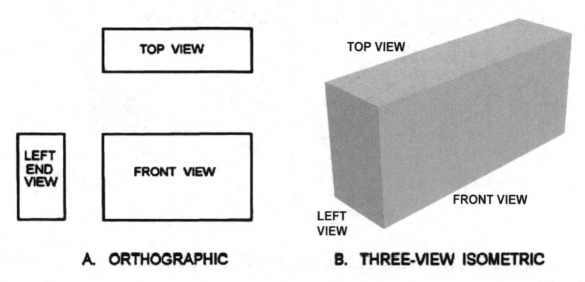

Figure 5.18 A. Orthographic views of an object (third-angle projection). B. Three-view isometric drawings of the same object.

corners are drawn in the same manner by locating the end of the radius on the straight line and then connecting the two points to form a triangle. The third point of the arc is actually the center of the triangle. Connect the three points with a freehand arc.

In isometric projections the direction of viewing is such that the three axes of space appear equally foreshortened. The displayed angles and the scale of foreshortening are universally known. However, in creating a final, isometric instrument drawing, a full-size scale—i.e., without the use of a foreshortening factor—is often employed to good effect because the resultant distortion is difficult to perceive.

Isometric drawing render a three-dimensional view of an object in which the two sets of horizontal lines are drawn at equal angles and all vertical lines are drawn vertically. In the resulting drawing all three angles are equally divided about a center point, and all three visible surfaces have equal emphasis. Orthographic techniques cannot be used in isometric drawings.

Any angle can be used to draw an isometric view, but the most common is 30 degrees because it can be drawn with a standard triangle and gives a fairly realistic view of an object. Today, CAD programs are the easiest way to draw isometric projections, but isometrics are also quick to draw manually and can be measured at any convenient scale. To manually draw in isometric, you will need a 30/60-degree set square.

There are four simple steps to manually draw a 5-inch box in isometric (Figure 5.19):

1. Draw the front vertical edge of the cube.
2. The sides of the box are drawn at 30 degrees to the horizontal to the required length.
3. Draw in the back verticals.
4. Drawn in top view with all lines drawn 30 degrees to the horizontal.

When you first start working with isometric techniques, use a simple box as a basic building block or guide to help you draw more complicated shapes. Figure 5.20 shows how to use such a simple box to accurately draw a more complicated L shape.

The first step is to lightly draw a guide box. This box is the size of the maximum dimensions. In this case, it measures 5 inches in length, 2.5 inch in width, and 5 inches in height. To achieve the L shape, we need to remove an area from this box. Draw a second box measuring 4 x 1 x 5 inches, the shape that needs to be removed from the first box to create the shape we require. For the finished shape, draw in the outline of the object using a heavier line. By using this technique complex shapes can be accurately drawn.

Circles in isometric do not appear circular. They appear skewed and are actually elliptical. There are several methods of constructing circles in isometric drawing. For many manual tasks the easiest method is to use an isometric circle template, which can be bought at most good art shops. These templates contain a number of isometric circles of various sizes.

Isometric circles can also be drawn manually using the following method:

1. Draw an isometric square and then draw in the diagonals, a vertical, and a line at 30 degrees from the midpoint of the sides, as illustrated in Figure 5.21.
2. Place your compass point on the intersection of the horizontal and the vertical lines and draw in a circle that touches the edges of the box
3. For the next section of the isometric circle place your compass point on the corner of the isometric square and draw in the arc as shown in the illustration.
4. Complete the circle by repeating the process for the other parts, using the appropriate techniques.

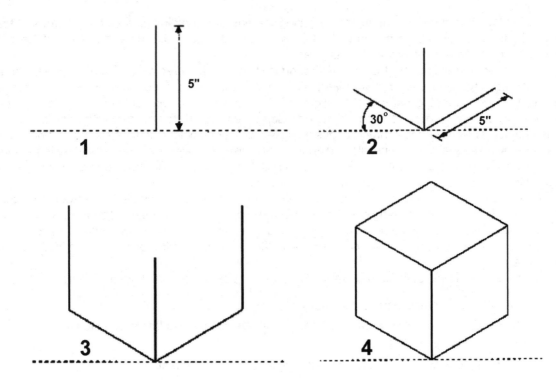

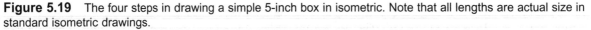

Figure 5.19   The four steps in drawing a simple 5-inch box in isometric. Note that all lengths are actual size in standard isometric drawings.

Dimetric Projection

Dimetric projection is an axonometric projection of an object placed in such a way that two of its axes make equal angles with the plane of projection and the third axis makes either a smaller or a greater angle. In dimetric projections, the directions of viewing are such that two of the three axes of space appear equally foreshortened, with the attendant scale and angles of presentation determined according to the angle of viewing; the scale of the third direction (vertical) is determined separately. Approximations are common in dimetric drawings. Figure 5.22 shows different arrangements for isometric, dimetric, and trimetric drawings.

Trimetric Projection

Trimetric projection is an axonometric projection of an object so placed that no two axes make equal angles with the plane of projection and each of the three principal axes and the lines parallel to them, re-

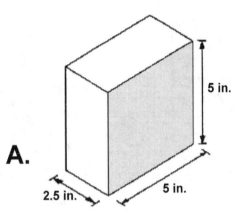

A.

5 in.

2.5 in.   5 in.

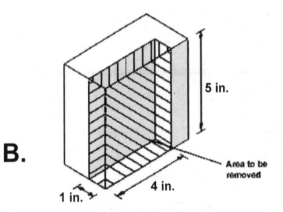

B.

5 in.

Area to be removed

1 in.   4 in.

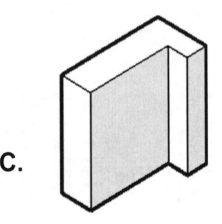

C.

Figure 5.20 Steps to draw a circle in isometric projection.

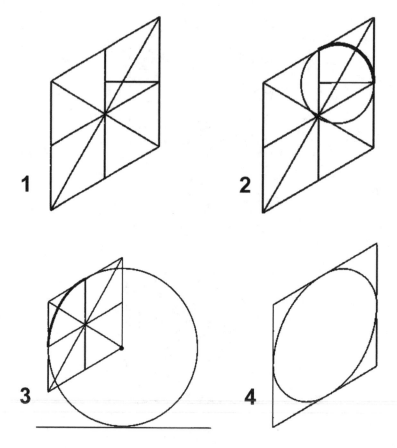

Figure 5.21   Steps to draw a circle in isometric projection.

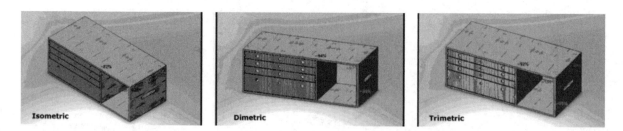

Figure 5.22   A comparison of isometric, dimetric and trimetric drawings (source: Wikipedia).

spectively, have different ratios of foreshortening (and are therefore drawn at different scales) when projected to the plane of projection. The wide angle choice gives the designer considerable flexibility and control of the pictorial view.

In trimetric projections, the direction of viewing is such that all of the three axes of space appear unequally foreshortened. The scale along each of the three axes and the angles among them are determined separately as dictated by the angle of viewing. Approximations in trimetric drawings are common.

Limitations of Axonometric Projection

Objects drawn with axonometric projection do not appear larger or smaller as they extend closer to or further away from the viewer. While advantageous for architectural drawings and sprite-based video games, this results in a perceived distortion, as, unlike perspective projection, it is not how our eyes or photography usually work. An additional problem in the case of isometric projection is that there are times when it becomes difficult to determine which face of the object is being observed. In the absence of proper shading, and with objects that are relatively perpendicular and similarly proportioned, it can become difficult to determine which is the top, bottom, or side of the object, since each face is given similar dimensions.

5.7   PICTORIAL DRAWINGS

Pictorial drawings are not often used for construction purposes. However, on some working drawings pictorial views are used to reveal information that orthographic views alone would be incapable of showing; other situations may require a pictorial drawing essentially to supplement a major view. Pictorial projection, unlike multiview projection, is designed to allow the viewer to see all three primary dimensions of the object in the projection. Pictorial architectural drawings and renderings are very easy to understand and are therefore used extensively to depict a three-dimensional view of an object and for explaining project designs to laypersons for sales-presentation purposes. They enable an inexperienced person to interpret drawings and quickly visualize the shape of individual parts or various components in complicated mechanisms. To convey as much information as possible, the view is oriented to show the sides with the most features. In many cases, orthographic (multiview) drawings provide information in a format that makes it difficult for laypersons to visualize the total project.

Orthographic/multiview drawings are typically dimensioned and are usually drawn to a specific scale (Figure 5.23). Although pictorial drawings may be dimensioned and drawn to scale, their main purpose is to give a three-dimensional representation of the building or object. As illustrators often take artistic liberties with scale and proportion, the reader should only use pictorial drawings for general reference. And although they are not usually dimensioned and exact scaling is not required, proportions are nevertheless expected to be maintained. When pictorial drawings are dimensioned and contain other specifications that are needed to produce the part or construct the object, they are considered to be working drawings.

Whereas a multiview drawing is designed to focus on only two of the three dimensions of the object, a pictorial drawing provides an overall view. The tradeoff is that a multiview drawing generally allows a less distorted view of the features in the two dimensions displayed while lacking a holistic view of the object (thus needing multiple views to fully describe the object).

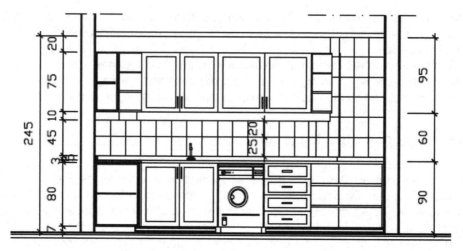

SECTION: AA

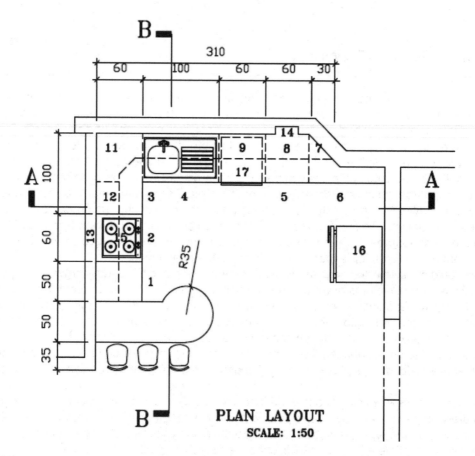

PLAN LAYOUT
SCALE: 1:50

Figure 5.23   Drawing showing an orthographic view (plan and section) of a kitchen design.

The same dimensioning rules that apply for an orthographic/multiview drawing also apply to a pictorial drawing. These include:

- Dimension and extension lines should be drawn parallel to the pictorial planes.
- When possible, dimensions are placed on visible features.
- Arrowheads lie in the same plane as extension and dimension lines.
- Notes and dimensions should be lettered parallel to the horizontal plane.

The three main types of pictorial drawings that are extensively used in architectural presentations are perspective drawings, isometric drawings, and oblique drawings.

The main difference between isometric and typical perspective drawings is that in the latter the lines recede to vanishing points. This gives the drawing a more realistic appearance but is technically inaccurate. Isometric drawings, on the other hand, show true dimensions. However, they create an optical illusion of distortion, mainly because the human eye is accustomed to seeing long object lines recede. For this reason, isometric drawings are primarily utilized for clarification of small construction details, since they depict their true size dimensions. In an oblique drawing two or more surfaces are shown at one time on one drawing. The front face of an object is drawn in the same way as the front view of an orthographic sketch.

5.8 PERSPECTIVE DRAWINGS

A good understanding of the principles of perspective is necessary to create an accurate and visually appealing piece of art. Perspective drawings are forms of pictorial drawing. Blueprint readers may not see a perspective drawing very often, but they will undoubtedly appreciate the descriptive information it offers. Perspective drawing is a system for representing three-dimensional space on a flat surface. It utilizes either one, two, or three points to where the receding lines will vanish. These vanishing points are placed along a horizontal line called a horizon line. In perspective drawings receding lines are no longer parallel to each other as in oblique or isometric drawings.

In perspective drawings, distant objects appear smaller but have the same shape and proportions as they would close up. In other words, as objects are further away, they become smaller and appear to vanish into the distance. The general principle behind perspective drawing is simple and shares many features with the way people actually perceive space and objects in it. It depends essentially on four interconnected criteria that will invariably affect the final image: the level of our eyes when viewing the scene or object, thus determining the horizon line; the distance from the picture plane to the object; the distance from the station point to the object and cone of vision; and the angle of the object to the picture plane.

Perspective drawings may be drawn as impressive artistic renderings to show landscapes or large structures. They can also be used to show a realistic representation of machine parts or layouts of architectural furnishings in a room. Perspective drawings are more difficult and time-consuming and are not used as extensively as they might be. They are used mainly for illustrative purposes, but their use as a descriptive drawing in manufacturing environments should not be overlooked.

In essence, there are three basic types of perspective drawings that are commonly used in architectural design and construction: one-point perspective, two-point perspective, and three-point perspective.

One-Point (Parallel) Perspective

The law of perspective is that parallel lines that lie in the same plane will appear to converge to a single point somewhere on the horizon (at the eye level), called the vanishing point (VP). This point is usually positioned within the view and gives objects an impression of depth. To draw a one-point perspective, simply draw a horizon line (HL) and draw a vanishing point anywhere on it. The horizon line may be located above, below, or at any other location on the drawing. Vertical and horizontal lines are drawn normally, and all receding lines are drawn to the vanishing point. Thus, the sides of an object diminish towards the vanishing point, whereas all vertical and horizontal lines are drawn with no perspective—i.e., face on (Figure 5.24).

A normal-view angle places the HL at a natural height as if the viewer were looking straight ahead without tilting the head up or down. Figure 5.25 shows two examples of normal-view one-point perspective. Altering the position of the VP changes the view of the object being drawn. For example, to look down at the top of the object, the vanishing point must be above the horizon line while to look up at the object, it must be below the horizon line. It is advisable to practice identifying the best locations for vanishing points to achieve the desired results. One-point perspective depicts a building or interior space with one side parallel to the picture plane (perpendicular to the observer's line of sight).

To set up a one-point perspective, connect the corners of the elevation to the vanishing point and mark off the depth through the lines of sight in the plan (Figure 5.26). One-point perspectives are often used to draw interiors, as they give an accurate depiction of the facing wall, in addition to both receding side walls. They are also typically used for roads, railroad tracks, interiors, and buildings viewed so that the front is directly facing the viewer. Objects that are made up of lines either directly parallel or perpendicular to the viewer's line of sight can be drawn with one-point perspective. However, it is of limited use, mainly because the perspective is too pronounced for small products, making them appear larger than they actually are.

Two-Point Perspective

Sometimes called angular perspective drawing; in this method only the vertical lines are drawn vertically. The horizontal, depth, and length receding lines are drawn to the vanishing points located on the horizon line. The front view is no longer true in shape but is now drawn in an isometric configuration. Again, the location of the horizon line and the vanishing points on the line will provide many different "looks" of the object.

In two-point perspective the sides of the object vanish to one of two vanishing points on the horizon. Vertical lines in the object have no perspective applied. Our distance from an object seen at an angle determines where the vanishing points lie on the horizon. Two-point perspective is a much more useful drawing system than the simpler one-point perspective and the more complex three-point perspective. Objects drawn in two-point perspective have a more natural look. Figure 5.27 illustrates a typical architectural application of two-point perspective for a department-store interior.

To set up a two-point perspective, connect the corner height line to the right and left vanishing points, and, with the lines of sight in the plan, mark off the depth of the object. The procedure for constructing a two-point perspective view is essentially the same as for one-point perspective except for the additional step of establishing two vanishing points. In two-point perspective you can make the object look big or small by altering the proximity of the vanishing points to the object (Figure 5.28).

Shade and shadow are often used in perspective drawings to give a better perception of the depth and form of a space or object. The drawing of shadows and reflections both follow the same immutable rules of perspective.

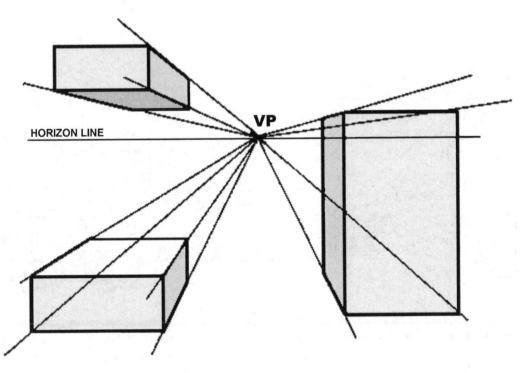

One-Point Perspective

Figure 5.24   A simple one-point perspective. Although it is possible to sketch products in one-point perspective, the perspective is often too aggressive to the eye, making objects appear bigger than they actually are.

The illustration in Figure 5.29 demonstrates how to draw a box in two-point perspective:

1. Put two vanishing points at opposite ends of the horizontal line.
2. Draw the front vertical of the box. Drawing the line below the horizontal will create a view that we are looking down on. To look at the object from below, draw the front vertical above the horizontal.
3. Draw lines from the top of the vertical that disappear to both of the vanishing points. Repeat the process for the bottom of the line.
4. To complete both sides, draw in the back verticals.
5. To draw the top of the box, draw lines from the back verticals to the opposite vanishing points.

Figure 5.25A A. Illustration of an interior using normal-view one-point perspective.

Three-Point Perspective

Three-point perspective is a development of two-point perspective and is usually used for buildings seen from above or below. Like two-point perspective it has two vanishing points somewhere on the horizon and a third somewhere above or below the horizon line toward which the verticals converge (Figure 5.30). This means that the object is either tilted to the picture plane or the spectator's central axis of vision is inclined upward or downward and the picture plane is tilted. Three-point perspectives usually indicate that the spectator is very close to the object or that the object is very large. It is best used for drawing tall objects such as buildings, although this form of perspective is not widely used in architectural presentations. Its best use may be to show a particular viewpoint of a tall object such as a skyscraper. This type of drawing is sometimes called an oblique perspective drawing, as the vertical lines are drawn to the third vanishing point not located on the horizon line.

Four-, five-, and six-point perspective drawings are more complex and challenging and require considerable understanding and skill to execute.

In general, most designers create drawings with a vanishing point far below the horizon so that the depth added to the verticals is only slight. In many cases the vanishing point isn't even on the paper. Learning how to apply vertical perspective will make your drawings more and more realistic.

Figure 5.25B A photograph of an interior space depicting a normal-view one-point perspective and showing the converging lines and horizon (source: Randy Sarafan).

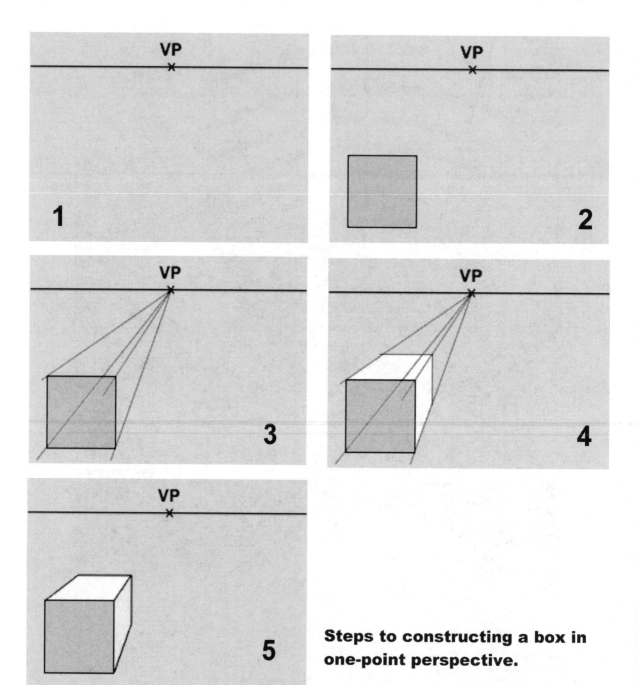

5 Steps to constructing a box in one-point perspective.

Figure 5.26   To construct a box in one-point perspective, draw a horizon line (HL) and place a vanishing point (VP) somewhere on this line; draw a square somewhere beneath the horizon—this will be the front of your box; draw four lines, one from each corner of the square that must pass through the vanishing point; and lastly, draw in the back vertical and horizontal to complete the box.

Figure 5.27   A typical architectural application of a two-point perspective used to show a department store interior.

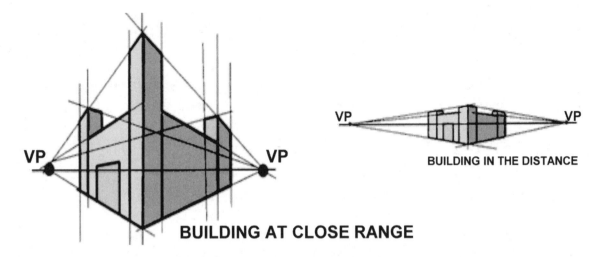

Figure 5.28   In two-point perspective you can make the object look big or small by altering the proximity of the vanishing points. In the image on the left the vanishing points are close to the building; the image on the right shows the vanishing points far away from the building.

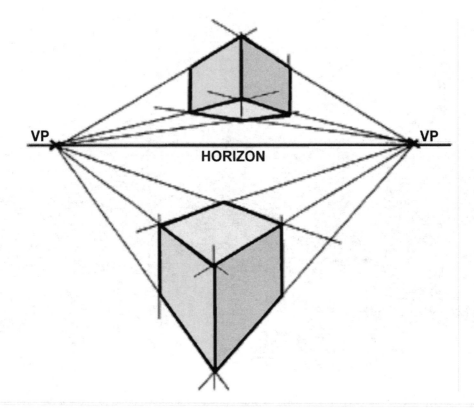

Figure 5.29   A simple two-point perspective of objects drawn above and below the horizon line.

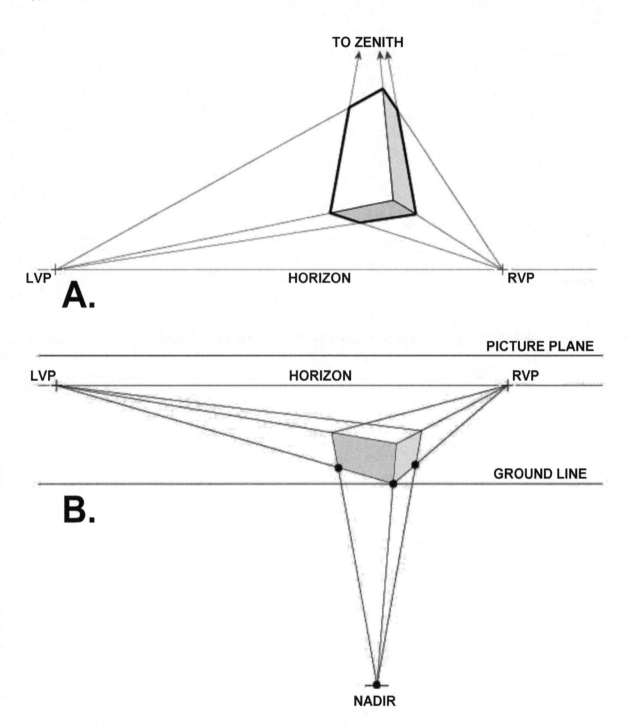

Figure 5.30A,B Two examples of a three-point perspective, one with the vanishing point above the horizon and the other with it below (source: Kevin Hulsey).

6

Layout of Construction Drawings

6.1 GENERAL OVERVIEW

For identification purposes, drawings associated with construction and the building trades can be categorized into four main types, preliminary drawings, presentation drawings, working drawings, and shop drawings.

Preliminary Drawings

These drawings are essentially intended to be concept design explorations and means of communication between the architect and the client. They are not intended to be used for construction but rather to interpret the client's needs and instructions, prepare functional studies, select materials, estimate preliminary costs, and solicit preliminary approval by civil authorities. They also form the basis for the final working drawings.

Presentation Drawings

These drawings are selling tools and normally consist of perspective views based on the preliminary design concept of the project. Presentation drawings are drawn to highlight the aesthetic qualities of a project and, in addition to perspectives, may include isometrics, colored elevations, and colored floor plans (Figure 6.1).

Construction or Working Drawings

The term "construction drawing" is generic in that it includes all the drawings needed by the various tradespeople to complete a building project. These drawings are prepared by the architect, engineer, and other specialists depending on the complexity of the project. Construction drawings are technical directions in graphic form, showing the size, quantity, location, and relationships of the building's components.

In construction drawings as much information as possible is presented graphically, or by means of pictures. Most construction drawings consist of orthographic views. General drawings consist of plans and elevations drawn at a relatively small scale. Detail drawings (discussed later in this chapter) consist of sections and details drawn at a relatively large scale. A plan view is a view of an object or area as it would appear if projected onto a horizontal plane passed through or held above the object area. The most common construction plans are plot plans (also called site plans), foundation plans, floor plans, and framing plans. A plot plan shows the contours, boundaries, roads, utilities, trees, structures, and other significant physical features on their sites. The locations of the proposed structures are indicated by appropriate outlines or floor plans. As an example, a plot may locate the corners of a proposed structure at a given distance from a reference or base line. Since the reference or base line can be located at the site, the plot plan provides essential data for those who will lay out the building lines. The plot can also have contour lines that show the elevations of existing and proposed earth surfaces and provide essential data for the graders and excavators.

The main functions of construction drawings include:

- Instruments for material take-offs: Labor, material, and equipment estimates are made from working drawings prior to construction.

- Instructions for construction: Working drawings show specific sizes, location, and relationships among all materials.

- Means for granting a building permit: Before construction can begin, the local building authority must review the working drawings to see that they meet the requirements of public safety in terms of structural soundness, fire, and other hazards. A building permit is issued to the builder only after approval of the drawings.

- Instruments for competitive bidding: In a free-enterprise system, working drawings allow potential contractors a uniform guide for preparing bids, thereby providing the owner with the most economical costs.

- Means of coordinating among the various trades: Working drawings are the basis of agreement between material suppliers and specialized trades.

- A permanent record for future remodeling or expansion or for legal use in the advent of a dispute: Working drawings eliminate remeasurements in case of future reconstruction. Drawings must be furnished in case of legal disputes. Building failures might possibly occur by natural, unavoidable causes, design errors, or neglect, but the drawings and design calculations are used as evidence and should be available for the life of the building.

- Basis for agreement between owner and tenant: In leasing all or portions of a building, the owner must use the working drawings in the contract agreement.

- A complement to the specifications: In obtaining written information from the specifications, contractors need the working drawings for interpretation. Information from one source is incomplete without the other.

Shop Drawings

These are technical drawings prepared by various suppliers participating in the construction. On many jobs the architect or designer must rely on specialists to furnish precise information about the components. For example, if complex cabinetwork is required, it must be built to exact sizes and specifications.

A shop drawing is needed to ensure that it will fit into the structure and that the structure will accommodate it. Approval of the shop drawings usually precedes the actual fabrication of the components. With shop drawings the architect or designer is able to check the quantity of other components that subcontractors propose to furnish.

The Construction Drawings Set

The transition from approved preliminary design/concept drawings to full-blown construction documents is very significant because it signifies the completion of one phase—that of making design decisions—and the beginning of a new phase—the production of construction documents, which is essential to the implementation of the project.

Construction drawings are sometimes referred to as "working drawings" or "production drawings." These drawings provide all the required information, both graphic and written, about the project. The information provided will be specific to every aspect of the proposed project. Preparing construction drawings represents the final step in the design process. The completed drawings become a "set" incorporating all the modifications made by the designer during the process of transition from the schematic-design to the working-drawings phase. Included in the construction drawing set will be detailed information regarding the building envelope, structural and mechanical systems, furniture, equipment, lighting, outlets, demolition, and so on. This information is explained through floor plans, interior or exterior elevations and sections, mechanical and electrical drawings, and detail drawings. Detailed specifications are also typically included.

The construction drawings and specifications are also used for pricing the project. Two or more general contractors are usually provided with the same set of documents to bid on. This facilitates fairness in that each contractor shares the same information and no one has any more or less information than his or her counterpart has when pricing.

Building Permits

The completed blueprints must also be submitted to the local building department to ensure that the proposed design is in compliance with all regulatory agency requirements, including those set by the zoning, fire, health, and other departments. A building permit will only be issued after the drawings are checked and approved by the various departments.

Almost all new construction (commercial, civic, industrial, residential, etc.) and renovation require a building permit. In order to obtain a permit, a complete set of construction drawings is required. Only in the case of minor changes to an existing building will a permit not be required. Examples of minor changes would include minor repair or painting. For residential projects, such as a renovation or addition to an existing building, a building permit will be required. All structural work must comply with applicable code standards.

A building permit is a document that states that approval from the local building department has been given to proceed with construction or demolition. This document is numbered and recorded at the local building department. The permit(s) must be posted in a visible location on the construction site. It is unlawful to start construction or demolition before a permit is issued. This document is necessary for all new construction, additions or renovations of both residential and commercial projects. An application is made at the building department in the city, town, or municipality in which the work is to take place.

A complete set of construction documents can be large—30 or more drawings, or small—10 to 15 drawings—depending on the size and type of project. A shopping mall would require many drawings, whereas a small residence would require fewer. Today almost all construction drawings are produced

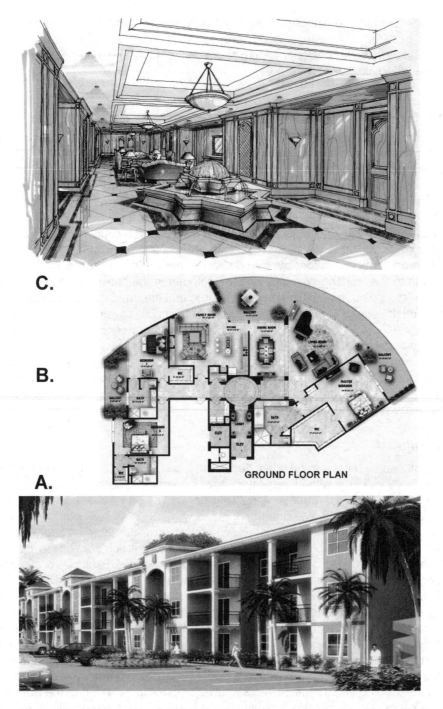

Figure 6.1 A and B. Computer-generated renderings of a building exterior and floor plan (source: Archiform Ltd.). C. Hand-drawn office-lobby interior with mahogany paneling (source: Kubba Design).

on computers. One reason is that CAD software is faster and results in greater drawing accuracy, consistency, correctability, and easier storage.

Construction drawings are considered to be legal documents, and everyone involved in the project—owner, architect/engineer, and general contractor—all use these drawings as their source of information. But in order to produce a comprehensive set of construction drawings, knowledge of design and building methods are necessary.

The types of drawings discussed in this chapter are essentially construction drawings, including architectural drawings, structural drawings, mechanical drawings, electrical diagrams, details, and shop drawings. Construction drawing is any drawing that furnishes the information required by the craftspeople to rough in equipment or erect a structure.

Cover Sheet

The first sheet in a set of working drawings is the cover sheet (Figure 6.2). This sheet is important because it lists the drawings that comprise the set (a drawing index) in the order that they appear. It normally lists the specific requirements of the building code having jurisdiction over the design of the project. A cover sheet should also list the project name and location, building permit information, key plan, and general notes. Names and contact information of all consultants should also be included. Other important information required includes the total square footage of the structure, the use group the structure will fall under, and the type of construction. Another important element on the cover sheet is the list of abbreviations or graphic symbols used in the set. Usually there is a section that contains general notes for the contractor, such as "Do not scale" or "All dimensions to be verified on site."

With larger projects, a second cover sheet is sometimes included that includes information not shown on the main cover sheet. Likewise, a location map is sometimes included to locate the project site in relation to nearby towns or highways.

Information presented in a set of working drawings, along with the specifications, should be complete so the craftspeople who use them will require no further information. Working drawings show the size, quantity, location, and relationship of the building parts. Generally, working drawings may be divided into three main categories: architectural, mechanical, and electrical. Regardless of the category, working drawings serve several functions.

They provide a basis for making material, labor, and equipment estimates before construction starts. They give instructions for construction, showing the sites and locations of the various parts. They provide a means of coordination between the different ratings. They complement the specifications; one source of information is incomplete without the other when drawings are used for construction work.

6.2   CIVIL DRAWINGS

The most obvious difference between civil and architectural drawings is the use of the engineer's scale. Like architectural drawings, civil drawings include symbols and graphics that convey intent with a minimum of words. Some of the symbols and graphics are unique to site plans, and others are very similar to those used on architectural plans. Some civil drawings offer a legend to decipher the symbols and graphics on a specific set of drawings. Site plans often contain several sheets, depending on the size and complexity of the project. They are usually numbered starting with a "C," such as sheet C-1, C-2, and so on. The term "site" is synonymous with plot or lot.

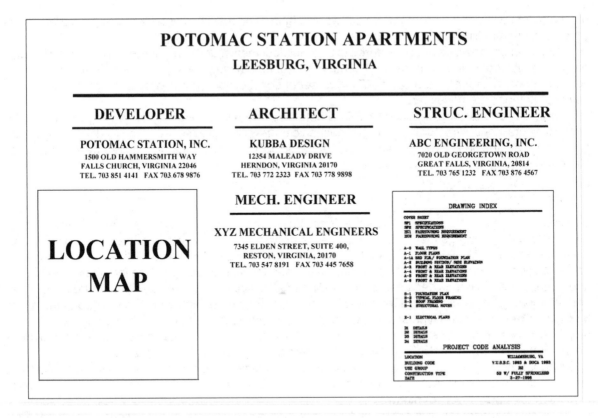

Figure 6.2 An example of a cover sheet used for a small- to medium-sized project.

The calculations that are required for submission on a plot plan are not complex. What is needed is a basic understanding of what you are calculating and why. Moreover, as mentioned earlier in the chapter, most drawings and calculations today are easily generated with the use of computers. Figure 6.3 is an example of a simple computer-generated site plan using AutoCAD software.

There are four basic calculations required for the plot plan: site area (lot size), lot coverage (bird's-eye view expressed as a percentage of the lot), gross floor area (combined size of parts of the structure), and floor-area ratio (FAR), a ratio of floor area to property size.

Site Plan

A site or plot plan is a scale drawing of a property that shows its size and configuration including the size and location of man-made features such as buildings, driveways, and walkways on the property. Plot plans show both what currently exists and what improvements are proposed (Figure 6.4). The main function of a basic site or plot plan is to determine the placement of the structure as it sits in reference to the boundaries of the construction site. Site plans clearly establish the building's dimensions, usually by the

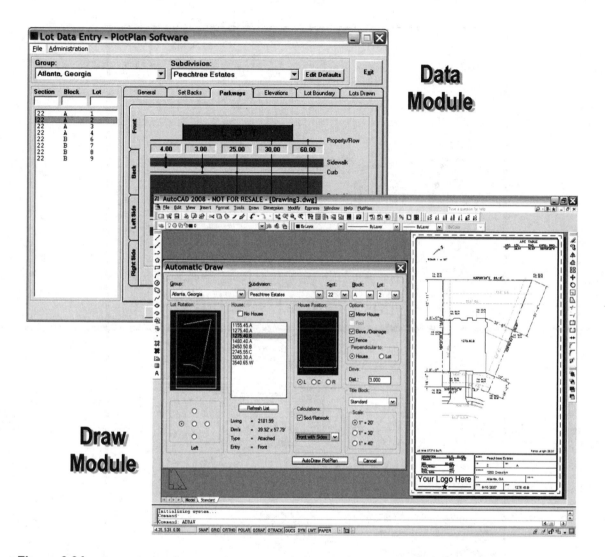

Figure 6.3A An example of a computer-generated site plan using AutoCAD software.

foundation's size and the distances to the respective property lines. Setback dimensions are shown in feet and hundredths of a foot, as opposed to feet and inches on architectural drawings. Thus, an architectural dimension reading 40 feet, 6 inches would be 40.5 feet on a site plan.

A site plan includes not only the project but also the surrounding area. Site plans should outline location of utility services, setback requirements, and easements. Topographical data is also sometimes indicated, specifying the slope of the terrain. The grades at fixed points are shown throughout the area to show the land slope before construction is started and the finished grade after construction is completed.

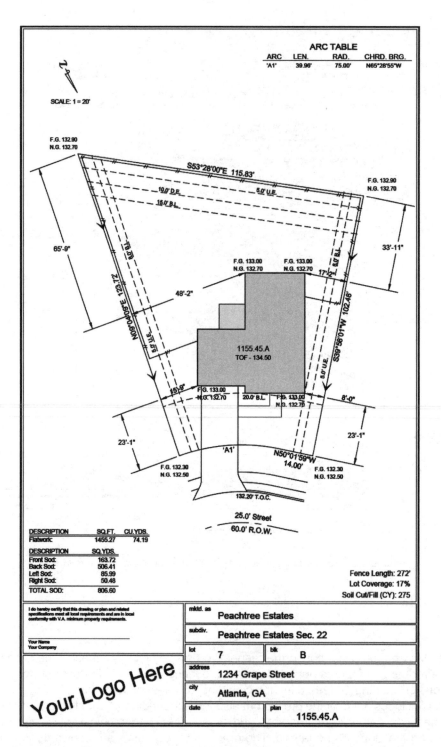

Figure 6.3B   An example of a computer-generated site plan using AutoCAD software.

Site plans are often required to accompany most applications in conjunction with a site-plan review submitted to the city or county in order to change how a particular property is used. For example, they are required for:

- Preapplication review of conceptual elements of multifamily, commercial, and industrial development

- Conditional-use permits

- Variances to zoning requirements

- Construction of new structures requiring a building permit

- A change of zone or a special zoning exception

A site plan should typically include:

- Legal description of the property based on a survey.

- Drawing scale: The site plan should be drawn to the most appropriate scale, for example, 1 inch = 10 feet,1 inch = 20 feet, or 1/4 inch = 1 foot.

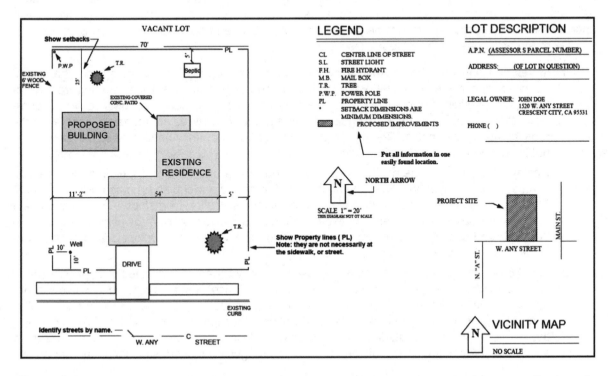

Figure 6.4   A hypothetical site plan showing proposed and existing structures, setbacks, property lines, north arrow, legal description, and vicinity map.

- An arrow indicating north, usually towards the top right-hand corner of the sheet.
- Property-line bearings and directions: For most additions, property lines need to be physically located. In many cases, a certificate of survey, signed by a licensed surveyor, is required.
- The distance between buildings and between buildings and property lines.
- The dimensions of the existing buildings.
- Location of adjacent streets and any easements.
- A clear indication of any proposed addition or alteration.
- Utilities.
- Other information that is relevant to the project.

Plat Map

A plat is a map drawn to scale (usually supplied by a land surveyor or civil engineer) of part of a city or township showing some specific area, such as a subdivision made up of several individual lots. A plat will often consist of many sites or plots. A plat delineates the divisions of a piece of land (property-line bearings, dimensions, streets, and existing easements) and represents the first of several stages in a site's development (Figure 6.5). City, town, or village plats chart subdivisions into blocks with streets and alleys. For additional clarification the blocks are split into individual lots, usually for the purpose of selling the described lots, usually termed subdivisions. In order for plats to become legally binding, they must be filed in local jurisdictions, such as a public-works department, urban-planning commission, or zoning board, which must typically review and approve them. Legal descriptions become part of the public record and can be reviewed at any time.

There are three basic types of legal descriptions:

1. Metes and bounds: This is a system that identifies a property by describing the shape and boundary dimensions of a unit of land using bearing angles and distances starting from a defined point of origin. The point of origin may be referenced to the corner of some section or quarter-section described by the rectangular survey system. The metes are measured in feet, yards, rods, or surveyor's chains. These legal descriptions are frequently used to describe land that is not located in a recorded subdivision.

2. Rectangular survey system: This system provides for a unit of land approximately 24 miles square, bounded by a baseline running east and west and a meridian running north and south. This 24-mile square is further divided into 6-mile squares called townships. A range is an east and west row of townships between two meridian lines 6 miles apart. A township is divided into 36 numbered sections, each 1 mile square. Farm, ranch, and undeveloped land is often described by this method.

3. Lot and block: This system is commonly used in many urban communities to legally describe small units of land because of its simplicity and convenience. A map is created in which a larger unit of land is subdivided into smaller units for the purpose of sale. The map is recorded after each lot has been surveyed by a metes-and-bounds description. Deeds then need only refer to the lot, block, and map book designation in order to describe the property. It is not necessary to state the survey bearings and distances or the rectangular survey description in the deed.

Lot lines are laid out by polar coordinates: that is, each line is described by its length plus the angle relative to true north or south. This is accomplished by the use of compass direction in degrees, minutes, and seconds. The lot line may read N6o 49' 29" W. The compass is divided into four quadrants, NW, NE, SW, and SE.

There are many reasons to plat:

- To designate roads and other rights of way
- To make sure that all property has access to a public right of way
- To create or vacate easements
- To dedicate land for other public uses, such as parks or areas needed for flood protection
- To ensure compliance with zoning

Demolition Plan

Many projects will be constructed on sites with existing features or exterior or interior elements that are not envisaged to be part of the final design. These will need to be removed or demolished prior to grad-

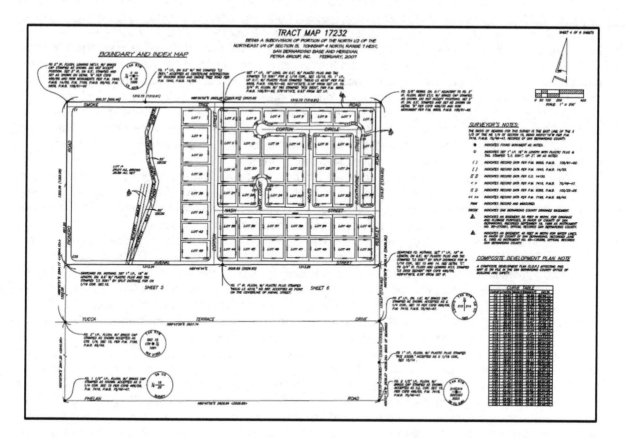

Figure 6.5 An example of a plat map. A plat represents the first of several stages in a site's development.

ing for construction and are shown on a demolition plan. Figure 6.5 is an example of a demolition plan for interior elements of a house. It shows the elements to be demolished and to be retained. Trees and other items that are to remain should be noted in the keynotes. Dotted lines indicate items (walls, fixtures, etc.) to be removed to make the space ready for the new design. A keynote legend is usually included on a demolition drawing sheet, showing each number and corresponding note.

Topography Map

The other important function of the site plan is to highlight the special surface conditions, or topography, of the lot. This will indicate to the builder the slope or flatness of the site. The topography of a particular lot may be indicated on the site plan (Figure 6.6). For some projects the topography needs to be shown separately for clarity, and a grading plan has to be used. Topographical information includes changes in the site's elevation, such as slopes, hills, valleys, and other variations in the surface. These changes in the surface conditions are shown on a site plan by means of a contour, which consists of a line connecting points of equal elevation. An elevation is a distance above or below a known point of reference, called a datum. This datum could be sea level or an arbitrary benchmark established for the particular project.

Architects normally adjust the existing contours of a site to accommodate the building construction and site-improvement requirements. Adjusting existing contours is one of the stages in the site-improvement process, in which the architect or designer requires a topography map to study the slope conditions that may impact the design. This map is usually prepared by a civil engineer and is meant to show in drawing form the existing contour lines and their accompanying numerical elevations. Normally, existing contour lines are illustrated by a dashed line, and new or proposed contours are normally shown as a solid line. The topography map can therefore be considered to be a plat map in which its broken lines and numbers indicate the grades, elevations, and contours of the site. The distance between the contour lines is at a constant vertical increment, or interval. Typically, an interval of five feet is used, but other intervals may be substituted, and one-foot intervals are not uncommon for site plans requiring greater detail (or where the change in elevation is more dramatic).

When reading a site plan, note that contours are continuous and often enclose large areas in comparison to the size of the building lot. This is why contours are often drawn from one edge to the other edge of the site plan. Contours do not intersect or merge except in the case of a vertical wall or plane. For example, a retaining wall shown in plan view would show two contours touching, and a cliff that overhangs would be the intersection of the contours. When contour lines are spaced far apart, the land is relatively flat or gently sloping. When the contour lines are close together, the land is much steeper.

A benchmark is normally required, which is basically a known reference point such as an elevation on the construction site. The benchmark is established in reference to the datum and is commonly noted on the site print with a physical description and its elevation relative to the datum. For example, "Northeast corner of catch basin rim - elev.1085'" might be found on a site plan. When individual elevations or grades are required for other site features, they are noted with a "+" and the grade. Grades differ from contours in that a grade registers accuracy to two decimal places, whereas a contour is shown as a whole number. A grading plan shows existing and proposed topography (Figure 6.7) and is used to delineate elevations and drainage patterns. Grading-plan requirements will vary from one location to the next, but a final grading plan will typically show the site boundaries, existing landform contours with a benchmark, existing site features, and proposed site structures. In some instances the grading plan may also show a cross section through the site at specified intervals or locations to more fully evaluate the surface topography.

The north arrow is used to show the direction of magnetic north as a reference for naming particular sides or areas of the project. Moreover, surveyors label the property lines in accordance with the directions normally found on a compass. This reference, in the form of an angle and its corresponding distance, is called the bearing of a line. The bearings of the encompassing property lines often provide the legal description of the building lot. Larger projects usually need several site plans to show the different scopes of related or similar work including drainage, utility plans, and landscaping plans.

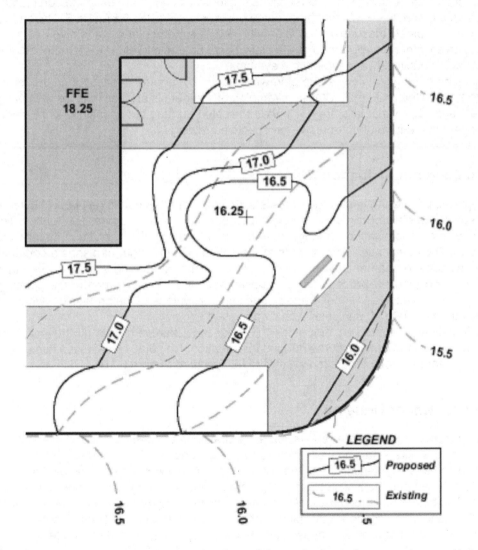

Figure 6.6 Example of a topography site plan showing existing and proposed contours (source: United States Air Force Landscape Design Guide).

Drainage and Utility Plans

Utility drawings show locations of the water, gas, sanitary sewer, and electric utilities that will service the building (Figure 6.7). The location of existing underground utilities is important so that they can be protected during excavation and construction. Drainage plans detail how surface water will be collected, channeled, and dispersed on- or offsite. Drainage and utility plans illustrate in plan view the size and type of pipes, their length, and their connections or terminations. The elevation of a particular pipe below the surface is given with respect to its invert. The invert of a pipe is the bottom trough through which the liquid flows. This is typically noted with the abbreviation for invert and an elevation, for example, "INV.123 feet."

Invert levels are shown at the intersections of pipes or other changes in the continuous run of piping, such as a manhole, sewer manhole, catch basin, etc. (Figures 6.8A and B). Inverts are normally provided only for piping that has pitch or a gravity flow. The method of sewage disposal is normally considered to be an important element of a site plan. There are several alternative methods of sewage disposal, including public sewers and private systems.

In addition to the plan view, a site plan will sometimes require clarification in the form of a detail, similar to an architectural detail. Classic examples of site details are sections through paving, precast structures, and curbing. Details are not limited to scaled drafting but occasionally for clarification purposes appear in the form of not-to-scale perspective drawings.

Landscaping and Irrigation Plan

Landscaping is usually the final stage of a site's development. Landscaping plans show the location of various species of plantings, ground cover, lawns, benches, garden areas, and fences. The plantings are noted with an abbreviation, typically three letters, along with the quantity of the particular species (Figure 6.9). The designation corresponds to a planting schedule that furnishes a complete list of the plantings by common name, Latin or species name, and quantity and size of each planting. Certain notes describing planting procedures or handling specifications accompany the planting schedule. Landscaping plans have additional graphics and symbols unique to the profession, which the blueprint reader should be aware of, particularly relating to plant symbols.

The irrigation plan often accompanies the landscape plan. It shows all the water lines, control valves, and types of watering fixtures needed for irrigation. Regional climatic conditions will impact the requirements for the type of irrigation system to be specified (Figure 6.10).

Site Improvement Drawings

Large or complex projects often require separate drawings to clarify proposed site improvements including curbing, walks, retaining walls, paving, fences, steps, benches, and flagpoles.

Paving and curbing plans indicate the various types of brick, concrete, and bituminous paving and curbing to be used and the limits of each. This information allows for calculation and measurement of paving and curbing. The legend symbols inform you where one material ends and another begins. No assumptions should be made by the plan reader. Details showing sections through the surface are used to differentiate between thicknesses and between material and the substrate below.

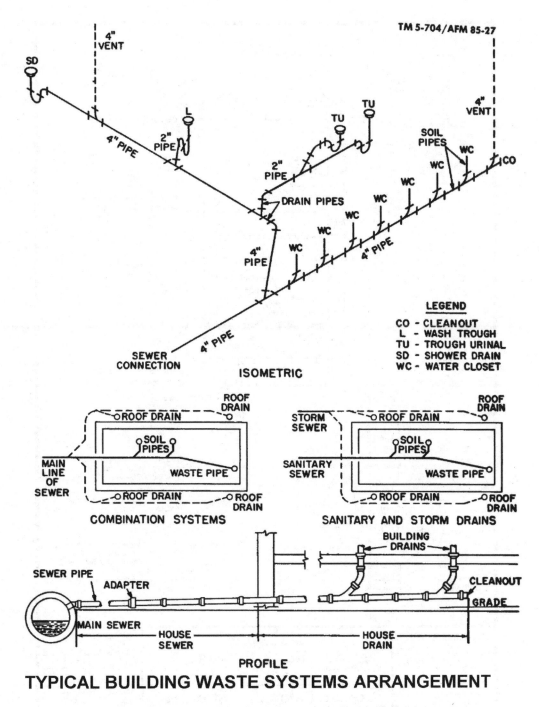

TYPICAL BUILDING WASTE SYSTEMS ARRANGEMENT

Figure 6.7 An isometric diagram of a typical building waste-system arrangement.

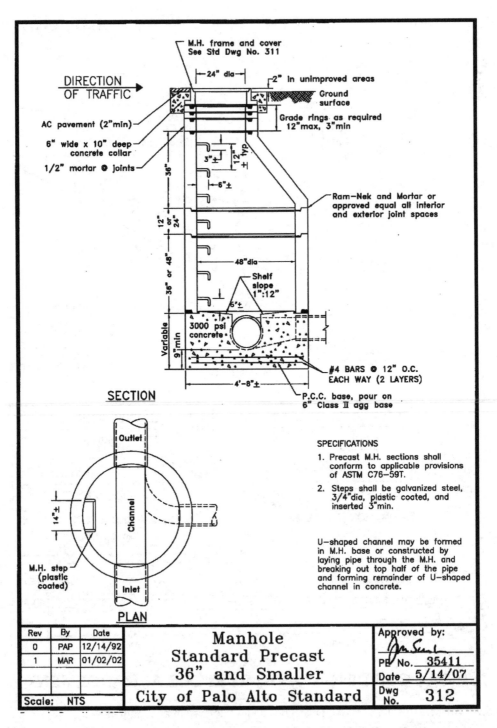

Figure 6.8A Standard precast manhole detail (source: City of Palo Alto, CA). Specification notes are placed in the bottom right-hand corner.

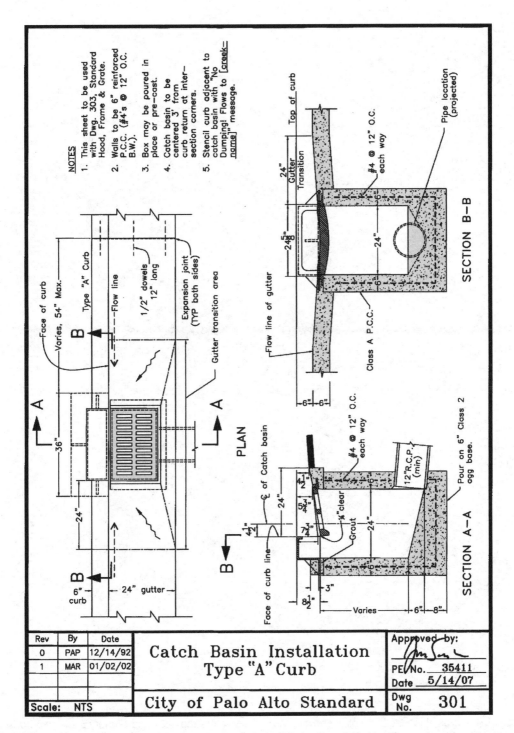

Figure 6.8B Catch-basin installation detail (source: City of Palo Alto, CA). Notes are placed on the top right-hand side of the sheet.

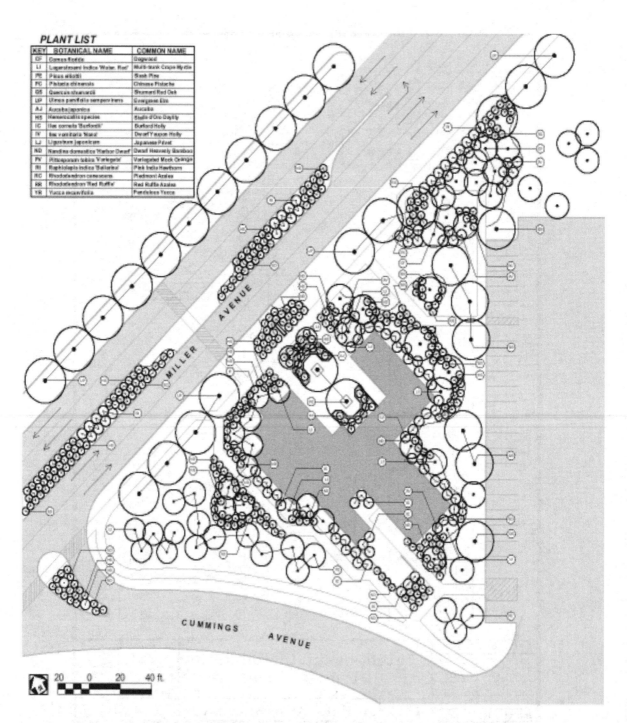

PLANT LIST

KEY	BOTANICAL NAME	COMMON NAME
CF	Cornus florida	Dogwood
LI	Lagerstroemia indica 'Watermelon Red'	Multi-trunk Crape Myrtle
PE	Pinus elliotii	Slash Pine
PC	Pistacia chinensis	Chinese Pistache
QS	Quercus shumardii	Shumard Red Oak
UP	Ulmus parvifolia sempervirens	Evergreen Elm
AJ	Aucuba japonica	Aucuba
HS	Hemerocallis species	Stella d'Oro Daylily
IC	Ilex cornuta 'Burfordii'	Burford Holly
IV	Ilex vomitoria 'Nana'	Dwarf Yaupon Holly
LJ	Ligustrum japonicum	Japanese Privet
ND	Nandina domestica 'Harbor Dwarf'	Dwarf Heavenly Bamboo
PV	Pittosporum tobira 'Variegata'	Variegated Mock Orange
RI	Raphiolepis indica 'Ballerina'	Pink India Hawthorn
RC	Rhododendron canescens	Piedmont Azalea
RR	Rhododendron 'Red Ruffle'	Red Ruffle Azalea
YR	Yucca recurvifolia	Pendulous Yucca

MILLER AVENUE

CUMMINGS AVENUE

20 0 20 40 ft.

Figure 6.9A Typical landscaping plan showing a plant list in the top left-hand corner. Scale and north point are placed in the lower left-hand corner. Leaders are used to identify plant type (source: United States Air Force Landscape Design Guide).

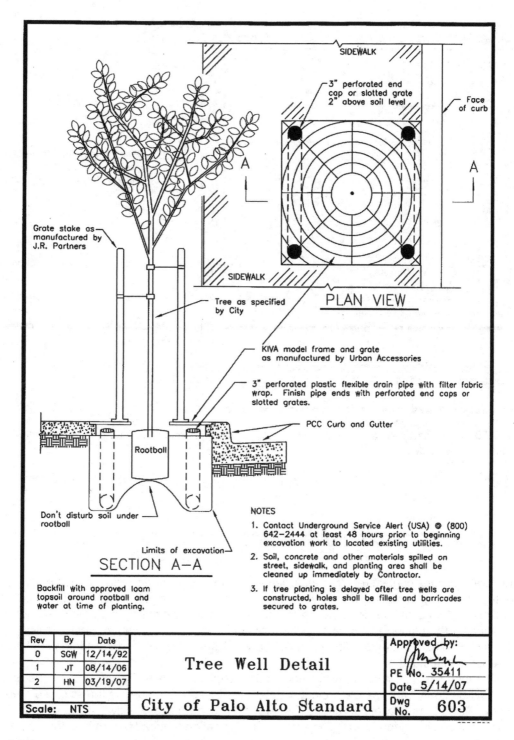

Figure 6.9B   Tree-well planting detail (source: City of Palo Alto, CA).

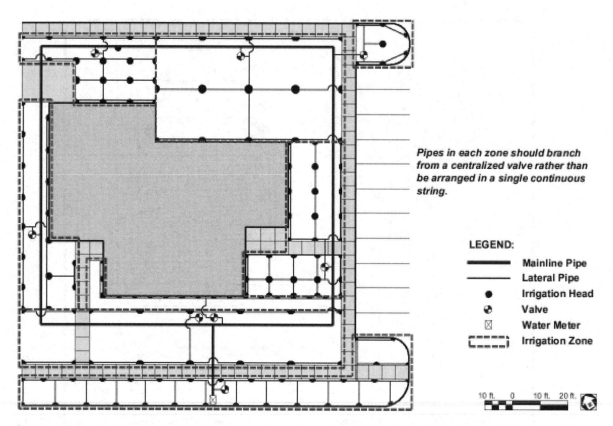

Pipes in each zone should branch from a centralized valve rather than be arranged in a single continuous string.

LEGEND:

▬▬▬	Mainline Pipe
———	Lateral Pipe
●	Irrigation Head
☉	Valve
⊠	Water Meter
⌐ ─ ─ ─ ┐	Irrigation Zone

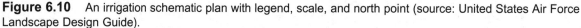

Figure 6.10   An irrigation schematic plan with legend, scale, and north point (source: United States Air Force Landscape Design Guide).

6.3 ARCHITECTURAL DRAWINGS

Architectural drawings contain required information on the size, material, and makeup of all main members of the structure, as well as their relative position and method of connection. In essence, they consist of all the drawings that describe the structural members of the building and their relationship to each other, including foundation plans, floor plans, framing plans, elevations, sections, millwork, details, schedules, and bills of materials. Architectural drawings are generally numbered sequentially with the prefix "A" for architectural.

The number of architectural drawings needed to effectively convey the scope of a construction contract and satisfactorily implement a project is determined by such factors as the size and nature of the structure and the complexity of operations. General plans consist of plan views, elevations, and sections of the structure and its various parts. The amount of information required determines the number and location of sections and elevations.

Plans

A plan is actually a part of the architectural drawing that represents a view of the project from above. The floor plan is the most common type of plan view. A floor plan is a two-dimensional view of a space, such as a room or building. It is a view of the space from above, as if the space were cut through horizontally at the windowsill level with the upper half removed. You are looking down at the floor. In general, a floor plan's main function is to identify and delineate the use of space. It identifies the locations and sizes of components such as rooms, bathrooms, doors, windows, stairs, elevators, means of egress, and room access (Figure 6.11). The floor plan will also show the locations of walls, partitions, doors, washrooms, and built-in furniture as well as dimensions and other pertinent information. When too much information is shown on a single plan, it becomes confusing, which is why very often, especially for complex projects, several different plans are required. These additional drawings may include demolition, partition, fixture, and floor-finish plans.

Architectural plans, when part of a working-drawing set, should be dimensioned to show actual length and width, thereby allowing the reader to calculate areas. Dimensions should be accurate, clear, and complete, showing both exterior and interior measurements of the space. For minor projects, a separate set of specifications is not always issued, depending on the financial constraints and on the assumption that the notes will suffice.

Plans are typically drawn to scale. The most common scales for floor plans (depending on size of project) are one-eighth inch = one foot (scale: 1/8 inch = 1 foot, 0 inches) and one-quarter inch = one foot (scale: 1/4 inch = 1 foot, 0 inches). The plan scale should always be noted on the drawing.

Elevations

Elevations are an important component of the construction-drawing set and the design and drafting process. Elevations are essentially views that show the exterior (or interior) of a building. They represent orthographic views of an interior or exterior wall. They are basically flat, two-dimensional views with only the height and width obvious. Exterior elevations provide a pictorial view of the exterior walls of a structure and indicate the material used (stone, stucco, brick, vinyl, etc.), the location of windows and doors, the roof slopes, and other elements visible from the exterior. Elevations are usually identified based on their location with respect to the headings of a compass (north, south, east, and west elevations). Alternatively, they may be labeled front, rear, right, and left elevations (Figure 6.12A). Four elevations are normally required to show the features of a building unless the building is of irregular shape, in which case additional elevations may be required.

The main function of exterior elevations is to provide a clear depiction of the façade treatment of the building and any changes in the surface materials within the plane of the elevation. They also show the location of exterior doors and windows (often using numbers or letters in circles to show types that correspond to information provided in the door and window schedule).

Elevations are typically drawn to the same scale as the floor plan. The scale of the elevation is noted either under or to the side of the title of the elevation or in the title block (Figure 6.12B). A common scale is one-quarter inch = one foot (scale: 1/4 inch = 1 foot, 0 inch), although a scale of one-eighth inch = one foot (scale: 1/8 inch = 1 foot, 0 inches) is used for larger buildings.

While floor plans show horizontal measurements of elements, elevations mainly provide vertical measurements with respect to a horizontal plane. These dimensions provide a vertical location of floor-to-floor heights, windowsill or head heights, floor-to-plate heights, roof heights, or a variety of dimensions from a fixed horizontal surface. These measurements can be used to calculate quantities of materials

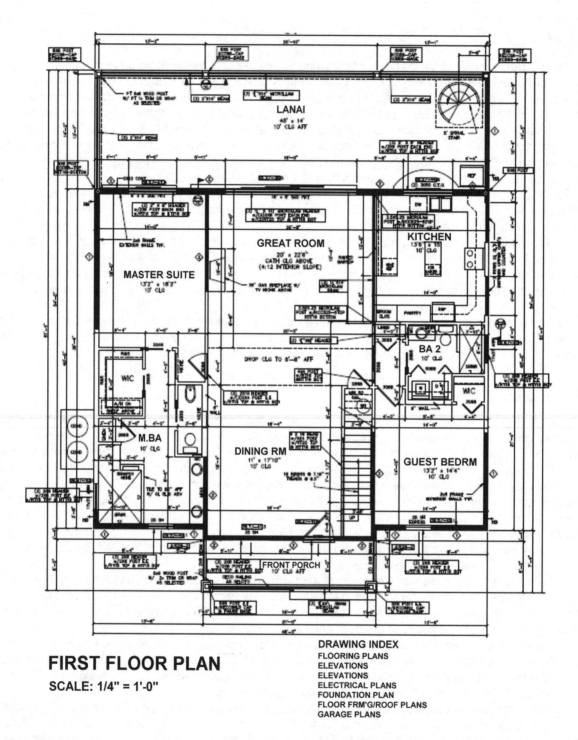

FIRST FLOOR PLAN

SCALE: 1/4" = 1'-0"

DRAWING INDEX
FLOORING PLANS
ELEVATIONS
ELEVATIONS
ELECTRICAL PLANS
FOUNDATION PLAN
FLOOR FRM'G/ROOF PLANS
GARAGE PLANS

Figure 6.11A Floor plans for a residence drawn to a scale of 1/4 inch = 1 foot, 0 inches. Additional notes are included on the original drawing.

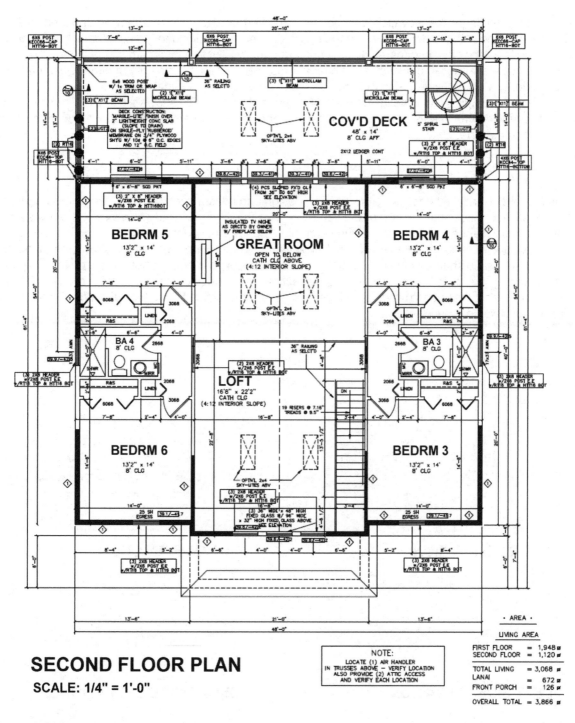

SECOND FLOOR PLAN

SCALE: 1/4" = 1'-0"

NOTE:
LOCATE (1) AIR HANDLER
IN TRUSSES ABOVE — VERIFY LOCATION
ALSO PROVIDE (2) ATTIC ACCESS
AND VERIFY EACH LOCATION

· AREA ·

LIVING AREA

FIRST FLOOR	= 1,948 ▫
SECOND FLOOR	= 1,120 ▫
TOTAL LIVING	= 3,068 ▫
LANAI	= 672 ▫
FRONT PORCH	= 126 ▫
OVERALL TOTAL	= 3,866 ▫

Figure 6.11B Floor plans for a residence drawn to a scale of 1/4 inch = 1 foot, 0 inches. Additional notes are included on the original drawing.

required. Sometimes the elevation dimensions are given as decimals (7.5 feet as opposed to 7 feet, 6 inches). Along with the dimensions on the elevations, notes are included to supplement and clarify information in a floor plan. In general, dimensions are usually kept to a minimum on elevations. Most consultants use elevations to depict sizes of major components, with the majority of dimensions placed on the sections, which provide greater clarity regarding construction materials and methods.

Interior elevations show the inside walls of a space. Figure 6.12C shows a kitchen elevation. Notice that the annotations take the form of specifications and are written at their appropriate location and not as notes to the side.

Sections

Sections are usually used to clarify the building design and construction process. Transverse and longitudinal sections are usually drawn at the same scale as the floor plan and show views of cross sections cut by vertical planes. A floor plan or foundation plan, cut by a horizontal plane, is a section as well as a plan view, but it is seldom called a section. They offer a view through a part of the structure not found on other drawings. To show as much construction information as possible, it is not uncommon for staggered (offset) cutting planes to be used in developing sections. To reduce the time and effort required for drafting and to simplify the construction drawings, it is common practice to use typical sections where exact duplications would otherwise occur.

FRONT ELEVATION   **REAR ELEVATION**

LEFT SIDE ELEVATION   **RIGHT SIDE ELEVATION**

Figure 6.12A Four exterior elevations of a house will normally accurately depict its features, including materials used and vertical heights.

FRONT ELEVATION C-11
SCALE: 1/4" = 1'-0"

Figure 6.12B Front elevation of a family residence. Notice that the roof material is not drawn but annotated. Likewise, the façade materials are only partially drawn (source: The Lessard Architectural Group, Inc.).

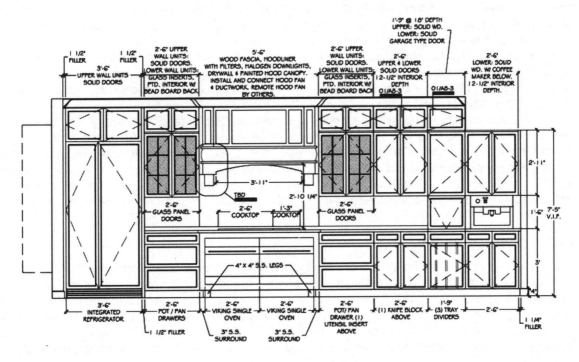

Figure 6.12C Elevation of a residential kitchen (source: JLC Studio).

Examples of building sections can be seen in Figures 6.13A and B. Figure 6.14A represents a section through a staircase drawn to a 3/8 inch = 1 foot, 0 inches scale. Several different sections may be incorporated into the drawings. Sections taken from a plan view are called cross-sections; those taken from an elevation are referred to as longitudinal or simply wall sections. Figure 6.14B shows a typical wall section. This type of section is commonly drawn at a scale of 3/4 inch = 1 foot, 0 inches. It is normally located in the structural division and provides information that is necessary to understand the structural arrangement, construction methods, and material composition of the walls of the building. Using sections in conjunction with floor plans and elevations allows the reader to get a better understanding of the project and how it is put together.

Sections are used as needed in each of the main divisions of construction drawings to show the types of construction required, the types and locations of materials used, and the method of assembling the building parts. Although they may be used in each of the divisions, the most common are the architectural and structural divisions. All sections are important to those responsible for constructing a building.

6.4   STRUCTURAL DRAWINGS

The structural drawings provide the reader with a view of the structural members of the building and how they will support and transmit its loads to the ground. Structural drawings (often referred to as "structurals") are sequentially numbered beginning with an "S," as in S-1, S-2, S-3, etc. They are normally located after the architectural drawings in a set of working drawings. For new construction, structural-engineering drawings will be needed for foundation and footing details, the structural frame design, beam sizes, and connections. In concrete structures, the structural drawings will indicate concrete forming details, dimensions of members, and reinforcing-steel requirements. If the structure is steel-framed, the size and type of steel framing will be indicated.

The benefit of structural drawings is that they provide information that is useful and can stand alone for subtrades such as framers and erectors. The structural drawings clearly indicate main building members and how they relate to the interior and exterior finishes without providing information that is not necessary for this stage of construction.

Structural drawings, like architectural drawings, start with the foundation plans, ground or first-floor plan, upper-floor plans, and the roof plan. The main difference is that only information pertinent to the structural systems is shown. For example, a second-floor structural plan would show the wood or steel framing and the configuration and spacing of load-bearing members but not doors or non-load-bearing partitions. Following the plan views are the sections and details in the same basic format as in the architectural drawings. Schedules are used to record such information as footings, columns, and trusses.

Types of Foundations

Houses and small frame buildings do not need complicated foundation systems. A simple inverted-T foundation is all that is normally needed to support the structure under normal conditions. Larger and more complex buildings impose a heavier burden on the foundation system and need to be carefully designed by structural engineers. Foundations for large commercial buildings perform the same functions as those for light-frame structures. The main difference in the foundations for a commercial building and that for a small residence is often the thickness of the concrete and the amount of reinforcing steel.

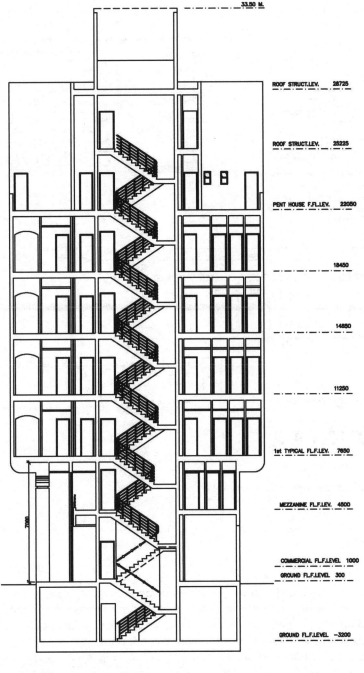

SECTION A-A

Figure 6.13A A. Cross-section of a multistory building displaying incomplete vertical dimensioning.

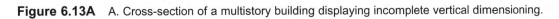

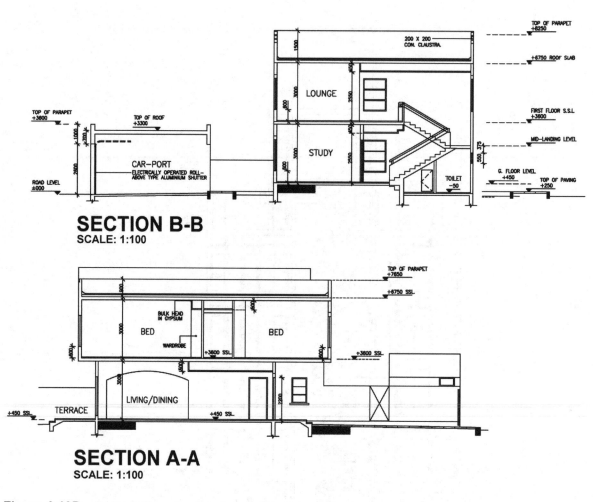

SECTION B-B
SCALE: 1:100

SECTION A-A
SCALE: 1:100

Figure 6.13B Sections through a two-story residence.

Foundations are usually constructed of continuous footings, pilings, or grade beams. Continuous or spread foundations are commonly used in residential and light-commercial construction. This type of foundation is based on a footing and wall. Concrete footings form the base of the foundation system and are used to displace the building loads over the soil. Piling foundations are typically used when conventional trenching equipment cannot be used safely or economically. Piling is a form of foundation system that uses columns to support the loads of the structure. They are rarely used for residences or low-rise commercial buildings. Grade beams are reinforced-concrete beams positioned at grade (ground) level below the stem wall to provide a bearing surface for the superstructure. They can be used in place of the foundation to provide added support for a foundation in unstable soil.

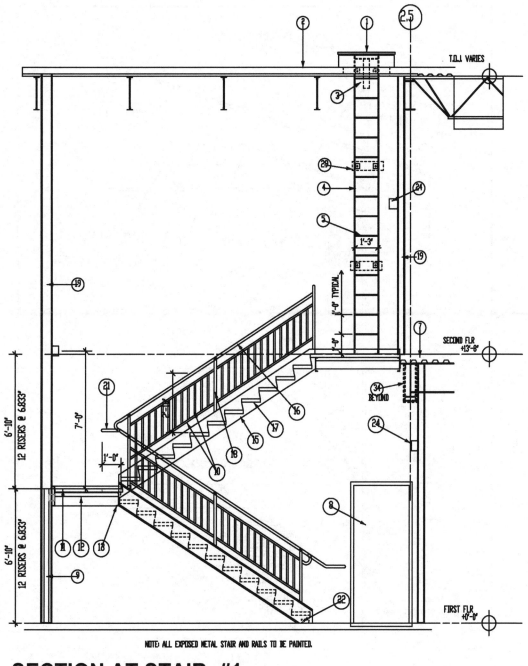

NOTE: ALL EXPOSED METAL STAIR AND RAILS TO BE PAINTED.

SECTION AT STAIR #1

SCALE: 3/8" = 1'-0"

Figure 6.14A   Section through a staircase. Scale is shown as 3/8 inch = 1 foot, 0 inches.

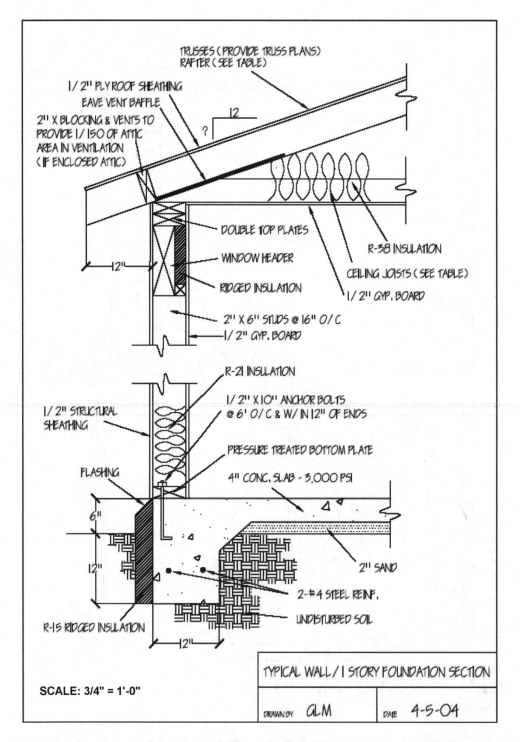

Figure 6.14B   Wall section through the foundations and roof at 3/4 inch = 1 foot, 0 inches scale.

Foundation Plan

A foundation plan (Figure 6.15) is a plan view of a structure projected onto an imaginary horizontal plane passing through at the level of the top of the foundation. A foundation sheet will indicate the size, thickness, and elevation of footings (footers), with notes regarding the placement of reinforcing bars (rebars). It will typically note locations for anchor bolts or weld-plate imbeds for structural steel and other elements. The footing schedule is normally found on the first sheet of the structural notes, which also includes notes regarding the reinforcing requirements and other written statements for structural strengths and testing requirements. To be able to properly interpret a foundation plan, you must first view the other plans, such as the sections and roofing plan.

The foundation plan shown in Figure 6.15 tells you that the main foundation of this structure will consist of an 8-inch masonry stem wall, centered on a 24-inch-wide by 12-inch-deep concrete footing reinforced with three #5 bars. The width and depth of the concrete footing will vary according to location. The drawing also shows a 4-inch-thick concrete slab reinforced with welded wire mesh. Besides the outside wall and footing, there will be two 30-inch-, 36-inch-, or 40-inch-square concrete footings reinforced with three #5 bars in each direction. From the drawing we can see that ½-inch-by-10-inch anchor bolts are used at 32-inch centers typically located on the building's perimeter.

Framing Plan

The framing plan will indicate the material used for framing the building and may include wood or metal studs, concrete-masonry units, or structural steel. Framing drawings include the basic skeletal structure of the building and are drawn to scale. Floor-joist locations, walls, and roof trusses are part of the overall detail of these plans. Generally, locations of each stud are not included, since the process is standard. However in some cases there are instructions for particular wall-construction methods.

Intermediate structural framing plans are used for multistory construction, where each level may require support columns, beams, joists, decking, and other elements. Structural drawings also typically incorporate numerous details relating to the structure. Figure 6.16 shows typical examples of structural details.

6.5   MECHANICAL DRAWINGS

The cover sheet for mechanical drawings should contain appropriate notes, legends (chart or table of symbols and abbreviations), and details. The mechanical plan specifies the design of or the modifications to the mechanical system, ductwork layout and dimensions, mechanical equipment location, damper locations, design air-delivery rates, diffuser locations, thermostat locations, and supplemental cooling systems if required. Mechanical plans are normally identified as M-1, M-2, M-3, etc. Some consultants prefer the heating, ventilating, and air-conditioning drawings, commonly referred to as the HVAC drawings, to be sequentially numbered and prefixed by the letter "H"; the plumbing drawings to be prefixed by the letter "P"; and the fire-protection drawings to be prefixed by the letters "FP." Most of the work shown on these types of drawings is in plan view. Because of the diagrammatic nature of mechanical drawings, the plan view offers the best illustration of the location and configuration of the work.

Due to the large amount of information required for mechanical work and the close proximity of piping, valves, and connections, the engineer utilizes a variety of symbols and abbreviations to convey the design intent. Examples of these symbols and their meaning can be found in Chapter 8.

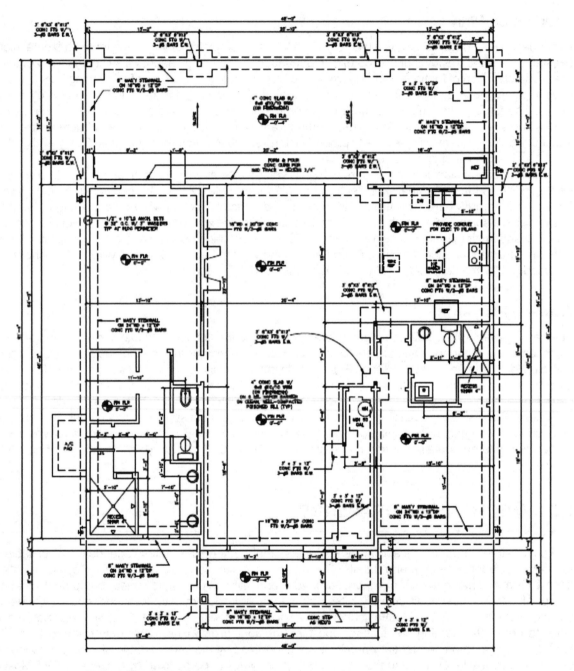

FOUNDATION PLAN

SCALE: 1/4" = 1'-0"

Figure 6.15   Foundation plan for a residence.

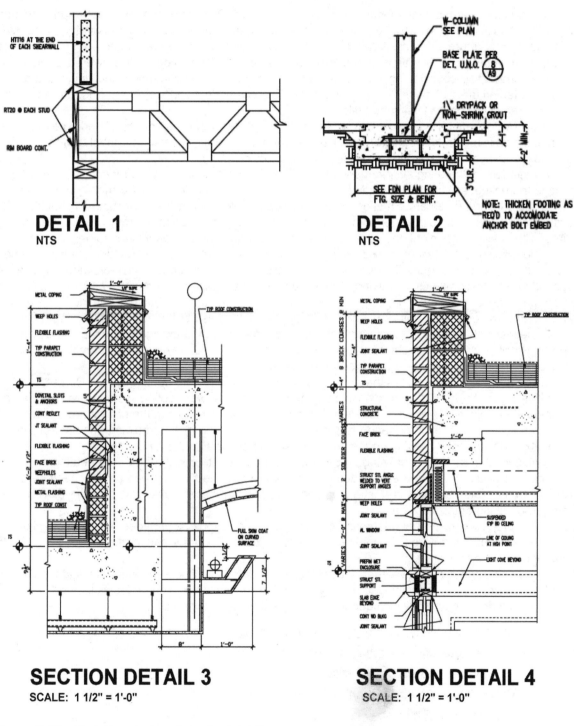

Figure 6.16 Typical examples of structural details.

Mechanical systems deal with the heating and cooling of buildings or spaces. The two primary methods of heating and cooling use air or water. In an all-air system hot or cold air is transported to the space with supply and return air ducts. A typical example is a residential forced-air furnace. The furnace uses gas or oil to heat the air. The air is forced through the ductwork by an electrically powered fan in the furnace. A separate air-conditioning unit is installed for cold air. For most commercial buildings, a large unit, often located on the roof, powers the all-air system. Supply-air ductwork, registers, and return-air grilles are required in all spaces within the building.

An all-water heating system uses a type of coil through which hot water is circulated. The most common example is the fin-tube radiator found in older homes, typically located in front of a window. Today the most common system is the radiant floor heating panel.

An all-electrical heating system uses electricity to heat elements within a radiator. The most common is the baseboard heater. It is used when a furnace is not installed. For example, many small cottages use baseboard heaters. Smaller, older commercial buildings rely on a baseboard installation. This system can also be found in larger commercial buildings as an addition to other systems. An electric radiator with a built-in fan might be located at an exterior entrance door to provide extra heat on the inside.

The mechanical drawings provide the client, the builder, and the permit department with the complete HVAC layout for the job. These drawings are typically part of the construction-drawing set. They are submitted with the construction drawings for a building-permit application (Figure 6.17). They are also part of the package for pricing the project. They are used for construction. All ducting, venting, exhaust fans, and heating and/or cooling units must be supplied and installed as per the approved drawings.

A mechanical-engineering consultant produces the mechanical drawings. Often the same person or company will produce the electrical and plumbing drawings. These drawings must comply with the various building codes including all provincial and local codes.

Generally, the engineer uses these plans and incorporates his/her ducting layout. Diffusers, return-air grilles, and exhaust fans are drawn in as symbols. Heating and/or cooling systems are specified and their location indicated. Legends, schedules, details, and notes specific to the project are added.

On small projects, all information required is covered on one or two drawing sheets. For large or complex projects, many drawing pages are necessary to cover all areas of the project.

Typically, the engineer's drawings must note the type, location, and number of heating and/or air-conditioning units. HVAC and electrical connections are specified, as well as any connections to gas lines or water systems. The thermostat type, location, and number are also noted. Figure 6.18 shows a refrigerant-piping detail diagram

Many projects require that heat-loss and heat-gain calculations be provided. Air-balancing information or air-distribution-device schedules are usually included. The information required depends on the type of project being built.

Many cities and towns have energy-conservation regulations. The engineer's drawings must abide by all codes and bylaws pertaining to the city, town, or province where the project is located.

The following are typically included in a set of mechanical drawings:

- Plans showing the size, type, and layout of ducting
- Diffusers, heat registers, return-air grilles, and dampers
- Turning vanes and ductwork insulation
- HVAC unit types, quantities, and location
- Thermostat types, quantities, and location
- Electrical, water, or gas connections

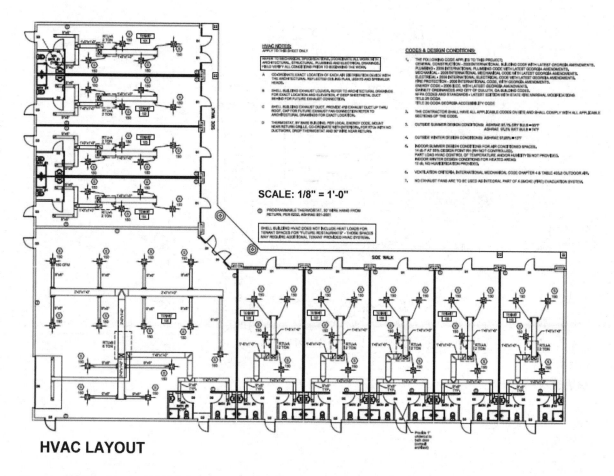

HVAC LAYOUT

Figure 6.17A Typical HVAC floor layout showing ducting drawn to a scale of 1/8 inch = 1 foot, 0 inches. HVAC notes and code and design conditions are included on the sheet.

- Ventilation and exhaust fans
- Symbol legend, general notes, and specific key notes
- Heating and/or cooling load summary

Other information, depending on the complexity of the project, may include:

- Connection to existing systems
- Demolition of part or all of existing systems
- Smoke detector and firestat for ducting
- Thermostat programming

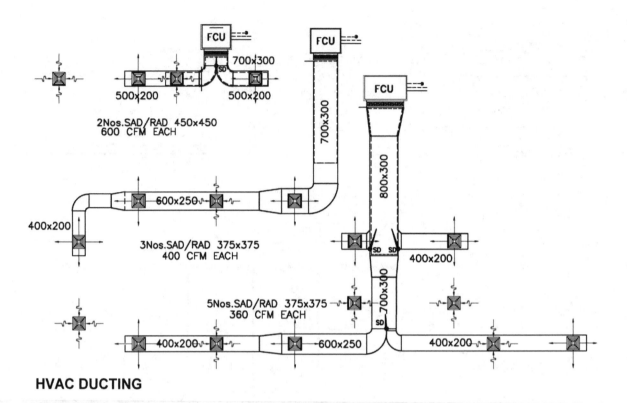

HVAC DUCTING

Figure 6.17B   Diagram showing HVAC ducting sizes, connections, and layout.

- Heat-loss and heat-gain calculations per area
- Round-duct, turning-vane, and lay-in-diffuser details
- Special conditions, such as seismic restraint codes

Engineer's drawings are required for all commercial projects involving HVAC work, including additions, renovations, or new construction. A permit is required prior to commencing any on-site work.

Drawings and permits are also needed for residential projects when any substantial work related to HVAC is to take place. For small projects, a licensed mechanical contractor can provide the information required to obtain a permit.

Concept and designs are the first stage of any project. When established, the next stage is construction drawings. Once the floor and reflected-ceiling plans are complete, they are passed to the engineer to produce the mechanical drawings. The engineer's drawings become part of the construction drawing set.

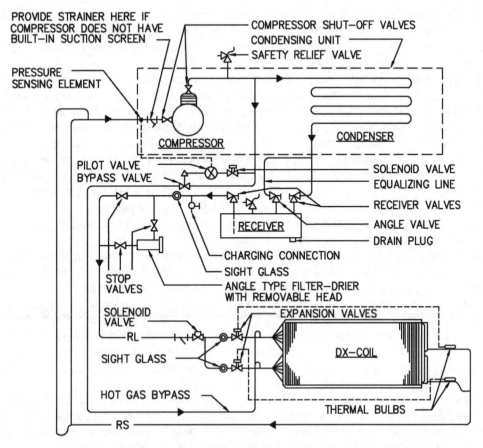

PROVIDE STRAINER HERE IF
COMPRESSOR DOES NOT HAVE
BUILT—IN SUCTION SCREEN

COMPRESSOR SHUT—OFF VALVES
CONDENSING UNIT
SAFETY RELIEF VALVE

PRESSURE
SENSING ELEMENT

COMPRESSOR

CONDENSER

PILOT VALVE
BYPASS VALVE

SOLENOID VALVE
EQUALIZING LINE
RECEIVER VALVES
ANGLE VALVE
DRAIN PLUG

RECEIVER

CHARGING CONNECTION
SIGHT GLASS
ANGLE TYPE FILTER—DRIER
WITH REMOVABLE HEAD

STOP
VALVES

SOLENOID
VALVE

RL

SIGHT GLASS

HOT GAS BYPASS

EXPANSION VALVES

DX—COIL

THERMAL BULBS

RS

REFRIGERANT PIPING DETAIL (AIR CONDITIONING)

Figure 6.18 Schematic diagram for a refrigerant-piping detail for air conditioning.

6.6 PLUMBING DRAWINGS

Plumbing drawings provide all pertinent information on the design of the plumbing system for a project, including line sizes and location, fixture location, isolation valves, storage-tank capacities, hot-water-heater capacities and locations, and drain locations and routing. Plumbing systems involve two major components, water supply and drainage. Water is supplied under pressure through pipes to plumbing fixtures. Drainage works by gravity: Drain pipes must slope downward. Vent pipes are required. A plumbing floor plan will typically show the location and type of plumbing fixtures, as well as the route pipes will be run (overhead or through walls) for potable water, drainage, waste, and vents. Plumbing drawings are usually numbered beginning with "P," as in P-1, P-2, etc.

The first component connected to a fixture is a trap. Traps are located at every fixture. A trap is the u-shaped pipe found below a sink. Some traps are part of the design of the fixture and are not visible, as in a toilet or double sink. The trap catches and holds a small quantity of water to provide a seal. This seal prevents gases from the sewage system entering the building.

From the trap sewage travels through drainage pipes in branch lines to a vertical stack. A soil stack carries waste from toilets. A waste stack carries the other waste from a sink, washing machine, or dishwasher. All drainage pipes must be connected to vents. Vents are open to the outside air. Vents allow built-up sewage gases to escape and pressure in the system to equalize. Figure 6.19 shows a schematic isometric of a two-bath plumbing system and the various connections and outlets needed.

Plumbing drawings are typically part of the construction-drawing set. In most cases, they are submitted with the construction drawings for a building-permit application. They are also part of the package for pricing the project for the client. They are used for construction. All related plumbing lines; drains, connections, and vents must be installed according to the approved drawings.

A mechanical-engineering company produces the drawings. They must comply with the National Plumbing Code and with national, provincial, and local codes.

Engineers produce their own drawings. They are based on plans provided by the interior designer or architect. These plans show the engineer the location of plumbing fixtures such as toilets, sinks, and water heaters in the design. Some projects require piping for equipment as well.

Generally, the engineer draws a plumbing plan and connection diagrams. Typical diagrams are of the water-supply system and the sanitary stack. Legends, schedules, and notes specific to the project are added. On small projects, there are usually only a few fixtures, a sink and a toilet. In this case, the required information is included on the mechanical-drawing sheets. For large or complex designs, the plan(s), diagrams, notes, etc., are on separate sheets. Several sheets may be required to cover all the information. Figure 6.20 shows a plumbing and sprinkler layout plan.

The engineer's drawings must provide information regarding the connections to the main water and sanitary sewer lines. The layout of any existing and new piping is indicated on the plan. The size for all lines for water, sanitary, and venting must be noted. The hookup to the water meter, new or existing, is covered, and the type, size, and location of the water heater are specified.

The following are typically included in a set of plumbing drawings:

- A plan with lines and symbols representing all piping
- Symbol legend, general notes, and specific key notes
- Fixture schedule, specifying the manufacturer and model for each item
- The sizes for all piping, cold/hot water, sanitary, vent lines, etc.
- Diagrams, such as water riser and sanitary stack
- Information regarding the water heater

Other information may be needed, depending on the complexity of the project:

- Details drawings, such as water heater, water-meter connection, or floor drains
- Diagrams or details referencing special equipment requirements
- Fire-protection notes
- Fire-sprinkler notes and symbols
- Special-air lines
- Natural-gas lines

A–2" re-vent. 3'-6" above floor
B–2" Vent, 6" Through Roof &
10" from cooler
C–3" Cleanout
D–1 1/2" Waste Line
E–Sanitary Tee
F–Fitting Double fixture
G–Combination Wye & eighth
Bend
H–2" Clothes WasherTrap 6"
to 10" above floor
I–2" Clothes Washer
Standpipe
J–1/1/2" Plumbing Vent 6"
above Roof
K–2" Cleanout
L–Sanitary Tee
M–2" waste Line
N–2" Waste Line
O–2"Cleanout
P–1 1/2" vent Thru Roof
Q–Sanitary Cross
R–Trap Arm
S–3" Waste Line
T–2" Shower Trap
U–Tub Trap
V–Cast Iron–90 deg Short
Swep
W–3"x2" Side Inlet "T"
X–2" Way Cleanout

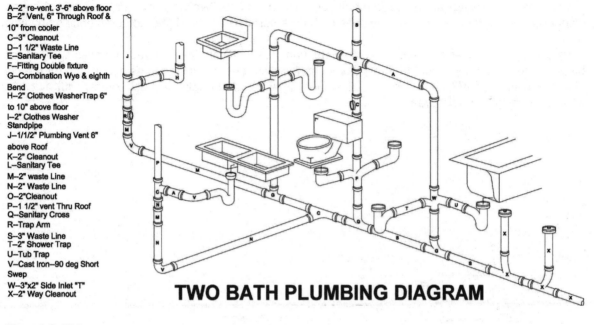

TWO BATH PLUMBING DIAGRAM

Figure 6.19A Isometric diagram of a two-bath plumbing system.

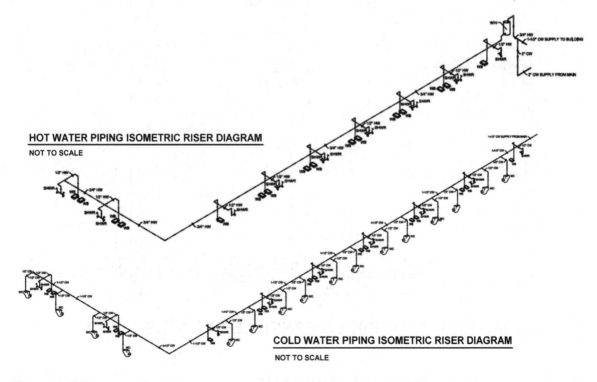

HOT WATER PIPING ISOMETRIC RISER DIAGRAM
NOT TO SCALE

COLD WATER PIPING ISOMETRIC RISER DIAGRAM
NOT TO SCALE

Figure 6.19B Isometric piping diagrams of hot- and cold-water riser systems.

Engineer's drawings are required for all commercial projects involving any plumbing work. This applies to additions, renovations, or new construction. A permit is required prior to commencing any work on site.

Building codes specify the number of toilets, urinals, and lavatories required in a building or space, based on the occupancy type. In many cases the facilities must be designed as accessible for the disabled as discussed in Chapter 11. The designer, architect, and engineer must comply with all codes when producing their final drawings.

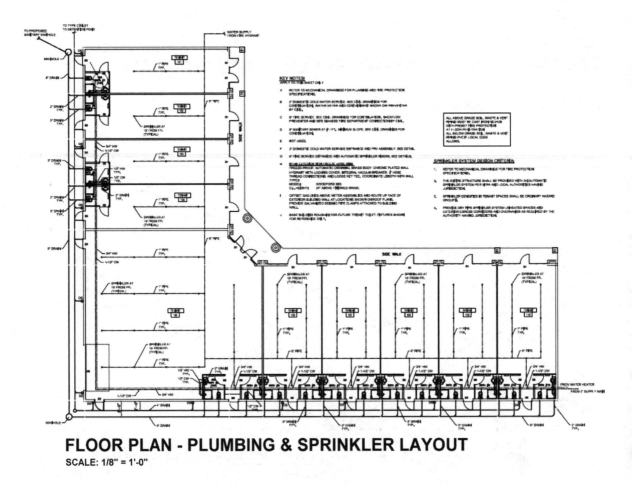

FLOOR PLAN - PLUMBING & SPRINKLER LAYOUT
SCALE: 1/8" = 1'-0"

Figure 6.20   Floor plan showing plumbing and sprinkler layout.

Drawings and permits are also needed for residential projects when substantial plumbing work is to take place. For small projects, a licensed plumber can submit the information required to obtain a permit.

A professional engineer also provides the required drawings and reports for a septic-tank installation. A sewage permit must be obtained. A septic tank is installed where a sanitary sewage connection to a municipal treatment facility is not possible.

Concept and designs are the first stage of any project. When established, the next stage is construction drawings. Once a floor plan is complete, it is passed over to the mechanical engineer to produce plumbing drawings. The drawings become part of the construction-drawing set.

In some residences and commercial structures, a separate plumbing plan is drawn to show fixtures, water-supply and waste-disposal lines, equipment, and other supply and disposal sources. These isometric drawings are much easier to understand and are invaluable to those responsible for preparing material estimates and to the craftspeople responsible for installing plumbing systems. The mechanical division of a set of construction drawings will include, in addition to plumbing plans and details, drawings for any heating, ventilation, and air-conditioning systems that a building might contain. Frequently, the drawing sheets in the mechanical division are identified by the designating letter M in the title block. However, remember that in the order of drawings, sheets containing heating, ventilation, and air-conditioning drawings will precede those for plumbing.

6.7   ELECTRICAL DRAWINGS

The final group of working drawings is usually the electrical drawings. Architects usually hire electrical consultants to design the electrical services in buildings (unless they have electrical engineers within the firm). The electrical drawings show the various electrical and communication systems of the building, and they provide the client, the builder, and the permit department with the complete power layout for the project. The electrical cover sheet indicates all electrical specs, notes, and panel schedules. This sheet includes the specification of supplemental electrical panels if required. These drawings are typically part of the construction-drawing set. They typically include electrical power and lighting plans, telecommunications, and any specialized wiring systems such as fire or security alarms.

Electrical drawings show the location of electrical circuits, panel boxes, and fixtures throughout the building, as well as switchgears, subpanels, and transformers when they are incorporated in the building. The electrical drawings are submitted with the construction drawings for a building-permit application. In some cases, they are submitted separately to obtain an electrical permit. Electrical plans are normally numbered E-1, E-2, E-3, etc.

Power Plan

The power plan is a drawing of the floor plan showing all required outlets, locating panels, receptacles, and the circuitry of power-utilizing equipment and special systems. The designer or architect will often draw a power plan and dimension these locations. This is important to the engineer, especially if the connection must be at a specific location or height or if it is floor-mounted or mounted within a fixture. Figure 6.21 shows two examples of electrical layouts, one for a residence and the other for a commercial installation.

If a power plan is not provided, then the engineer draws it. This document indicates all outlets and circuitry, electrical-distribution system, riser location, routing of service, design voltage and amperage, and transformer size and location. It is an engineering drawing separate from the architectural set.

The engineer draws in the circuiting for every power receptacle. The circuit tie to the electrical panel is noted. A legend provides a description for each symbol used on the plan. Conduit size, special power specifications, and notes are included. A project with special equipment or systems requires additional legends, wiring schedules, or diagrams.

Electrical systems provide power for lighting, outlets, and equipment. The local power company supplies electricity to the building, including the meter. A licensed electrician can install an electrical panel of the appropriate type and size.

Circuit breakers in the panel trip off if a circuit is overloaded. It is often an indication that there is a problem with an appliance or other equipment or that too many items are plugged into one outlet source. Outlets in wet areas must be grounded. Ground-fault-interrupter (GFI) outlets are required in bathrooms, kitchens, or outdoor areas.

Special power outlets, called dedicated circuits, are placed on their own circuit. They are used for sensitive equipment such as computers or equipment that requires voltage greater than 120 volts, including outlets for electric ranges, large copy machines, or other special equipment.

An electrical engineer designs the system for a commercial or large residential project. An electrical contractor can design the system for smaller residential projects. The drawings are part of the package for pricing the project for the client and are used for construction. The electrician must wire all outlets, lights, and panels according to the approved drawings.

The National Fire Protection Association publishes the National Electrical Code, which specifies the design of safe electrical systems. Electrical engineers and electricians should know this code, as it is an accepted standard for electrical installations. They should also be familiar with any state or local codes that apply.

Reflected Ceiling Plan

A reflected ceiling plan (RCP) is a drawing of a room or space looking down at the interior ceiling. The designer or architect produces this plan to graphically show the ceiling treatment, ceiling grid, and the placement of all light fixtures as well as light fixtures to be relocated or removed. The plan indicates the type of ceiling (acoustical tile, gypsum board, etc) and the ceiling heights. The location of all light fixtures, speakers, special lighting, ceiling outlets, and switch locations is indicated and labeled. A ceiling-fixture legend is also normally included to provide a description for each symbol.

The engineer creates the electrical drawing for the RCP. It is an engineering drawing separate from the architectural set. The engineer prepares his/her own drawings based on floor and reflected ceiling plans provided by the interior designer or architect. The plans indicate to the engineer the location of light fixtures, special ceiling features, toilet rooms, and any equipment requiring special venting.

The engineer's drawing shows the circuiting and switching for each item on the ceiling. The circuit connection to the panel is labeled. Conduit size when required, legends, and general and/ or specific notes are provided on this drawing.

The RCP is basically a view of the ceiling from above. It is as if you were floating above the ceiling and looking down at it (Figure 6.22). This view will show the location of light fixtures, drywall or T-bar ceiling patterns, and any items that may be suspended from the ceiling. Figure 6.22 also shows a lighting layout plan.

Many projects will require electrical and mechanical drawings. The interior designer or architect will provide the electrical and/or mechanical engineer with their reflected-ceiling-plan design. The engineer will add the required information.

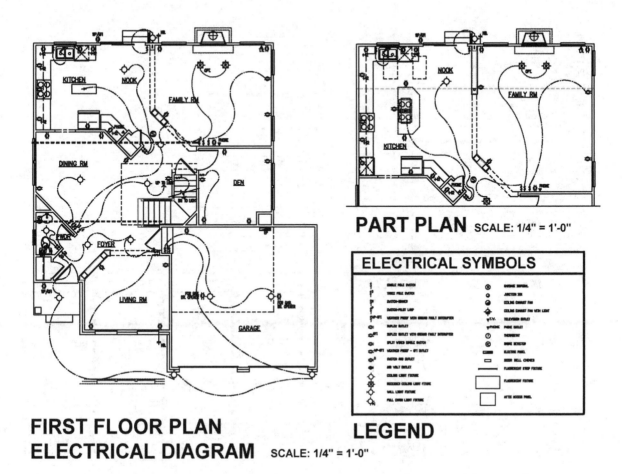

Figure 6.21A   Electrical diagram for a small residence showing location of switches and fixtures.

A reflected ceiling plan is drawn to scale. This means that the plan is measured proportionally to a size that will fit on a drawing sheet. A reflected ceiling plan will most often be at the same scale as the floor plan. A common scale is one-quarter inch = one foot (scale: 1/4 inch = 1 foot, 0 inches). The reflected ceiling plan will be titled with the scale noted at the bottom.

In a reflected ceiling plan each light fixture has an identification letter. A light-fixture legend is included on the actual drawing sheet. In the legend, each fixture is listed with its letter and a specification. The numbers in the hexagon shape are key notes. They describe items on the reflected ceiling plan. For example: number 2 would list the specification for the T-bar ceiling. Ceiling heights are noted in the oval shape.

The engineer's drawings must specify the type, location, and number of panels. On large or complex projects a circuit-breaker layout is included. This can take the form of a legend or diagram which typically includes number of panels and ampere loads.

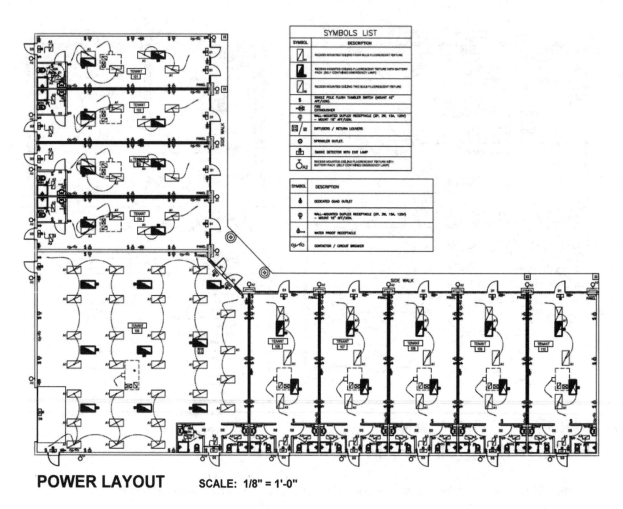

POWER LAYOUT SCALE: 1/8" = 1'-0"

Figure 6.21B Power-plan layout for a commercial installation drawn to a scale of 1/8 inch = 1 foot, 0 inches. Both plans display electrical symbols and legends.

The engineer must provide a load summary of the total connected load (amps/watts) for all items shown on the power and RCP drawings. This ensures that the main service is adequate. Many cities and towns have energy-conservation regulations regarding electrical loads. The engineer's drawings must abide by all codes and bylaws pertaining to the city, town, or province where the project is located.

The following are typically included in a set of electrical drawings:

- Type and location of outlet, (duplex, dedicated, isolated ground, GFI, etc.)
- Size and type of conduits (data, communication, phone)
- Volts of switches, wiring, and circuitry
- Lamps and model numbers of light fixtures

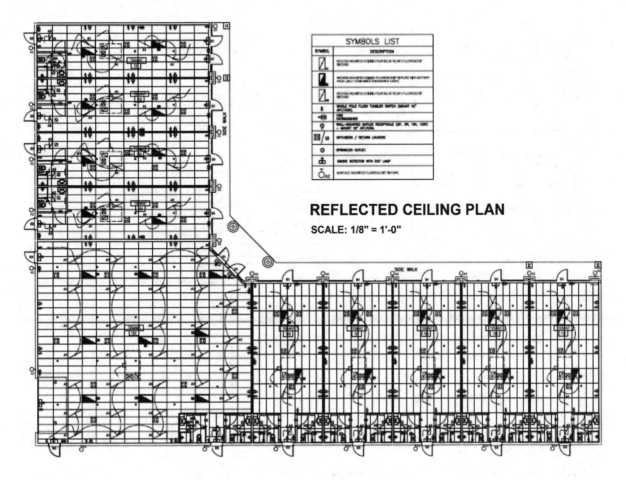

Figure 6.22A Reflected ceiling plan drawn to a scale of 1/8 inch = 1 foot, 0 inches.

Other information may be needed, depending on the complexity of the project:

- Direct connections (junction boxes, etc.)
- Emergency lighting and exit signs
- Alarm and security systems
- Fire-alarm systems
- Sound systems, speakers, monitors, and cameras
- Special equipment (kitchen, entertainment)
- Special technical devices (computers, gauges, medical, etc.)
- Special wiring (signs, heating, saws)

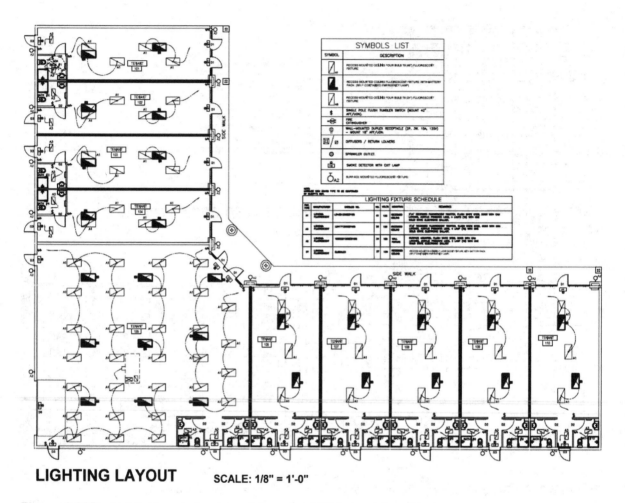

LIGHTING LAYOUT   **SCALE: 1/8" = 1'-0"**

Figure 6.22B   Lighting layout plan drawn to a scale of 1/8 inch = 1 foot, 0 inches.

Engineer's drawings are required for all commercial projects involving electrical work, including additions, renovations, and new construction. A permit is required prior to commencing any on-site work. Drawings and permits are also needed for residential projects when substantial electrical work is to take place. For small projects, a licensed electrician can provide the information required to obtain an electrical permit.

Concept and designs are the first stage of any project. When established, the next stage is construction drawings. Once the floor and reflected ceiling plans are complete, they are passed to the engineer to produce the electrical drawings. The engineer's drawings become part of the construction-drawing set.

6.8 MISCELLANEOUS DRAWINGS

Millwork

Millwork refers to custom, shop-built wood components for interior-finish construction for both residential and commercial work. These are items such as custom wood chair rails, bases, built-in bookcases, paneling, doors, cash units, display fixtures, and so on.

The project interior designer or architect produces the millwork drawings. These drawings provide information about each custom-designed piece. Styles, dimensions, types of wood, finishes, and desired details are drawn and noted.

Millwork items are primarily built of wood, although glass and metal parts may also be included within a fixture. For example, a cabinet for entertainment equipment can have a metal swivel device for the television and a metal rack for DVD storage. A display case may have a wood base with a glass showcase on top.

Wood may include solid wood, wood veneer, MDF, particle board, or plywood. Information on the drawing indicates the specific type of wood—maple, pine, oak, etc.—and whether it is solid or veneer. The finish is also noted—stained, lacquered, or laminate, for example.

On smaller projects, millwork drawings are typically pages within the construction-drawing set. On larger projects it is common practice to create a separate set of drawings. As a separate set the pricing and construction process is simplified. The company working on this portion of the project has all items clearly laid out.

A separate set of drawings also works well for chain stores. The head office will have the custom fixtures designed and drawn. They will then contract out large quantities of standard fixtures directly to a woodworking shop. This enables them to mass-produce items for their stores, such as cash units or display cases. By producing more than one at a time they are able to negotiate a better overall price.

Typically, the general contractor subcontracts the millwork portion of a project to a woodworking company. The contracted company will produce shop drawings when a request is noted on the design drawings. Shop drawings show exact construction methods in detail, including the finish. They are submitted to the interior designer or architect for approval before building starts. This ensures that items are built as intended for the project. Approved shop drawings are then used for building.

Figures 6.23A,B and C show examples of millwork drawings. The drawings have been reduced to fit on the page and are meant only to provide an overview and not to show specific information.

The drawings will typically show a plan view of the item. Front, side, and rear elevations are used when required to explain the shape. Sections are used when necessary to provide information for various segments. All are dimensioned. Specific materials and finishes are noted.

Shop-fabricated units are preassembled in the shop and shipped to their location in one piece. Field-installed components such as baseboards are shipped in lengths to the job site. They are put in place and fitted on site, using metal fastenings such as screws or blind nailing in combination with the joint.

Details

Architectural details are essentially enlarged drawings of specific construction assemblies and are normally provided by the architect or structural engineer. Their main purpose is to offer greater clarification and understanding where required to implement a project. Contractors frequently request additional construction details during the execution stage. When an area of construction is drawn to a larger scale in

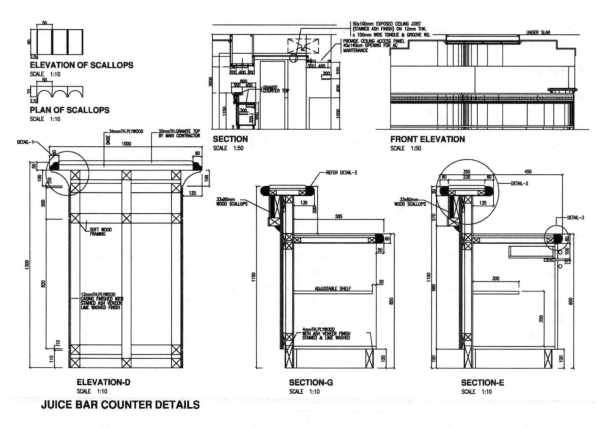

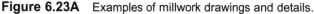

JUICE BAR COUNTER DETAILS

Figure 6.23A   Examples of millwork drawings and details.

order to clearly show the materials, dimensions, method of building, desired joint or attachment, and so on, these enlargements are referred to as details.

Details are most commonly drawn as sections. It is as if a slice is made through a specific area and the inner components made visible. Detail drawings are one of the most important sources of information available to the contractor about specific parts of the construction. A detail contains both graphic and written information. There are many, many types of details. A drawing sheet will often include several details. The complexity of the project will determine which areas need to be shown at a larger scale. Details are not limited to architectural drawings but can be used in structural and site plans and, to a lesser degree, in mechanical or electrical plans.

Details are always drawn to scale. A typical scale for a detail is three inches = one foot (scale: 3" = 1'0"). The scale for each detail will vary depending on how much information is required to make the construction clear to the builder. Each detail will be titled with the scale noted below it. Figure 6.24 shows different types of details found in a typical construction set.

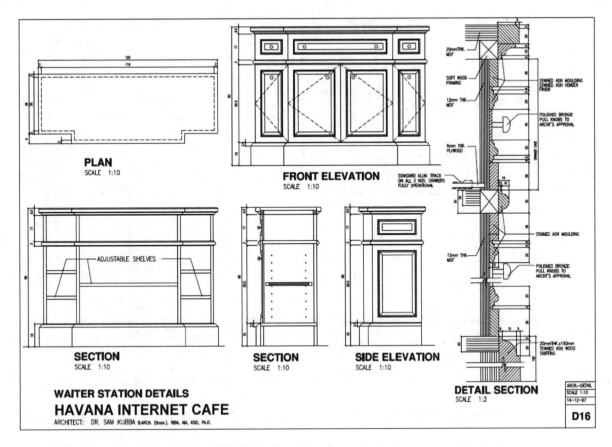

PLAN
SCALE 1:10

FRONT ELEVATION
SCALE 1:10

ADJUSTABLE SHELVES

SECTION
SCALE 1:10

SECTION
SCALE 1:10

SIDE ELEVATION
SCALE 1:10

DETAIL SECTION
SCALE 1:2

WAITER STATION DETAILS
HAVANA INTERNET CAFE
ARCHITECT: DR. SAM KUBBA B.ARCH. (Hons.), RIBA, AIA, ASID, Ph.D.

Figure 6.23B Examples of millwork drawings and details.

Shop Drawings

Very often the consultant will request shop drawings for certain components during the construction process. These are normally prepared by the contractor, subcontractor, manufacturer, or fabricator, depending on the element being manufactured. Shop drawings are typically required for prefabricated components such as windows, precast components, elevators, structural steel, trusses, or millwork. The drawings represent the manufacturer's or contractor's interpretation of the consultant's drawing and are thus expected to show more specific detail than normally provided by the construction documents. They are drawn to explain the fabrication and/or installation of the required product. The primary emphasis of a shop drawing always relates to the particular product to be manufactured or its installation and excludes information regarding other products unless integration with another product is necessary.

Shop drawings should include information for the consultant to compare to the specifications and contract documents. They should also include dimensions, manufacturing conventions, and special fab-

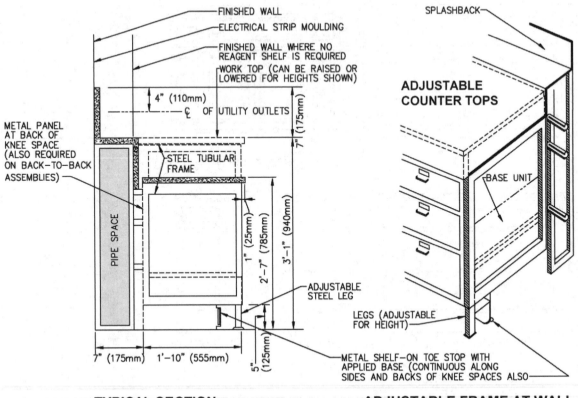

FINISHED WALL

ELECTRICAL STRIP MOULDING

FINISHED WALL WHERE NO
REAGENT SHELF IS REQUIRED

WORK TOP (CAN BE RAISED OR
LOWERED FOR HEIGHTS SHOWN)

4" (110mm)

℄ OF UTILITY OUTLETS

7" (175mm)

**ADJUSTABLE
COUNTER TOPS**

METAL PANEL
AT BACK OF
KNEE SPACE
(ALSO REQUIRED
ON BACK-TO-BACK
ASSEMBLIES)

STEEL TUBULAR
FRAME

PIPE SPACE

BASE UNIT

1" (25mm)

2'-7" (785mm)

3'-1" (940mm)

ADJUSTABLE
STEEL LEG

LEGS (ADJUSTABLE
FOR HEIGHT)

7" (175mm)   1'-10" (555mm)

5" (125mm)

METAL SHELF—ON TOE STOP WITH
APPLIED BASE (CONTINUOUS ALONG
SIDES AND BACKS OF KNEE SPACES ALSO

SPLASHBACK

TYPICAL SECTION   **ADJUSTABLE FRAME AT WALL**

NOTE: GRAPHIC REPRESENTATION ONLY. CONSULT MANUFACTURER'S LITERATURE FOR ACTUAL DIMENSIONS,
CLEARANCE AND UTILITY REQUIREMENTS.

Figure 6.23C   Examples of millwork drawings and details.

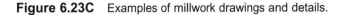

rication instructions and should address the appearance, performance, and prescriptive descriptions in the specifications and construction drawings. Shop drawings are designed to assist the consultant in gaining approval of the product and thus should be precise, clear, and as complete as possible. Shop drawings are normally accompanied by samples, catalogs, and any other pertinent information. Any proposed changes should be clearly shown on the shop drawings for the consultant's approval. Of note, shop drawings are not typically produced by the consultant under the contract with the owner.

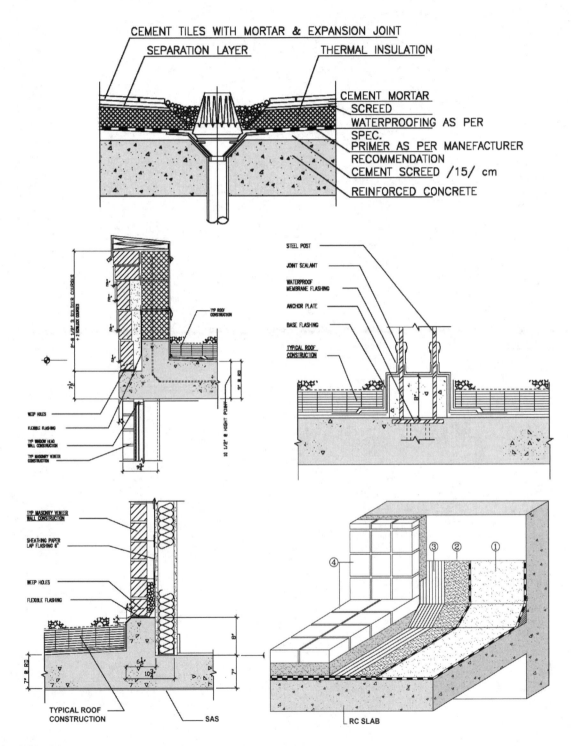

CEMENT TILES WITH MORTAR & EXPANSION JOINT
SEPARATION LAYER
THERMAL INSULATION

CEMENT MORTAR
SCREED
WATERPROOFING AS PER SPEC.
PRIMER AS PER MANEFACTURER RECOMMENDATION
CEMENT SCREED /15/ cm
REINFORCED CONCRETE

STEEL POST
JOINT SEALANT
WATERPROOF MEMBRANE FLASHING
ANCHOR PLATE
BASE FLASHING
TYPICAL ROOF CONSTRUCTION

WEEP HOLES
FLEXIBLE FLASHING
TYP WINDOW HEAD WALL CONSTRUCTION
TYP MASONRY VENEER CONSTRUCTION
TYP ROOF CONSTRUCTION

TYP MASONRY VENEER WALL CONSTRUCTION
SHEATHING PAPER LAP FLASHING 6"
WEEP HOLES
FLEXIBLE FLASHING
TYPICAL ROOF CONSTRUCTION
SAS
RC SLAB

Figure 6.24 Examples of various types of construction details.

7

Understanding
Industrial Blueprints

7.1   GENERAL

Industrial drawings may often necessitate more description and detail than some other types of working drawings, mainly because of the close tolerances and finished surface requirements. In this chapter we will cover some of the more common terms and symbols that the blueprint reader must be familiar with in order to read machine drawings. Some of the basic mechanisms usually found in detail and assembly drawings of machines are also presented.

One of the first steps in learning to read machine drawings is to become familiar with key terms, symbols, and conventions in general use in the industry. Although today's CAD packages make the production of industrial drawings much easier, it is still imperative to follow industry standards and conventions.

Tolerance

Tolerance represents the total amount a dimension may vary. It is basically defined as the difference between the upper and lower limits. Working to absolute or exact basic dimensions is impractical and unnecessary in most instances; therefore, the designer calculates in addition to the basic dimensions an allowable variation. Work must therefore be implemented within the limits of accuracy specified on the drawing. A clear understanding of tolerance and allowance can go a long way toward preventing small but potentially critical errors.

Tolerance is shown on a drawing as ± (plus or minus) a certain amount, either as a fraction or decimal. Limits are the maximum and/or minimum values prescribed for a specific dimension, while tolerance represents the total amount by which a specific dimension may vary. Tolerances may be shown on drawings in a number of different ways. Figure 7.1 shows three examples: A. The unilateral method, which is used when variation from the design size is permissible in one direction only; B. The bilateral method, where the dimension figure shows the variation in either direction that is acceptable; and C. The limit-dimensioning method, where the maximum and minimum measurements are both stated. Figure

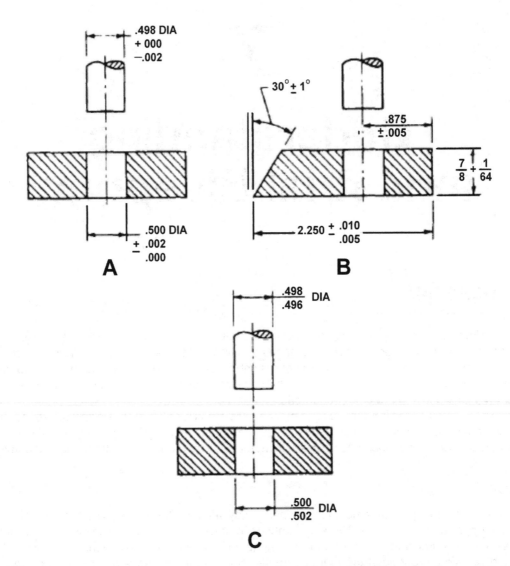

Figure 7.1   Three methods of indicating tolerances: A. Unilateral method, B. Bilateral method, and C. Limit dimensioning method (source: Blueprint Reading and Sketching, Navedtra 14040).

7.2 illustrates a typical method used to show tolerances for holes and shafts. Surfaces being toleranced have geometrical characteristics such as roundness or perpendicularity to another surface. Figure 7.3 demonstrates typical symbols used in lieu of or in conjunction with notes to state the geometric characteristics being toleranced.

If tolerances are not actually specified on a drawing, certain assumptions can be made regarding the anticipated accuracy by applying the following principles: For dimensions ending in a fraction of an inch, such as 1/8, 1/16, 1/32, or 1/64, the required accuracy will be to the nearest 1/64 inch. When the dimension is given in decimal form, the following principles should be followed: If a dimension is given as 2.000 inches, the accuracy expected is ±0.005 inch; if the dimension is given as 2.00 inches, the accuracy expected is ±0.010 inch. The ±0.005 is called in shop terms "plus or minus five thousandths of an inch." The ±0.010 is called "plus or minus ten thousandths of an inch."

Fillets and Rounds

Fillets are concave metal corner (inside) surfaces. In a cast, a fillet normally increases the strength of a metal corner because a rounded corner cools more evenly than a sharp corner, thereby reducing the possibility of a break. Rounds or radii are edges or outside corners that have been rounded to prevent chipping and to avoid sharp cutting edges. Figure 7.4 shows an example of fillets and rounds.

Slots and Slides

Slots and slides are used to mate two specifically shaped pieces of material and securely hold them together, while allowing them to move or slide. Figure 7.5 shows two common types: the tee slot and the dovetail slot. A tee-slot arrangement is usually used on a milling-machine table, while a dovetail slot is often used on the cross-slide assembly of an engine lathe.

Keys, Keyseats, and Keyways

These terms denote several types of small wedgelike metal objects designed to fit mating slots in a shaft and the hub of a gear or pulley to prevent slippage and provide a positive drive between them. Parts can be keyed together by means of a flat piece of steel that is partly seated in a recessed shaft or groove called a keyseat. Figure 7.6 shows three types of keys, a keyseat, and a keyway. A keyseat is a slot or groove on the outside of a part into which the key fits. A keyway is a slot or groove within a cylinder, tube, or pipe. A key fitted into a keyseat will slide into the keyway and prevent movement of the parts. Drawings usually include a height dimension that states the measurement from the circumference to the depth to which the keyway and keyseat are machined.

Coordinate Measuring Machines

Coordinate measuring machines (CMM) use software and sensors to compare computer prototype drawings to the finished prototype. Tool and die makers must use accurate prototypes to tool a production line that will produce identical quality components.

Casting

It is the production of metal components by pouring molten metal into moulds and allowing it to solidify. It can also be a metal component produced by casting.

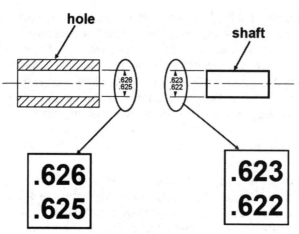

Figure 7.2 A typical method used to show tolerances for holes and shafts (source: College of Engineering, Ohio State University).

⌒	**FLATNESS & STRAIGHTNESS**
∠	**ANGULARITY**
⊥	**PERPENDICULARITY**
‖	**PARALLELISM**
⊙	**CONCENTRICITY**
⊕	**TRUE POSITION**
○	**ROUNDNESS**
≡	**SYMMETRY**
Ⓜ	**MAXIMUM MATERIAL CONDITION (MMC)**
Ⓢ	**REGARDLESS OF FEATURE SIZE (RFS)**
-A-	**DATUM IDENTIFYING SYMBOL**

Figure 7.3 Examples of geometric symbols used in industrial drawings.

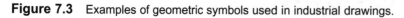

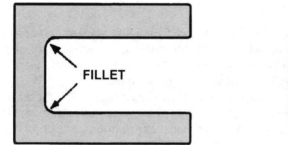
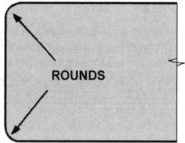

Figure 7.4   Fillets and rounds.

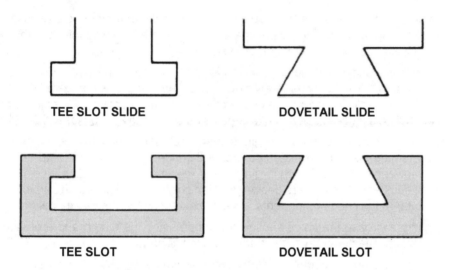

Figure 7.5   Examples of slot and slide arrangements.

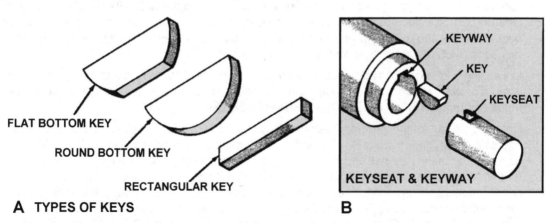

A TYPES OF KEYS   **B**

Figure 7.6   Diagram showing three types of keys, a keyseat and a keyway (source: Blueprint Reading and Sketching, Navedtra 14040).

Forging

Forging is a process of shaping metal while it is hot and pliable by suitably applying compressive force. Usually the compressive force takes the form of hammer blows using a power hammer or a press. To achieve good forgings, a proper lubricant is necessary. The lubricant helps prevent the work piece sticking to the die and also acts as a thermal insulator to help reduce die wear.

Forging refines the grain structure and improves the physical properties of the metal. With proper design, the grain flow can be oriented in the direction of principal stresses that are encountered in actual use. Grain flow is the direction of the pattern that the crystals take during plastic deformation. Physical properties (such as strength, ductility, and toughness) are much better in a forging than in the base metal, whose crystals are randomly oriented. Forgings yield parts that have a high strength-to-weight ratio. Common forging processes include:

- Open-die or hand forgings are a traditional and antiquated manufacturing process based on repeated blows in an open die, where the operator manipulates the work piece in the die. The finished product is a rough approximation of the die.

- Press forging is the shaping of metal between dies on a mechanical or hydraulic press. The action is that of kneading the metal by relatively slow application of force as compared with the action of hammering. This results in uniform material properties and is necessary for large weight forgings. Parts made with this process can be quite large as much as 125 kg (260 lb) and 3m (10 feet) long.

- Impression-die forgings or precision forgings are further refinements of blocker forgings (blocker forging is a forging that approximates the general shape of the final part with relatively generous finish allowance and radii. Such forgings are sometimes specified to reduce die costs where only a small number of forgings are desired and the cost of machining each part to its final shape is not exorbitant). Finished components more closely resemble the die impression.

- Upset forgings increase the cross section by compressing the length and are used in making heads on bolts and fasteners, valves, and other similar parts.

- In roll forging (also known as draw forging) a round or flat bar stock is placed between die rollers, which reduces the cross section and increases the length to form parts such as axles, leaf springs etc. Draw forging also involves pulling a hollow tube through a series of hardened steel dies of gradually decreasing diameter. Before each step of the drawing process, the tube is pointed at one end to fit through the next smaller die and is then gripped by automatic jaws attached to a rotating drawing machine.

- In swaging a tube or rod is forced inside a die and the diameter is reduced as the cylindrical object is fed. The die hammers the diameter and causes the metal to flow inward, causing the outer diameter of the tube or the rod to take the shape of the die.

- Net and near net shape forging represents a number of recent developments of the conventional impression die forging process. Net and near net shape forgings are distinguished by geometric features that are thinner and more detailed, varying parting line locations, virtual elimination of draft, and closer dimensional tolerances. The resulting product benefits are much fewer machining operations (in many cases, the only machining operations required are drilling of attachment holes), reduced weight and lower costs for raw materials and energy. The processes are quite costly in terms of tooling and the capital expenditure required. Thus, these processes can only be justified for current processes that are very wasteful where the material savings will pay for the significant increase in tooling costs.

Die

A die is a tool used to form or stamp out metal parts or to cut external threads. Dies can be simple objects or made of a series of jigs and fixtures to ensure that the die makes contact with the metal stock at the correct place and angle.

Tempering

Tempering is a process of heat treatment to change the physical characteristics of ferrous alloys. The object of tempering or drawing is to reduce the brittleness in hardened steel and to remove the internal strains caused by the sudden cooling in the quenching bath. The tempering process consists of heating the steel by various means to a certain temperature and then cooling it. The rate of cooling usually has no effect on the metal structure during tempering. Therefore, the metal is usually allowed to cool in still air. When steel is in a fully hardened condition, its structure consists largely of martensite. On reheating the steel to a temperature of about 300 to 750 degrees F, a softer and tougher structure known as troostite is formed. If the steel is reheated to a temperature of 750 to 1290 degrees F, a structure known as a sorbite is formed, which, while lacking the strength of troostite, has far greater ductility. High-speed steel is one of the few metals that will become harder instead of softer after it is tempered.

Drill

A drill is a pointed tool that is rotated to cut holes in material. Figure 7.7A shows a typical drill with its related nomenclature and two holes,one 0.75 deep and the other through, as well as the operations of upright drilling machines.

Boring

To bore is to enlarge and finish the surface of a cylindrical hole by the action of a rotating boring bar (cutting tool) or by the action of a stationary tool pressed (fed) against the surface as the part is rotated. Boring produces a hole with a continuous inside diameter or a tapered or contoured diameter. When a straight and smooth hole is required that is too large or odd-sized for drills or reamers, a boring tool can be utilized to bore any size hole into which the tool holder will fit by inserting it into the drilling machine. A boring bar with a tool bit installed is used for boring on larger drilling machines. To bore accurately, the setup must be rigid, the machine sturdy, and a power feed used.

Reaming

Reaming consists of enlarging a hole. It is similar to boring but is more precise and is implemented after boring or drilling. Reaming is performed with a drill machine or lathe; it is difficult if not impossible,to drill a hole to an exact standard diameter. Where extra accuracy is specified, the holes should first be drilled slightly undersized and then reamed to size. The majority of hand and machine reamers have a slight chamfer at the tip to aid in alignment and starting (Figure 7.8).

Tapping

Tapping is the process of cutting a thread in a drilled hole and cutting internal threads. Tapping can be accomplished on a lathe or drilling machine. On a drilling machine it is done by selecting and drilling the

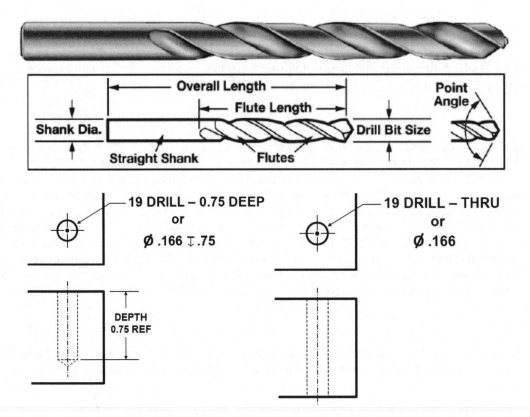

Figure 7.7A   A typical drill with the terminology related to it and two holes—one 0.75 deep and the other through.

tap drill size, then using the drilling-machine chuck to hold and align the tap while it is turned by hand. There are typically three types of taps used: a tapered tap to facilitate initial thread cutting; an intermediate type, used to progress the thread after it has been started; and a "bottoming" thread, which is used to obtain the full thread depth when cutting partway through the piece.

Finish Marks

Many metal surfaces must be finished with machine tools for various reasons. The acceptable roughness of a surface depends upon how the part will be used. Sometimes only certain surfaces of a part need to be finished while others are not. A modified symbol (check mark) with a number or numbers above it is used to show these surfaces and to specify the degree of finish required. The proportions of the surface roughness symbol are shown in Figure 7.9. On small drawings the symbol is proportionately smaller.

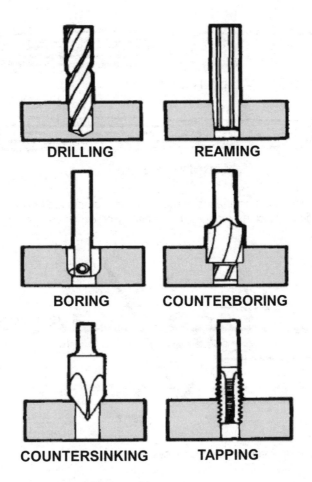

DRILLING   **REAMING**

BORING   **COUNTERBORING**

COUNTERSINKING   **TAPPING**

Figure 7.7B Operations of the upright drilling machine.

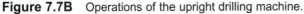

The number in the angle of the check mark, in this case "02," tells the machinist the degree of finish the surface should have. This number is the root-mean-square value of the surface roughness height in millionths of an inch. In other words, it is a measurement of the depth of the scratches made by the machining or abrading process.

Wherever possible, the surface roughness symbol is drawn touching the line representing the surface to which it refers. If space is limited, the symbol may be placed on an extension line on that surface or on the tail of a leader with an arrow touching that surface, as shown in Figure 7.9.

When a part is to be finished to the same roughness all over, a note on the drawing will include the direction "finish all over" along the finish mark and the proper number. When a part is to be finished all over but a few surfaces vary in roughness, the surface roughness symbol number or numbers are applied to the lines representing these surfaces, and a note on the drawing will include the surface roughness symbol for the rest of the surfaces.

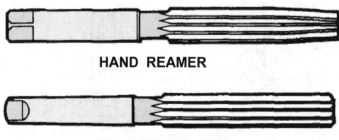

HAND REAMER

MACHINE REAMER

Figure 7.8   A hand reamer and a machine reamer. Solid hand reamers are typically used when a greater accuracy in hole size is desired. Machine reamers can generally be expected to produce clean holes when used properly.

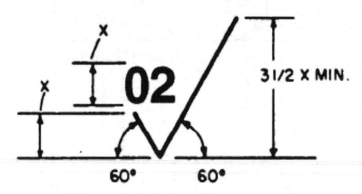

Figure 7.9A   The proportions of the surface roughness symbol.

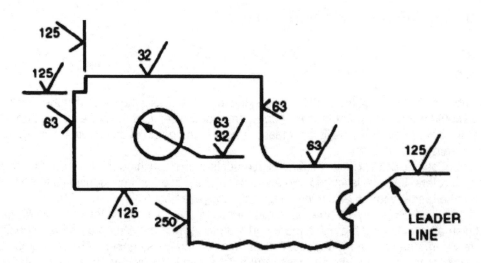

Figure 7.9B   The method of placing surface roughness symbols.

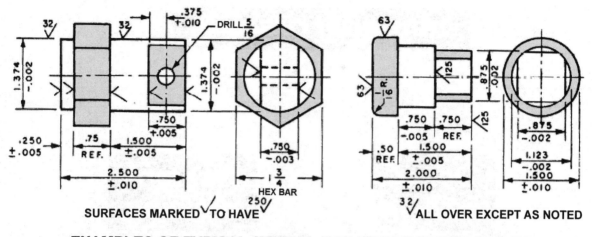

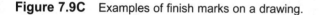

EXAMPLES OF TYPICAL SYMBOL USE IN MACHINE DRAWINGS

Figure 7.9C Examples of finish marks on a drawing.

7.2 INDUSTRIAL DRAWINGS

A working drawing is a drawing that includes all the information required to successfully and accurately execute a job. A detail drawing is essentially a working drawing that incorporates much more specific information including the size and shape of the project, the materials to be used, finishing information, and degree of accuracy needed. An assembly drawing, on the other hand, may incorporate very little detail (Figure 7.10). The purpose of this type of technical drawing is mainly to show how the machine is to be assembled.

There are several types of detail drawings used in various industrial settings. They convey the information and instructions for manufacturing the part, including the object's shape, size, and specifications. The detail drawing should provide all the information needed to produce the particular part (Figure 7.11). In addition to the part production, the detail drawing can be used when bidding a job or as a master drawing from which other drawings are produced. The information on a detail drawing is specific to the part to be produced, typically drawn one part per page or sheet. Included in the information should be the pertinent shape, size, specifications, and notes. The notes and/or title-block information would normally include the scale, tolerances, surface texture, and specified material. Other data may be required that stipulate finish specification, moisture contents, color code, and other specifications requested by the customer.

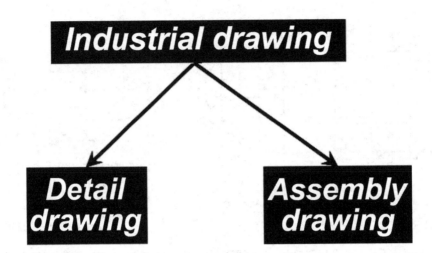

Figure 7.10 Two categories of industrial drawings are detail drawings and assembly drawings.

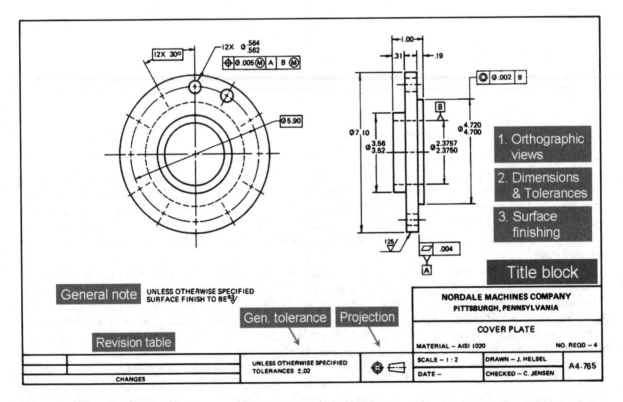

Figure 7.11 A detail drawing showing some of the items that need to be included such as orthographic views, dimensions and tolerances, surface finishes, general notes, revision table, and title block.

A casting drawing details the size and location of separation lines, called parting lines. When the mold separates, the parting line is left on the casting. The other information typically found on a casting detail drawing may be a pattern number, sand requirements, part numbers, customer name, draft, shrinkage, finish, ribs and other support webbing, and the material used in the casting/mold.

Wood patterns, made by patternmakers in the general male or female shape of the finished part are configured into the sand mold and removed, leaving an impression to be filled with molten metal. When cooled, this metal casting is removed and the process is continued. These castings are then machined into items such as engine blocks, cast-iron tools, or other metal castings we use every day. Bearing blocks, machinery bases and parts, and many household items have castings produced in this manner.

7.3 MACHINIST DRAWINGS

Machinists generally use precision machine tools such as lathes, boring machines, milling machines, and cylindrical or surface grinders to shape materials such as steel, brass, iron, bronze, aluminum, titanium, and plastics and to manufacture components to meet customers' specifications.

A machining detail drawing provides all the information needed to manufacture a specific part and is used to machine the casting into a finished component. Typically only one part is detailed in each drawing. Machining details are generally used when machining a rough part into a finished part; a machining detail drawing will specify the surfaces to be machined, bolt holes and locations, reference points, geometric dimensioning and tolerancing (GD & T), and other machined areas. Machining drawings contain critical information to the traditional or CNC (computer numerically controlled) machinist, which would include angles, reference points, surface finish, etc. Once complete, this machined part should match and fit other machined parts as defined in an assembly/detail drawing. Parts that would not normally need to be drawn are standard parts—those that may be purchased from an external source more economically than it would cost to manufacture. Such parts may include screws, nuts, bolts, keys, and pins. While they do not need to be drawn, they nevertheless need to be included as part of the information on each sheet. The blueprint reader must clearly understand the shape, size, material, and surface finish of a part, which shop operations are necessary, and the limits of accuracy that must be observed from the detail drawing. Figure 7.12 is an example of a typical detail drawing.

Normally, detail drawings contain information that is classifiable into three groups:

1. Shape description: describes and explains or portrays the shape of the component
2. Size depiction: shows the size and location of features of the component
3. Specifications: relates to such items as material and finish

Detail machine drawings should include all or most of the following information:

- Views of the component as necessary to allow visualization
- Material used to manufacture the component
- Dimensions
- General notes and specific manufacturing information
- Identification of the project name, the part, and the part number
- Name or initials of who worked on or with the drawing
- Any engineering changes and related information

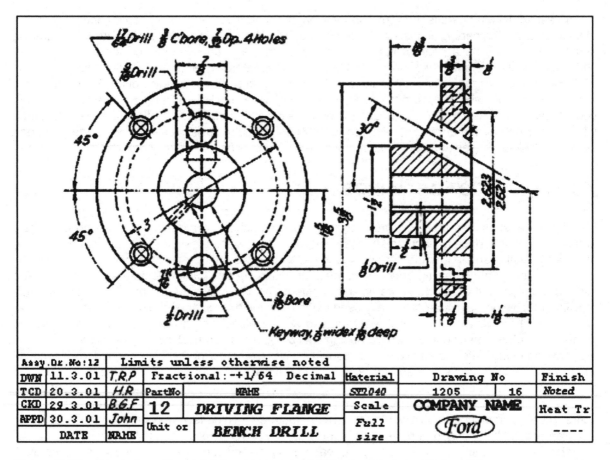

Figure 7.12 A typical machine detail drawing (source: Middle East Technical University).

A complete assembly drawing is a presentation of the product or structure put together, showing the various components in their operational positions. The separate components come to the assembly department after their manufacturing processes are finished, and here they are put together according to the assembly drawings.

Many products consist of more than one part or component. A bill of materials (BOM) or list of component parts is often included on an assembly drawing to facilitate the assembly, as well as necessary dimensions and component labeling (Figure 7.13). A three-dimensional picture of a complete assembled unit would help the reader to understand the final shape of the assembly. Front, side, and top views may be critical to communicate dimensions or shapes to the reader. If the assembly drawing is actually one of several subassemblies, the print should indicate this in the title block or bill of materials. Movements of components in an assembly detail drawing should be indicated with the use of phantom lines.

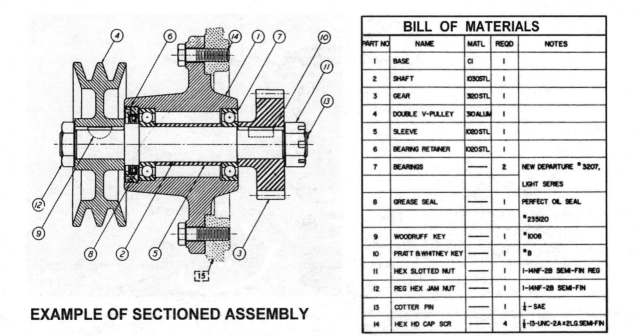

EXAMPLE OF SECTIONED ASSEMBLY

PART NO	NAME	MATL	REQD	NOTES
	BILL OF MATERIALS			
1	BASE	CI	1	
2	SHAFT	1030 STL	1	
3	GEAR	320 STL	1	
4	DOUBLE V-PULLEY	310 ALUM	1	
5	SLEEVE	1020 STL	1	
6	BEARING RETAINER	1020 STL	1	
7	BEARINGS	——	2	NEW DEPARTURE #3207, LIGHT SERIES
8	GREASE SEAL	——	1	PERFECT OIL SEAL #235120
9	WOODRUFF KEY	——	1	#1006
10	PRATT & WHITNEY KEY	——	1	#B
11	HEX SLOTTED NUT	——	1	1-14NF-2B SEMI-FIN REG
12	REG HEX JAM NUT	——	1	1-14NF-2B SEMI-FIN
13	COTTER PIN	——	1	¼ - SAE
14	HEX HD CAP SCR	——	4	⅜-13-UNC-2A×2LG.SEMI-FIN

Figure 7.13   A sectioned assembly drawing with a bill-of-materials table (source: College of Engineering, Ohio State University).

There are various types and versions of assembly drawings including:

• Layout assembly drawings initially used in development of a new product.

• Exploded assembly drawings pictorially showing parts laid out in their correct order of assembly, found in machinery catalogs designed for homeowners or suppliers for ordering parts (Figure 7.14).

• Diagram assembly drawings use conventional symbols and are used to show the approximate location and/or sequence of the components to be assembled or disassembled.

• Working assembly drawings are fully dimensioned and noted. When applied to very simple products, they can act as alternatives to detail drawings.

• Installation assembly drawings are used to show how to install large components of equipment.

As mentioned earlier, an assembly drawing is a drawing of various parts of a machine or structure in their relative working positions. An assembly drawing essentially conveys the completed shape of the product as well as its overall dimensions, relative position of the different parts, and the functional relationship of its components. When all the parts are produced using their respective machining detail draw-

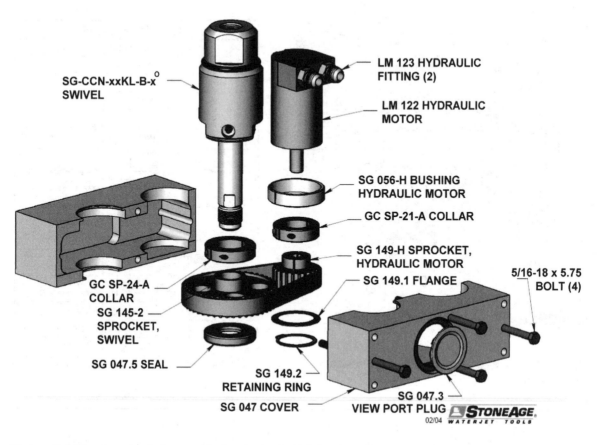

SG-CCN-xxKL-B-x$^\circ$
SWIVEL

LM 123 HYDRAULIC
FITTING (2)

LM 122 HYDRAULIC
MOTOR

SG 056-H BUSHING
HYDRAULIC MOTOR

GC SP-21-A COLLAR

SG 149-H SPROCKET,
HYDRAULIC MOTOR

SG 149.1 FLANGE

5/16-18 x 5.75
BOLT (4)

GC SP-24-A
COLLAR

SG 145-2
SPROCKET,
SWIVEL

SG 047.5 SEAL

SG 149.2
RETAINING RING

SG 047 COVER

SG 047.3
VIEW PORT PLUG

STONEAGE
WATERJET TOOLS
02/04

Figure 7.14A An exploded assembly drawing typical of drawings found in machinery catalogs showing the various components laid out in their correct order of assembly (source: StoneAge, Inc.).

ings, an assembly drawing provides the information the print reader needs to assemble the components. The bill of materials (BOM), which is basically a tabulated list, may be placed either on the assembly drawing or on a separate sheet. The list gives critical information such as part numbers, names, quantities, material detail-drawing number, and sometimes stock sizes of raw material, etc. The term "bill of materials" is usually used in structural and architectural drawing, whereas the term "part list" is used in machine-drawing practice.

A three-dimensional picture of a complete assembled unit facilitates the reader's ability to visualize the final shape of the assembly (Figure 7.15). Front, side, and top views may be necessary to communicate dimensions or shapes to the reader. If this assembly drawing is actually one of several subassembly drawings, the print should indicate this in the title block or the BOM.

Likewise, a maintenance technician would normally need assembly drawings at the work site to evaluate the best sequence for the dismantling of specific machinery, to locate parts that must be ac-

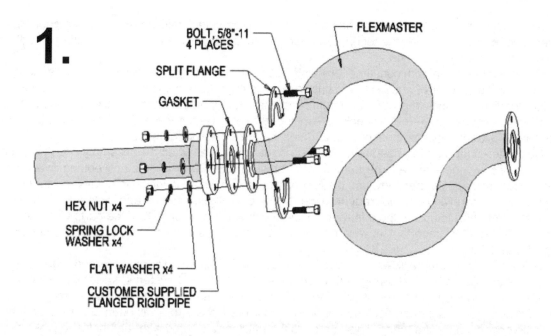

1.

BOLT, 5/8"-11
4 PLACES

FLEXMASTER

SPLIT FLANGE

GASKET

HEX NUT x4

SPRING LOCK
WASHER x4

FLAT WASHER x4

CUSTOMER SUPPLIED
FLANGED RIGID PIPE

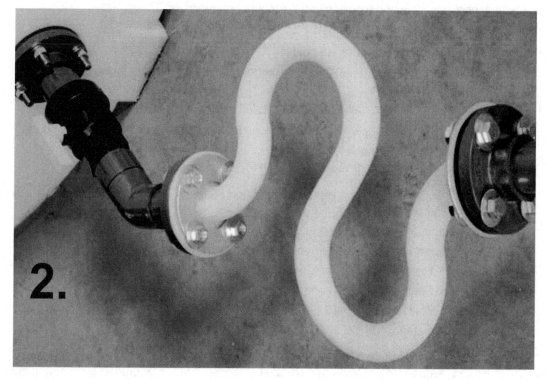

2.

Figure 7.14B An exploded assembly drawing and photograph of a Flexmaster expansion joint (source: Snyder Industries, Inc.).

cessed or attachment bolts of those to be removed, and to provide detail information on components disassembled for repair. Finally, the technician needs to accurately discern the correct alignment of components upon reassembly.

Customers who deal with consumer products such as electronic goods will also normally require the use of exploded CAD drawings to help in understanding the relationships among assembled parts. Exploded drawings are indispensable for a number of manufacturing industries. While generating assembly drawings, critical interference checks are included to ensure that the entire assembly is integrated, thus saving a tremendous amount of time and costs in the prototyping phase.

Computer-aided drafting has proved to be a huge timesaver when an assembly drawing is being produced. Today, there are a large number of sophisticated CAD programs and equipments, and the vast majority of manufacturers now use these programs to recover high initial production costs. Although many assembly drawings do not require dimensions, overall dimensions and distances between the centers or from part to part of the different pieces to clarify the relationship of the parts with each other can be included. Most important, however, an assembly drawing should be easy to read and not be overloaded with detail.

Utilizing CAD programs likewise allows individual component details to be merged together to create an assembly or working drawing of the component(s). With CAD systems, three-dimensional (3-D) models can be created that makes it possible to superimpose images and to graphically measure clearances. When parts have been designed or drawn incorrectly, the errors will often stand out so that appropriate correction can be made. This improves the efficiency of the drafter and helps to make the details in the final print accurate and the resultant parts function properly.

Information normally required for general assembly drawings includes:

- Parts to be drawn in their operating position
- Part list (or bill of materials) including item number, descriptive name, material, and quantity required per unit of machine
- Leader lines with balloons drawn around part numbers
- Machining and assembly operations and critical dimensions related to the operation of the machine

Steps in creating an assembly drawing include the following:

1. Analyze the geometry and dimensions of the various parts in order to understand the assembly steps and overall shape of the object.
2. Select an appropriate view of the object.
3. Choose the major components—components that require assembly of several parts .
4. Draw a view of the major components according to a selected viewing direction.
5. Add detail views of the remaining components at their working positions.
6. Add balloons, notes, and dimensions as required.
7. Create a bill of materials (BOM).

Assembly drawings can require one, two, three, or more views, although they should be kept to the minimum necessary. A good viewing direction should be chosen that represents all (or most) of the parts assembled in their working position.

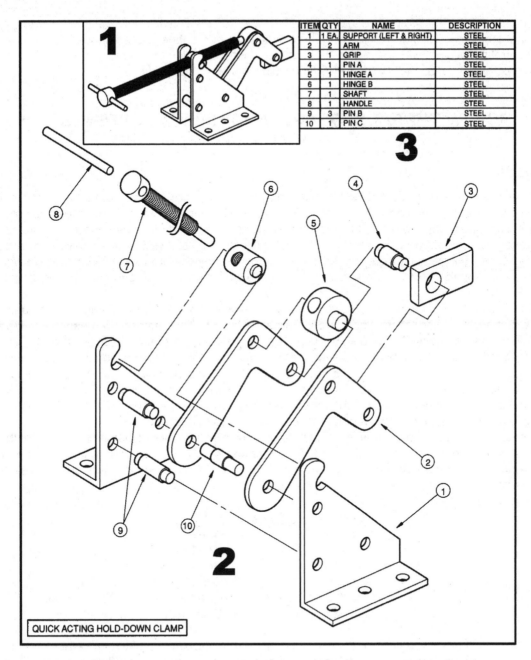

ITEM	QTY	NAME	DESCRIPTION
1	1 EA.	SUPPORT (LEFT & RIGHT)	STEEL
2	2	ARM	STEEL
3	1	GRIP	STEEL
4	1	PIN A	STEEL
5	1	HINGE A	STEEL
6	1	HINGE B	STEEL
7	1	SHAFT	STEEL
8	1	HANDLE	STEEL
9	3	PIN B	STEEL
10	1	PIN C	STEEL

QUICK ACTING HOLD-DOWN CLAMP

1. PICTORIAL   2. EXPLODED   3. BILL OF MATERIALS

Figure 7.15   Diagram showing how multiple parts fit together and a bill of material and pictorial view of the assembled object (source: College of Engineering, The Ohio State University).

In mating parts the two main considerations are the surface finishing and the tolerance (especially size and geometry). The surface finishing means the level of roughness of a surface. Its main purpose is to control the accuracy in positioning and tightness between the mating parts. The other objective is to reduce the friction, especially for parts that move relative to other parts.

7.4   SCREW THREADS, GEARS, AND HELICAL SPRINGS

Threads

Machine screws are extensively used for securing parts. There are many different types and sizes of machine screws, nuts, and bolts. Moreover, drafters use different methods to show thread on drawings. Figure 7.16 shows a thread profile in section and a common method of drawing threads. To save time, the drafter uses symbols that are not drawn to scale. The drawing shows the dimensions of the threaded part, but other information may be placed in "notes" almost anywhere on the drawing but most often in the upper left corner. Figure 7.17 gives an example of a typical note showing the thread designator "1/4-20x1, RHMS."

The first number of the note, 1/4, is the nominal size, which is the outside diameter. The number after the first dash, 20, means that there are 20 threads per inch. The number 1 represents the length of the screw, and the letters RHMS identify the head type (round-head machine screw). It would normally also have letters identifying the thread series (e.g., UNC if it were a unified national coarse thread) and a number to identify the class of thread (e.g., "3") and tolerance, commonly called the fit. If it is a left-hand thread, a dash and the letters LH will follow the class of thread. Threads without the LH designation are right-hand threads. Figure 7.18 shows right-hand and left-hand screws. Figure 7.19 is another example of a screw type and the general nomenclature used to describe it.

Specifications necessary for the manufacture of screws include thread diameter, number of threads per inch, thread series, and class of thread. The two most widely used screw-thread series are unified or national form threads, which are called national coarse or NC, and national fine, or NF, threads. NF threads have more threads per inch of screw length.

<u>Common NC Screw-Thread Sizes</u>

o2-56	o1/4-20
o4-40	o3/8-16
o6-32	o1/2-13
o8-32	o5/8-11
o10-24	o3/4-10

Diameter = $(N*.013) + .060$ (inches)

Classes of threads are distinguished from each other by the amount of tolerance and/or allowance specified. Classes of thread were formerly known as class of fit, a term that will probably remain in use for many years. External threads or bolts are designated with the suffix "A," internal or nut threads with the suffix "B." Figure 7.19 shows different types of screws and screw heads.

The unified and American (national) thread forms designate classifications for fit to ensure that mated threaded parts fit to the tolerances specified. The unified-screw-thread form has specified a number of classes of threads:

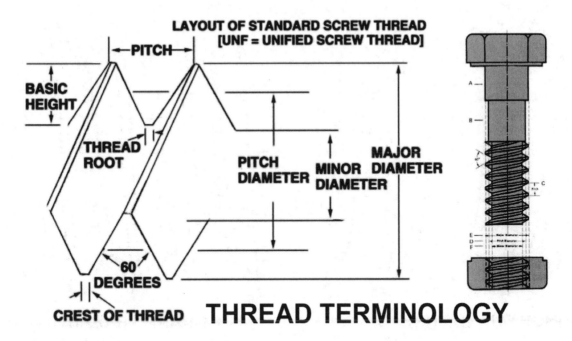

Figure 7.16 A typical screw thread with profile section and the terminology associated with its use in today's industry.

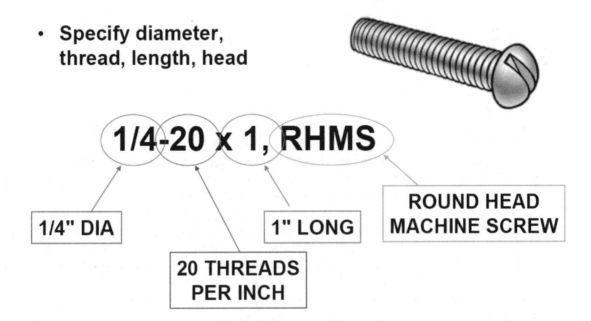

Figure 7.17 A typical note showing the thread designator "1/4-20x1, RHMS." It specifies the diameter, thread, length, and head.

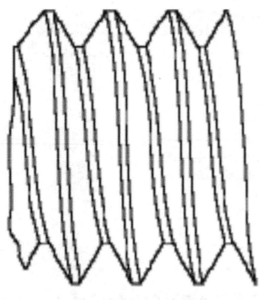   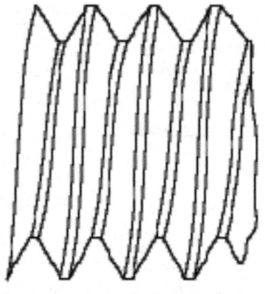

RIGHT-HAND THREAD   **LEFT-HAND THREAD**

Figure 7.18   Right- and left-handed thread screw types.

- Class 1A and 1B: for work of rough commercial quality where loose fit, quick assembly, and rapid production are important and shake or play is not objectionable.

- Class 2A and 2B: the recognized standard for normal production of the great bulk of commercial bolts, nuts, and screws. Classes 2A and 2B provide a small amount of play to prevent galling and seizure in assembly and use and sufficient clearance for some plating.

- Class 3A and 3B: these have no allowance and 75 percent of the tolerance of Classes 2A and 2B. A screw and nut in this class may vary from a fit having no play to one with a small amount of play. This fit represents an exceptionally high quality of commercially threaded product and is recommended only in cases where the high cost of precision tools and continual checking are warranted.

- Class 4: this is a theoretical rather than practical class and is now obsolete.

- Class 5: for a wrench fit, used principally for studs and their mating tapped holes or a force fit requiring the application of a high torque for semipermanent assembly. It is also a selective fit if initial assembly by hand is required. It is not as yet adaptable to quantity production.

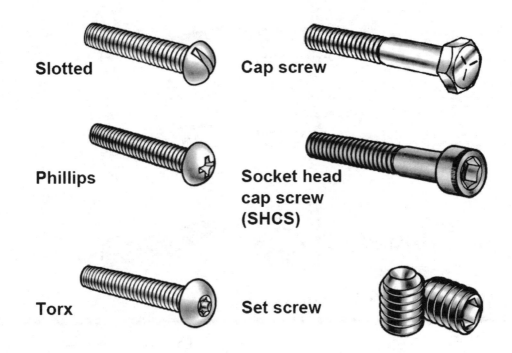

Slotted

Cap screw

Phillips

Socket head cap screw (SHCS)

Torx

Set screw

Figure 7.19A   Different types of screws and screw heads.

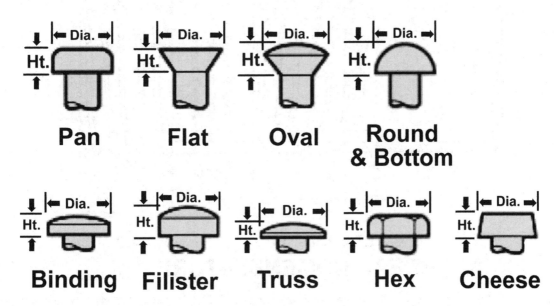

Pan   Flat   Oval   Round & Bottom

Binding   Filister   Truss   Hex   Cheese

Figure 7.19B   Different types of screws and screw heads.

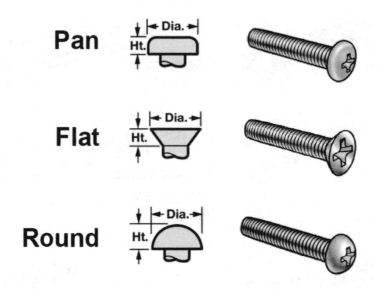

Figure 7.19C   Different types of screws and screw heads.

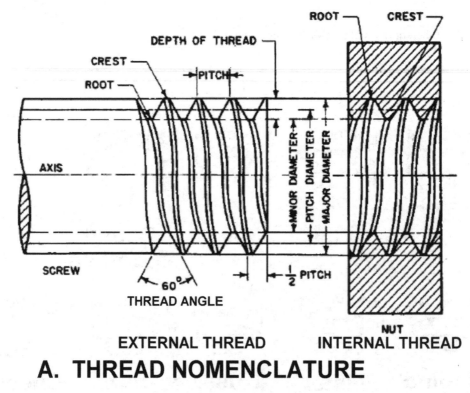

A. THREAD NOMENCLATURE

Figure 7.20   A typical screw thread with the terminology and nomenclature associated with its use in today's industry.

Handedness

Almost all threaded fasteners tighten when the head or nut is rotated clockwise. That is, as a viewer turns a nut clockwise it moves away from her. Such a fastener is said to have a right-hand thread; all screw fasteners are assumed to be right-hand unless otherwise specified. Left-hand threads are usually found only on rotating machinery. For example, the axles of bicycle pedals screw into threaded holes in the cranks. In a pair of pedals one will have a right-hand thread and the other a left-hand thread. That way the rotation of the pedals doesn't tend to unscrew their axles. To designate a left-hand thread, the letters "LH" are placed after the class of fit: 3/8-16 UNC 2B LH (21).

Helix

A helix denotes the curve formed on any cylinder by a straight line in a plane that is wrapped around the cylinder with a forward progression. Common objects formed like a helix are a spring, a screw, and a spiral staircase.

Helices can be either right-handed or left-handed. With the line of sight being the helical axis, if clockwise movement of the helix corresponds to axial movement away from the observer, then it is a right-handed helix. If counterclockwise movement corresponds to axial movement away from the observer, it is a left-handed helix.

External Thread

An external thread, also known as a male thread, is a thread on the outside of a cylinder or cone. An example is the thread of a bolt.

Internal Thread

Also known as a female thread, an internal thread is a thread on the inside of a hollow cylinder or bore. An example is the thread inside a nut.

Major and Minor Diameters

A major diameter is the largest diameter of an external or internal screw thread. A minor diameter is the smallest diameter of an external or internal screw thread; it is also known as the "root diameter."

Cut Thread

Screw threads are cut or chased (to cut thread in a lathe rather than with a die); the unthreaded portion of the shank will be equal to the major diameter of the shank.

Axis

The axis is the center line running lengthwise through a screw—the line, real or imaginary, passing through the center of an object about which it could rotate, or a point of reference.

Crest

The crest is the top or outer surface of the thread joining the two sides (also called "flat"). It is the surface of the thread corresponding to the major diameter of an external thread and the minor diameter of an internal thread.

Root

The root is the surface of the thread corresponding to the minor diameter of an external thread and the major diameter of an internal thread. It represents the bottom or inner surface joining the sides of two adjacent threads.

Depth

The depth is the distance from the root of a thread to the crest, measured perpendicularly to the axis.

Pitch

The pitch is the distance from any point on the screw thread to a corresponding point on the next thread measured parallel to the axis. For example, nuts and bolts need to have the same pitch and diameter if they are to be used together.

Lead

The lead is the distance a screw thread advances in one complete turn, measured parallel to the axis. On a single-thread screw the lead is equal to the pitch; on a double-thread screw the lead is twice the pitch; on a triple-thread screw the lead is three times the pitch.

Gears

Gears are enhanced assembly relationships and work by determining a relationship (or ratio) between adjacent parts, such as gears and pulleys. Gears efficiently solve motion between related components without the need to define physical contact relationships. They can be used to describe rotary to rotary (gears), rotary to linear (rack and pinion), or linear to linear (telescopic or hydraulic cylinder) relationships. When gears are drawn on machine drawings, the drafter usually draws only enough gear teeth to identify the necessary dimensions. Figure 7.21 illustrates important gear nomenclature. The terms in the figure are explained in the following list:

Pitch Diameter

Pitch diameter (PD) is the diameter of the pitch circle (or line), which equals the number of teeth on the gear divided by the diametral pitch. It represents the diameter of a thread at an imaginary point where the width of the groove and the width of the thread are equal (i.e., the simple effective diameter of screw thread). It is approximately halfway between the major and minor diameters. This is the critical dimension of threading, as it determines the fit of the thread (not used for metric threads).

Diametral Pitch

Diametral pitch (DP) is the ratio of the number of teeth on a gear to the number of inches of pitch diameter or the number of teeth to each inch of pitch diameter (i.e,. the number of teeth on the gear divided by the pitch diameter). Diametral pitch is usually referred to as simply pitch.

Number of Teeth

The number of teeth (N) is the diametral pitch multiplied by the diameter of the pitch circle (DP x PD).

Addendum Circle

The addendum circle (A) is the circle over the tops of the teeth.

Outside Diameter

The outside diameter (OD) is the diameter of the addendum circle.

Circular Pitch

The circular pitch (CP) is the length of the arc of the pitch circle between the centers or corresponding points of adjacent teeth.

Addendum

The addendum (A) is the height of the tooth above the pitch circle or the radial distance between the pitch circle and the top of the tooth.

Dedendum

The dedendum (D) is the depth or that portion of a gear tooth from the pitch circle to the root circle of gear or the length of the portion of the tooth from the pitch circle to the base of the tooth.

Chordal Pitch

The chordal pitch is the distance from center to center of teeth measured along a straight line or chord of the pitch circle.

Root Diameter

The root diameter (RD) is the diameter of the circle at the root of the teeth.

Clearance

The clearance is the distance between the bottom of a tooth and the top of a mating tooth or the distance by which one object clears another or the clear space between them.

Whole Depth

The whole depth (WD) is the distance from the top of the tooth to the bottom, including the clearance.

Face

The face is the working surface of the tooth over the pitch line.

Thickness

The thickness is the width of the tooth, taken as a chord of the pitch circle.

Pitch Circle

The pitch circle is the circle with the pitch diameter, the line (circle) of contact between two meshing gears.

Working Depth

The working depth is the greatest depth to which a tooth of one gear extends into the tooth space of another gear.

Rack

A rack may be compared to a spur gear that has been straightened out. The linear pitch of the rack teeth must equal the circular pitch of the mating gear.

Helical Springs

The helical spring, in which wire is wrapped in a coil that resembles a screw thread, is probably the most commonly used mechanical spring. There are three classifications of helical springs: compression, extension, and torsion. Drawings seldom show a true presentation of the helical shape; instead, they usually show springs with straight lines. Figure 7.22A shows several methods of spring representation including both helical and straight-line drawings. Figure 7.22B shows the use of typical helical spring nomenclature.

A spring can be designed to carry, pull, or push loads. Twisted helical (torsion) springs are used in engine starters and hinges. Helical tension and compression springs have numerous uses, notably automobile suspension systems, gun-recoil mechanisms, and closing.

In manufacturing coiled helical springs, the spring wire is wrapped around a mandrel that serves as a support; when seamless tubing is being extruded, a long mandrel forms the internal diameter.

Helical springs, often called spiral springs, are probably the most common type of spring. They may be used in compression, extension, tension, or torsion. A spring used in compression tends to shorten in action, while a tension spring lengthens in action. Torsion springs, which transmit a twist instead of a direct pull, operate by a coiling or an uncoiling action. In addition to straight helical springs, cone, double-cone, keg, and volute springs are classified as helical. These types of springs are usually used in compression.

7.5   RECOMMENDED PRACTICE

With industrial drawings and blueprint reading, as with other types of graphical representation, it is always advisable to follow the best industry practices current at the time:

1. Draw one part to one sheet of paper. If this is not practical, apply enough spacing between parts and draw all parts using the same scale. Otherwise, the scale should be clearly noted under each part's drawing. Standard parts such as bolts, nuts, pins, and bearings do not require detail drawings.
2. Always use decimals.
3. Lettering should be all in capitals.
4. Select a front view that best describes the part.

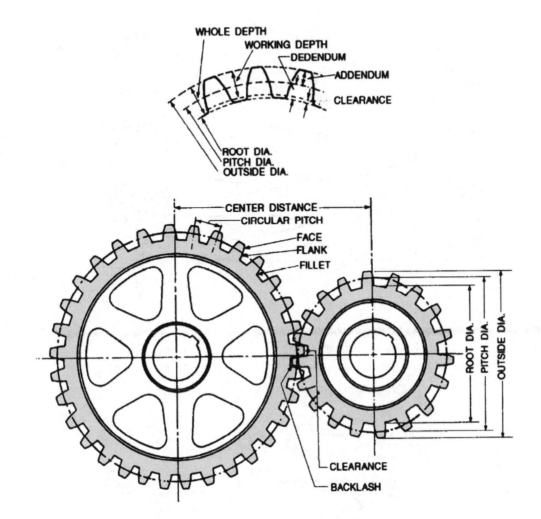

Figure 7.21 Gear nomenclature used in machine blueprints.

5. Omit hidden lines unless absolutely necessary to describe the shape of the object.

6. Consider data and dimensioning schemes based on feature relationships.

7. Place dimensions between views where possible.

8. Where possible, minimize mathematical equations.

9. Do not dimension hidden lines.

10. Check that the product is designed for manufacturability and that an inspection process is in place to check dimension accuracy.

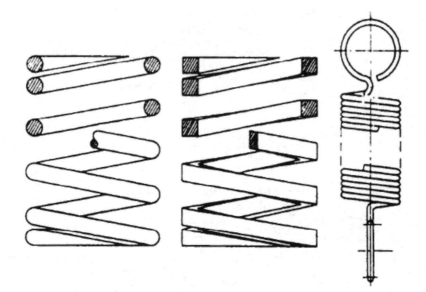

Figure 7.22A   Common types of helical springs.

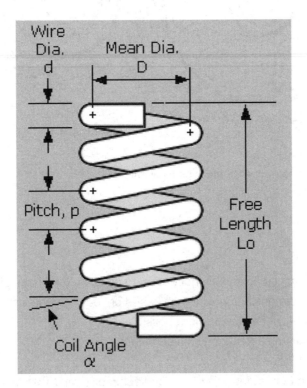

Figure 7.22B   Typical helical spring nomenclature.

11. Avoid duplicating dimensions; use reference dimensions if necessary to duplicate.
12. Dimension lines should not cross other dimension lines.
13. Dimension lines should not cross extension lines.
14. No dimensions should appear on the body of a part. Offset 0.38 inch from the object outline.
15. Place all dimensions for the same feature in one view if possible.
16. Extension lines can cross extension lines.
17. Use center marks in view(s) only where a feature is dimensioned.
18. Use centerlines and center marks in views only if a feature is being dimensioned or referenced; otherwise omit.
19. When multiples of the same feature exist in a view, dimension only one of the features and label as "NumberX DIM," meaning that the feature exists in that view a certain number of times. For example, "4X .250" implies that in the view, there are four like dimensions for the dimensioned feature.
20. Minimize the use of centerlines between holes; they add little value and clutter the object being drawn.

7.6 STANDARDS

The main international standards organizations for geometrical dimensioning and tolerancing are:

- ASME
- ANSI
- ISO

For blueprint production, whether the print is drawn by a human hand or by computer-aided drawing (CAD) equipment, the American industry has adopted a new standard, ANSI Y14.5M. It standardizes the production of prints from the simplest on-site handmade job to single or multiple-run items produced in a machine shop with computer-aided manufacturing (CAM) software.

There exist standards and practices for creating technical drawings of mechanical parts and assemblies. The governing agency mainly responsible for setting the standards in the United States is ASME. There are a number of documents that cover various aspects of mechanical drawings, including:

- ASME Y14.100 -2004 Engineering Drawing Practices
- ASME Y14.4M -1989 Pictorial Drawing
- ASME Y14.3M -Multi- and Sectional View Drawings
- ASME Y14.1-1995 Decimal Inch Drawing Sheet Size and Format
- ASME Y14.5M-1994 Geometric Dimensioning and Tolerancing
- ASME Y14.13M-1981 Mechanical Spring Representation

It is important that these standards are implemented and followed by the drafter to ensure that the drawings are correctly interpreted by others.

8

The Meaning of Symbols

8.1 GENERAL INFORMATION

Blueprint drawings—as applied to the building-construction industry—are generally used to show how a building, object, or system is to be constructed, implemented, modified, or repaired. One of the main functions of graphic symbols on construction drawings is to reference other drawings within the set. For example, a circle drawn around an area of a drawing with an extension to a number would indicate that this portion of the drawing has been drawn to a larger scale to provide more information than would be possible at the existing scale (Figure 8.1). In the preparation of working drawings for the building-construction industry, architects and engineers have devised systems of abbreviations, symbols, and keynotes to simplify the work of those preparing the drawings and to keep the size and bulk of the construction documents to an acceptable, comprehensible minimum. Drawing simple building components without the use of symbols would indeed be a tiresome task. Visualizing and reading construction drawings therefore necessitate a knowledge of symbols and abbreviations used in the construction industry and of their proper use in representing materials and other components and their locations (Figure 8.2). Symbols may vary slightly from one locale to another.

The majority of architects and engineers today use symbols adopted by the American Institute of Architects (AIA) and the American National Standards Institute (ANSI). However, designers and drafters continue to modify some of these symbols to suit their own particular needs for the types of projects they are normally commissioned to design. For this reason, most drawings have a symbol list or legend drawn and lettered either on each set of working drawings or in the written specifications. Modified symbols are normally selected by the consultant because they are easier to draw and interpret and are sufficient for most applications.

In order for this system of symbols to work, each drawing within the set has its own unique number. This is usually a combination of numbers: the number for the individual drawing as well as the page or sheet number on which the specific drawing appears. Individual drawings may be referenced many times throughout a set of construction drawings. The symbols discussed in this chapter are not all-inclusive by any means, but they are the ones that the builder or designer is likely to encounter in most general building-construction applications.

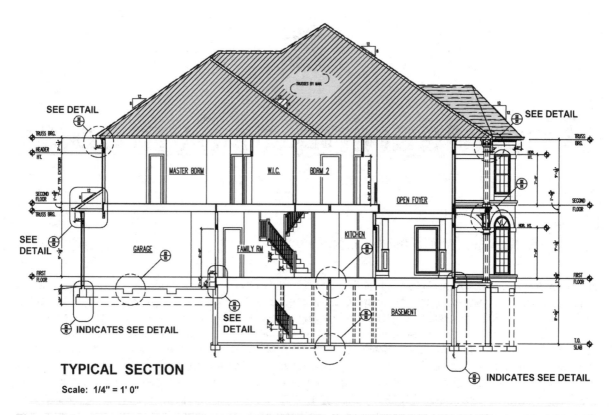

SEE DETAIL

SEE DETAIL

TRUSS BRG.

HEADER HT.

MASTER BDRM   W.I.C.   BDRM 2

TRUSS BRG.

OPEN FOYER

SECOND FLOOR

TRUSS BRG.

SEE DETAIL

GARAGE

FAMILY RM

KITCHEN

SECOND FLOOR

FIRST FLOOR

FIRST FLOOR

SEE DETAIL

BASEMENT

INDICATES SEE DETAIL

T.O. SLAB

TYPICAL SECTION

Scale: 1/4" = 1' 0"

INDICATES SEE DETAIL

Figure 8.1 A computer-generated section of a residential structure showing different elements with encircled portions. These portions are detailed to a larger scale at another location within the set. Notice that the highlighted portions do not necessarily have to be circles but can take on other shapes.

Graphic symbols are often used on building plans to show elements such as gas and water service lines and window types as well as to list drawing notes and identify finishes and revisions. The same graphic can be used for more than one purpose. For example, the same symbol is used for every revision; it is the number within the graphic that carries specific information. Trade-specific symbols are included for the electrical, HVAC, and plumbing trades.

One of the most important symbols to use right at the beginning of a new job is the directional symbol. This symbol, which is usually an arrow labeled "N" for north, enables the reader of a construction drawing to orient it. However, there are numerous variations of the "North" point symbol, depending on the designer's fancy (Figure 8.2). Whichever symbol is decided upon, a drawing is properly oriented when it is held so that the north arrow shown on the drawing is pointing toward north. The drawing must be properly oriented so the reader can relate the information on it to the surrounding area.

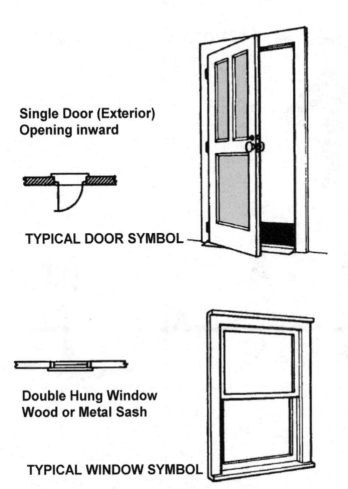

Single Door (Exterior)
Opening inward

TYPICAL DOOR SYMBOL

Double Hung Window
Wood or Metal Sash

TYPICAL WINDOW SYMBOL

Figure 8.2   Examples of typical door and window symbols shown in both plan (as it may appear on a blueprint) and in pictorial form. Reading working drawings necessitates the ability to read, and understand and visualize the various symbols.

The following figures show some of the most common standard symbols; during the course of your work you will likely see many other types as well. For various reasons, some of the symbols on a drawing may not be standard. Many times you will figure out what a symbol means by analyzing it and thinking about what it looks like. The legend on a drawing should show any nonstandard symbols and their meanings. Occasionally a symbol for a particular component or device may have been specifically created by the architect or engineer who developed the drawing.

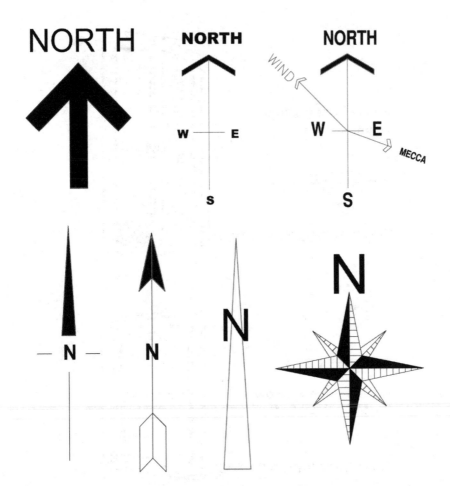

Figure 8.3   Different examples of north-point symbols. Architects and designers often like to design their own symbols for particular items.

8.2   ARCHITECTURAL GRAPHIC SYMBOLS

Reading blueprints requires a good understanding of line types. For example, on a plot plan some of the information shown will include property lines, rights-of-way, easements, topographic features, and a north arrow. Line types are discussed in greater detail in Chapter 3. Figure 8.4A and B show various door, window, and wall symbols used in general construction.

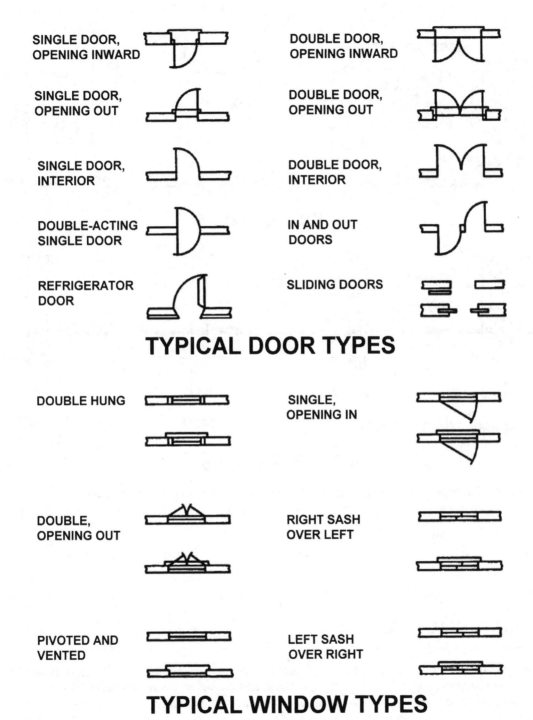

SINGLE DOOR, OPENING INWARD

DOUBLE DOOR, OPENING INWARD

SINGLE DOOR, OPENING OUT

DOUBLE DOOR, OPENING OUT

SINGLE DOOR, INTERIOR

DOUBLE DOOR, INTERIOR

DOUBLE-ACTING SINGLE DOOR

IN AND OUT DOORS

REFRIGERATOR DOOR

SLIDING DOORS

TYPICAL DOOR TYPES

DOUBLE HUNG

SINGLE, OPENING IN

DOUBLE, OPENING OUT

RIGHT SASH OVER LEFT

PIVOTED AND VENTED

LEFT SASH OVER RIGHT

TYPICAL WINDOW TYPES

Figure 8.4A Examples of different common door and window symbols shown in plan form.

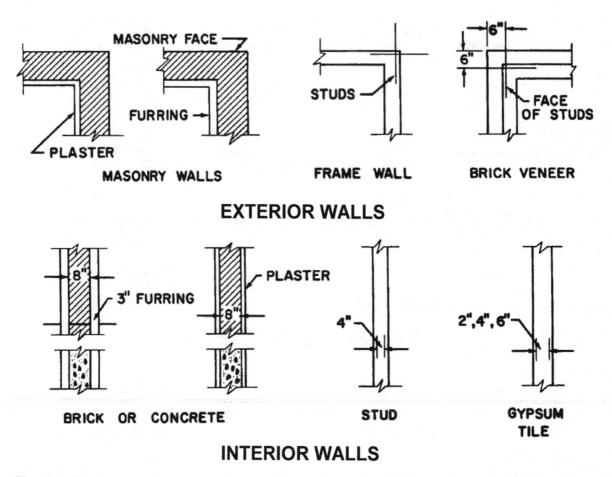

Figure 8.4B   Examples of different exterior and interior wall symbols used in general construction.

8.3   MATERIAL SYMBOLS

Material symbols are used to represent materials or contents on floor plans, elevations, and detail drawings. Different symbols may be used to represent the same item on these drawings. The outline of the drawing may be filled in with a material symbol to show what the object is made of. Many materials are represented by one symbol in elevation and another symbol in section. Examples of this are concrete block and brick. Other materials look pretty much the same when viewed from any direction, so their symbols are drawn the same in sections and elevations as seen in Figure 8.5.

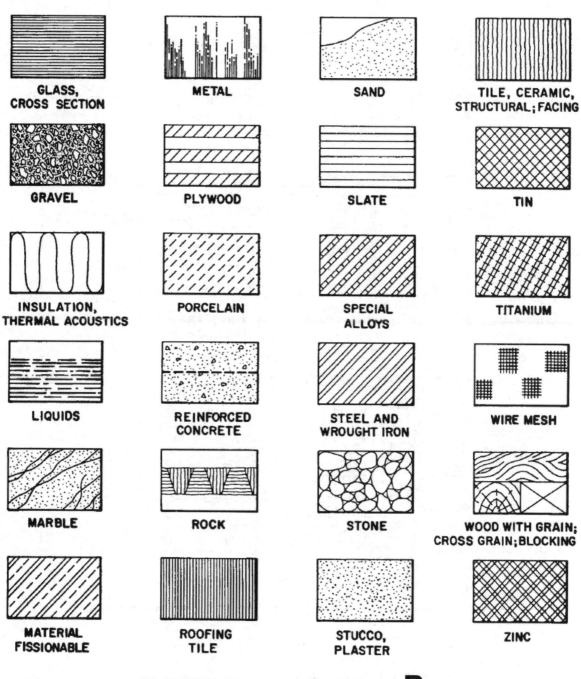

MATERIAL SYMBOLS B

Figure 8.5 Examples of graphic symbols of materials used by architects and engineers when preparing blueprints.

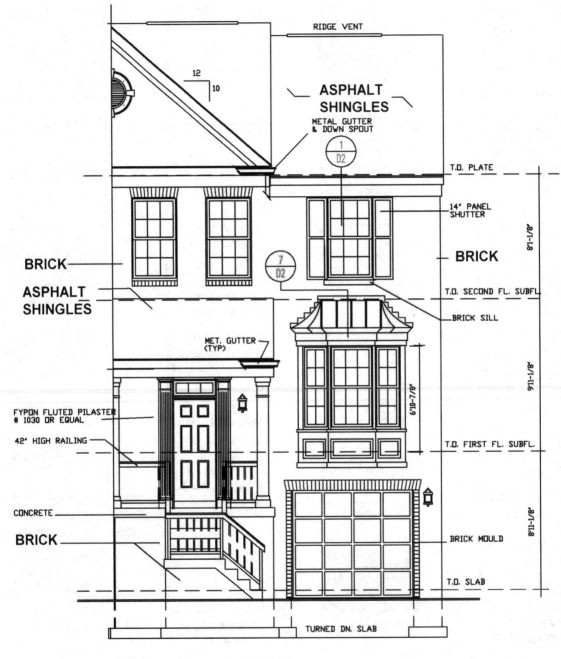

RIDGE VENT

ASPHALT SHINGLES

METAL GUTTER & DOWN SPOUT

12
10

BRICK

ASPHALT SHINGLES

MET. GUTTER (TYP)

FYPON FLUTED PILASTER # 1030 OR EQUAL

42" HIGH RAILING

CONCRETE

BRICK

T.O. PLATE

14" PANEL SHUTTER

BRICK

T.O. SECOND FL. SUBFL

BRICK SILL

T.O. FIRST FL. SUBFL.

BRICK MOULD

T.O. SLAB

TURNED DN. SLAB

8'11-1/8"

9'11-1/8"

6'10-7/8"

8'11-1/8"

FRONT ELEVATION

Figure 8.6 Drawing of a part elevation of a townhouse showing the use of notes for the brick on the facade and asphalt roof shingles rather than partial drawing of the symbols.

When a large area is made up of one material, it is common to only draw the symbol in a part of the area. Some drafters simplify this even further by using a note to indicate what material is used and omitting the symbol altogether (Figure 8.6).

Throughout the remainder of this text, material symbols are presented as they appear in plan and elevation views. Many symbols are designed to approximate the actual appearance of material. This is especially true on elevation drawings, as shown in Figure 8.6. Because of the complexity and space required, many symbols do not have any graphic relationship to the items they represent. These must be memorized if drawings are to be consistently interpreted. When material symbols are similar, always look for a notation, different view, detail, or specification for information about the material.

8.4 ELECTRICAL SYMBOLS

Electrical drawings—like other types of building-construction drawings—must be prepared by competent electrical drafters in a given time period to avoid unnecessary delays. Symbols are used on electrical drawings to simplify the drafting work for both the drafters and the workers interpreting the drawings. It should be noted that electrical symbols are not standardized throughout the industry, which is one reason why electrical drawings typically have a symbol legend or list.

Many electrical symbols are used to show the desired lighting arrangement. Switch symbols are usually placed perpendicular to the wall and read from the right side or bottom of the sheet. Figure 8.7 shows two lists of electrical symbols currently in use on most electrical drawings. The lists represent a good set of electrical symbols in that they are easy to draw, easily interpreted by workers, and sufficient for most applications.

The electrical system for a small property would typically include wiring as well as devices such as switches, receptacles, light fixtures, and appliances. Wiring is indicated by lines that show how devices are connected. These lines are not shown in their actual position. They simply indicate which switches control which lights. Outlets (receptacles) and switches are usually shown in their approximate positions. Major fixtures and appliances are shown in their actual positions. In Figures 8.8, 8.9, and 8.10 we see examples of receptacle symbols, alarm symbols, and other electric symbols.

Some of the symbols listed contain abbreviations, such as TV for television outlet and WP for weatherproof. Others are simplified pictographs, such as the symbol for an electric meter. One way to develop the ability to interpret a drawing and its symbols is by first learning the basic form of the different symbols. This is because many symbols are similar (square, circle, etc.), and the addition of a line, letter, or number determines the specific meaning of that symbol.

8.5 PLUMBING SYMBOLS

One must differentiate between industrial and residential or commercial piping. Industrial piping is generally designed to carry liquids and gases used in manufacturing processes. In heavy construction steel pipes have welded or threaded fittings and connections. Piping used in commercial and residential applications is generally termed "plumbing" and is designed to carry fresh water, liquid and solid wastes, and gas. These pipes can be made of plastic, copper, galvanized steel, or cast iron. In preparing a plumbing drawing, all pipe fittings, fixtures, valves, and other components are represented by symbols

ELECTRICAL SYMBOLS

⏀	CONVENIENCE OUTLET - SET 12" AFF U.N.O.
⏀ A	CONVENIENCE OUTLET -42" AFF
⏀ H	CONV. OUTLET -MOUNTED HORIZ. IN BACKSPLASH
⏀ 220	CONV. OUTLET -220 VOLT
⏀	CONVENIENCE OUTLET - 1/2 SWITCHED 12" A.F.F.
⏀ WP/GFI	CONV. OUTLET- WATER PROOF- GRND FAULT INTERUPT
⬥ GDO	GARAGE DOOR OPENER
$ $₃ $₄	SWITCH, 3-WAY, 4-WAY- SET 38" AFF U.N.O.
$₂	STACKED SWITCH
−○−	LIGHT FIXTURE
├○−	LIGHT FIXTURE
⊡R	RECESSED LIGHT FIXTURE
▭	FLOURESCENT LIGHT
▭ CHIMES	DOORBELL CHIMES
DB ▣−	DOORBELL @ 32" A.F.F.
Ⓔ	ELEC. METER
Ⓖ	GAS METER
PHONE ▼	TELEPHONE OUTLET + 12" A.F.F.
T.V. ▼	T.V. OUTLET + 12" A.F.F.
⏀ SD	SMOKE DETECTOR

◑ VENT TO EXT.	EXHAUST FAN
−◑− FAN / LITE COMBO VENT TO EXT	EXHAUST FAN / LIGHT COMBO
▭ EP	ELECTRIC PANEL
▢○	COMPRESSOR
−○−	CLNG FAN
V̂AC	CENTRAL VACCUUM OUTLET
Ⓢ−	SECURITY PANEL
◇ INTRCM	INTERCOM
▢	NON FUSED DISCONNECT SWITCH
▭ KEY PAD	CONTROL CONSOLE
◆	SECURITY SENSORS
▽△	ALARM HORN
■ MD	MOTION DETECTOR
◉	JUNCTION BOX @ DISH WASHER
Ⓣ	THERMOSTAT
T	DOORBELL TRANSFORMER
WP	WEATHER PROOF

Figure 8.7　List of general electrical symbols in current use.

such as those listed in Figures 8.11A,B,C. The use of these symbols simplifies considerably the preparation of piping drawings and conserves a considerable amount of time and effort.

　　In residential applications, plumbing information can often be provided on the floor plans in the form of symbols for fixtures and notes (for pipe sizes, slope, etc.) for clarification. Other plumbing items and information to be shown on the floor plans include water connections, floor drains, vent pipes, and sewer connections. Floor drains are shown in the approximate location on the plans and sections, with notes indicating the size, type, and slope to drain. Sewer and water-service lines are located in the position in which these utilities enter the building. Service lines are normally indicated on the site plan. Vent pipes

ELECTRICAL SYMBOLS

SYMBOLS	DESCRIPTION
O	LIGHT FITTING CEILING MOUNTED
⊗	STANDARD RECESSED DOWN LIGHT FITTING 'X' FOR WATERPROOF DIFFUSER, 'W WEATHER PROOF
Ⓝ	BULK HEAD FITTING WITH NUMERALS (N)
♦₅O	LOW VOLTAGE SPOT LIGHT FITTING
⊕	EXTRACT FAN (EXHAUST FAN)
⊗	CEILING FAN
Ⓡ	REGULATOR
⊢O	WALL MOUNTED LIGHT FITTING "X" FOR WEATHER PROOF
▬▬▬	FLOURESCENT LIGHT FITTING 1X36 W
▬▬▬	FLOURESCENT LIGHT FITTING 2X36 W
☐	MODULAR FLUORESCENT LIGHT FITTING 2' X 2'
⬠	EMERGENCY LIGHT FITTING WITH EXIT SIGN
⊿	EMERGENCY LIGHT FITTING WITHOUT EXIT SIGN
⬠	SHAVER SUPPLY UNIT
⊣⊢	T.V. SIGNAL OUTLET
◆	CHANDILIER, DONTE A-0.3KW, B-0.5KW, C-1KW, D-2KW E-3KW, F-4KW
⊙	BELL PUSH
∕	15/20A SINGLE POLE LIGHTING SWITCH (NO. OF GANG SHOWN AS INDICATED) 5A INDICATED WHERE SPECIFIED
⋎	15A SINGLE POLE 2 WAY LIGHTING SWITCH
∕	5A SINGLE POLE 2 WAY LIGHTING SWITCH
∮	15A SINGLE POLE INTERMEDIATE LIGHTING SWITCH
⋎	TIME DELAY SWITCH
∮	20A D.P. SWITCH (W.H. WATER HEATER)
⊞⊞	BELL INDICATOR
⌷	SPUR UNIT (P-DENOTE PILOT LIGHT, F-DENOTE FLEX O/L)
⌷	SWITCHED SPUR UNIT (P-PILOT LIGHT, F-FLEX O/L)
▣	DIMMER UNIT
▣	JUNCTION BOX FOR LIGHTING
▣	NEON SIGN POINT
⚐	13A SINGLE SWITCHED SOCKET OUTLET
⚐	13A TWIN SWITCHED SOCKET OUTLET
⚐	5A UNSWITCHED SOCKET OUTLET FOR TABLE LAMP CONNECTED IN LIGHTING CIRCUIT
⚐	20A SWITCH FUSED SPUR WITH FLEX OUTLET
⚐	13A SINGLE SWITCHED SOCKET OUTLET (WEATHER PROOF)
⊠	WINDOW A/C UNIT

SYMBOLS	DESCRIPTION
⚐	13A SINGLE SOCKET OUTLET
⚐	3 PHASE +N+E SOCKET OUTLET WITH MATCHING PLUG
⚐	15A SINGLE SWITCH SOCKET OUTLET
⚐	15A SINGLE SOCKET OUTLET
▬	45A CONNECTOR
▢ᴴ	LOAD BREAK SWITCH
▣	THERMOSTAT
▣ᴴ	45A COOKER CONTROL UNIT WITH 13A S.S.O. (WITH NEON INDICATOR)
⚡	CENTRAL T.V. ANTENNA
▣	AUDIO/VIDEO DOOR PHONE
▣	RADIO/T.V. JUNCTION BOX
▣	MAIN TELEPHONE JUNCTION BOX
▣	TELEPHONE JUNCTION BOX
▣	MOTOR CONTROL CENTER
⋇	MCCB (MOULDED CASE CIRCUIT BREAKER)
⌿	MINIATURE CIRCUIT BREAKER
⌿	SWITCHFUSE OR FUSESWITCH
▨	H R C FUSES
▣	KILOWATT HOUR METER
▭	MCB CONSUMER UNIT
▬	SUB MAIN DISTRIBUTION BOARD
▭	MAIN M.V. CUBICLE PANEL
◁	TELEPHONE OUTLET (ITSALAT APPROVED)
◀	INTERCOM HAND SET
◀	DOOR PHONE OUTLET
⑧	SMOKE DETECTOR - IONISATION
⊕	OPTICAL DETECTOR
⑧	HEAT DETECTOR (F-DENOTE FIXED TEMPERATURE)
▣	BREAK GLASS UNIT
▥	FIRE ALARM MAIN PANEL
▣	FIRE ALARM REPEATER PANEL
⌂	FIRE ALARM BELL (INDICATE WHETHER 8" OR 10")

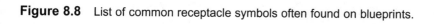

Figure 8.8 List of common receptacle symbols often found on blueprints.

SYMBOLS - FIRE ALARM SYSTEM

DESCRIPTION

$\boxed{F}_{SD}$ SMOKE DETECTOR DUCT TYPE.

$\boxed{OSY}$ VALVE SUPERVISORY SWITCH.

$\boxed{DH}$ ELECTROMAGNETIC TYPE DOOR HOLDER OUTLET.

$\boxed{FATC}$ FIRE ALARM TERMINAL CABINET.

$\boxed{PIV}$ POST INDICATOR VALVE.

FAT_{2-4-6} FIRE ALARM TRANSMITTER (BASE LOOP) NUMERALS DENOTE CODE.

FTT_{2-4-6} FIRE ALARM TROUBLE TRANSMITTER (BASE LOOP) NUMERALS DENOTE CODE.

■ CITY FIRE ALARM MASTER STATION MTD 5'-6" [1676mm] AFF UNLESS NOTED.

$\boxed{FAR}$ FIRE ALARM RECORDER.

$\boxed{BAT}$ FIRE ALARM BATTERIES.

$\boxed{CRG}$ BATTERY CHANGER.

$\boxed{FACC}$ FIRE ALARM CENTRAL CONSOLE.

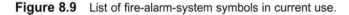

Figure 8.9   List of fire-alarm-system symbols in current use.

SYMBOLS - TELEVISION SYSTEM

DESCRIPTION

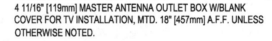

4 11/16" [119mm] MASTER ANTENNA OUTLET BOX W/BLANK COVER FOR TV INSTALLATION, MTD. 18" [457mm] A.F.F. UNLESS OTHERWISE NOTED.

OUTLET FOR TV CAMERA (CCTV SYSTEM), MTD 18" [457mm] A.F.F.

OUTLET FOR TV MONITOR (CCTV SYSTEM).

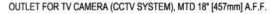 CCTV CONDUIT RUN, MINIMUM 1" [25mm] CONDUIT.

SYMBOLS - RADIO SYSTEM

DESCRIPTION

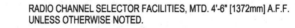

RADIO CHANNEL SELECTOR FACILITIES, MTD. 4'-6" [1372mm] A.F.F. UNLESS OTHERWISE NOTED.

SPEAKER PROGRAM SELECTOR SWITCH & VOLUME CONTROL MTD. 4'-6" [1372mm] A.F.F. UNLESS OTHERWISE NOTED.

SPEAKER OUTLET, MTD. 7'-6" [2286mm] A.F.F. UNLESS OTHERWISE NOTED.

RADIO CONDUIT RUN, MINIMUM 3/4" [19mm] CONDUIT.

CEILING MICROPHONE SPEAKER.

NOTE: ALSO SEE VA BARRIER FREE DESIGN GUIDE PG-18-13.

Figure 8.10   Symbols for typical television and radio systems.

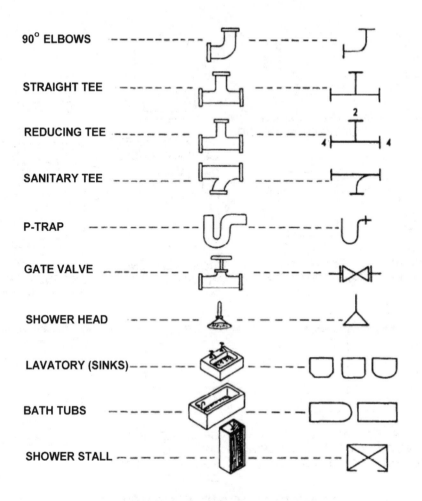

90° ELBOWS	
STRAIGHT TEE	
REDUCING TEE	
SANITARY TEE	
P-TRAP	
GATE VALVE	
SHOWER HEAD	
LAVATORY (SINKS)	
BATH TUBS	
SHOWER STALL	

COMMON PLUMBING SYMBOLS

Figure 8.11A Examples of common plumbing fittings and symbols.

(which allow for a continuous flow of air through the building) are drawn in the wall according to their location on the plan, with a note indicating the material and size.

When plumbing drawings are prepared, they are not usually shown on the same sheet as the architectural floor plan. The main plumbing items shown on the floor plans are the fixtures. Drainage, vent, and water systems are drawn with thick lines using symbols, abbreviations, and notes.

90° ELBOW
REDUCING

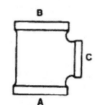

90° STREET ELBOW
REDUCING ON MALE END

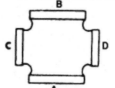

TEE WITH ONE END
OF RUN REDUCED

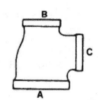

TEE WITH ONE END OF
RUN & OUTLET REDUCED

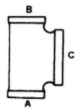

TEE (BULL HEAD)
BOTH ENDS OF RUN
REDUCED

TEE WITH
REDUCED OUTLET

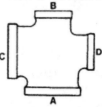

CROSS REDUCING
ON BOTH OUTLETS

CROSS REDUCING ON
ONE END OF RUN AND
ON ONE OUTLET

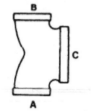

DOUBLE BRANCH
ELBOW REDUCING ON
BOTH ENDS OF RUN

45° Y-BEND
(LATERAL) REDUCING
ON OUTLET ONLY

CROSS REDUCING
ON ONE END OF
RUN AND ON BOTH
OUTLETS

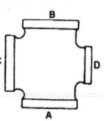

CROSS REDUCING ON
ONE OUTLET ONLY

HOW TO READ PLUMBING FITTINGS

Figure 8.11B   Examples of different types of plumbing connections.

8.6   HEATING, VENTILATION, AND AIR-CONDITIONING SYMBOLS

The main function of HVAC drawings is to show the location of the heating, cooling, and air-conditioning units as well as the piping and ducting diagrams. Graphical symbols on HVAC drawings are similar in pattern to those used for plumbing. In Figures 8.12A and B we see examples of typical HVAC symbols in current use in the United States and Europe.

It should be noted that on some HVAC drawings one sees two types of symbols. The main duct may consist of a rectangle to represent the actual size of the ductwork—including reductions. However, the

ITEM	SYMBOL	SAMPLE APPLICATION	ILLUSTRATION
PIPE	SINGLE LINE IN SHAPE OF PIPE, USUALLY WITH NOMINAL SIZE NOTED		
JOINT - FLANGED	DOUBLE LINE		
SCREWED	SINGLE LINE		
BELL & SPIGOT	CURVED LINE		
OUTLET TURNED UP	CIRCLE & DOT		
OUTLET TURNED DOWN	SEMICIRCLE		
REDUCING OR ENLARGING FITTING	NOMINAL SIZE		
REDUCER CONCENTRIC	TRIANGLE		
ECCENTRIC	TRIANGLE		
UNION SCREWED	LINE		
FLANGED	LINE		

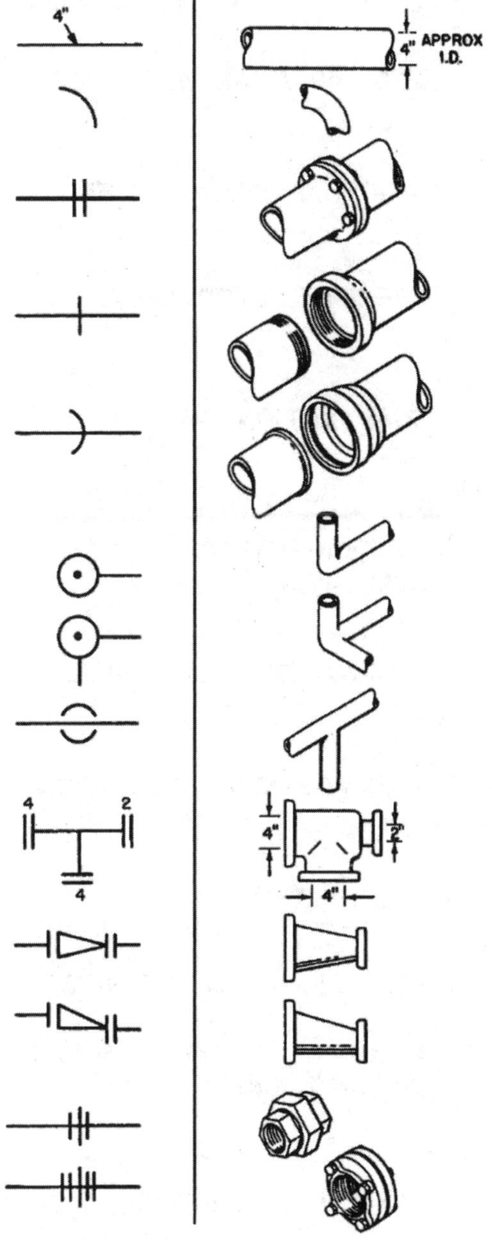

PIPE FITTING SYMBOLS

Figure 8.11C Examples of typical pipe fitting symbols showing sample applications and pictorial representations of the fitting.

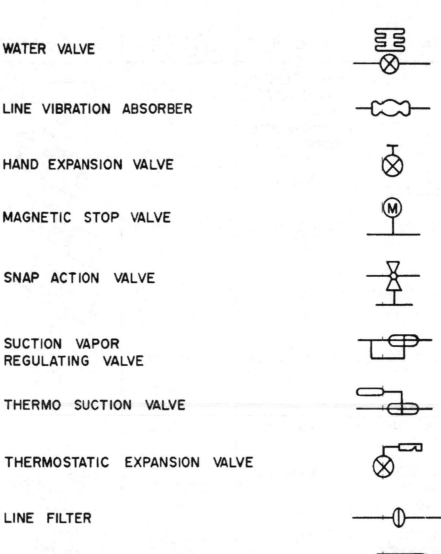

WATER VALVE

LINE VIBRATION ABSORBER

HAND EXPANSION VALVE

MAGNETIC STOP VALVE

SNAP ACTION VALVE

SUCTION VAPOR
REGULATING VALVE

THERMO SUCTION VALVE

THERMOSTATIC EXPANSION VALVE

LINE FILTER

LINE FILTER & STRAINER

NATURAL CONVECTION FINNED-TYPE
COOLING UNIT

FORCED CONVECTION COOLING UNIT

AIR-CONDITIONING SYMBOLS

Figure 8.12A   Common air conditioning symbols in general use.

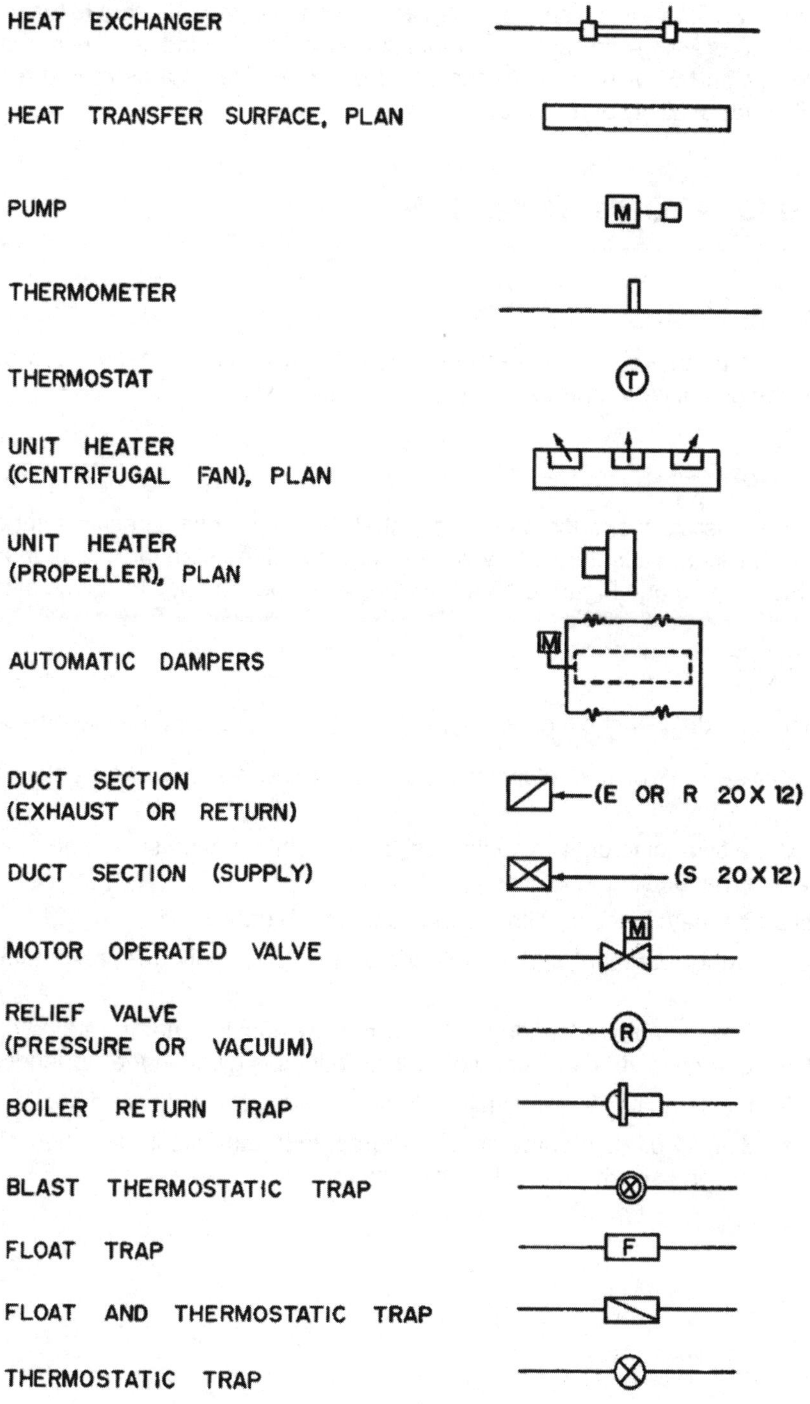

HEAT EXCHANGER	
HEAT TRANSFER SURFACE, PLAN	
PUMP	
THERMOMETER	
THERMOSTAT	
UNIT HEATER (CENTRIFUGAL FAN), PLAN	
UNIT HEATER (PROPELLER), PLAN	
AUTOMATIC DAMPERS	
DUCT SECTION (EXHAUST OR RETURN)	(E OR R 20 X 12)
DUCT SECTION (SUPPLY)	(S 20 X 12)
MOTOR OPERATED VALVE	
RELIEF VALVE (PRESSURE OR VACUUM)	
BOILER RETURN TRAP	
BLAST THERMOSTATIC TRAP	
FLOAT TRAP	
FLOAT AND THERMOSTATIC TRAP	
THERMOSTATIC TRAP	

Figure 8.12B   Examples of heating symbols in general use.

branches are shown by a single line with the dimensions of each noted adjacent to the lines. Both methods are used on HVAC drawings throughout the building-construction industry, either singly or in combination (Figure 8.13A and B). A number of custom HVAC CADD programs are currently available to help improve drafting productivity and efficiency.

8.7   MISCELLANEOUS SYMBOLS

Component Symbols

Component symbols represent such items as furniture, fixtures, and appliances. Many component symbols are similar to the actual item represented in plan/elevation view.

Specialist Symbols

There are many specialist trades, such as welding, that have their own specific symbols. Figure 8.14 shows examples of basic and supplementary weld symbols and the standard location of elements of a welding symbol. Figure 8.15 and Figure 8.16 are further examples of specialist type symbols.

Indexing Symbols

To summarize, when reading symbols on a blueprint, the following should be remembered:

- Placement of symbols on floor plans and elevations may be approximate unless they are dimensioned.
- Do not scale a symbol to determine its size. Specific information should be obtained from relevant details, schedules, or specifications.
- A symbol's size may vary according to the scale of the drawing.
- Symbols drawn to represent surface materials may cover only a representative portion of an area.
- Abbreviations are often used in place of a graphic symbol in order to reduce drawing time and space. A list of architectural/construction abbreviations is given in the Appendix.
- Symbols for the same items are usually different in plan, elevation, and section drawings.
- Some symbols may use the same or similar graphic treatments in elevation. Check notations, detail drawings, and specifications for identification.

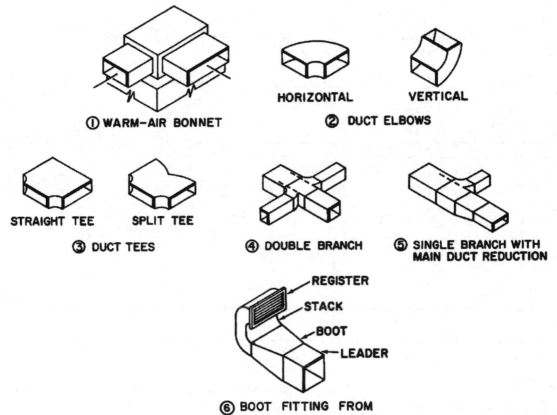

① WARM−AIR BONNET

HORIZONTAL   VERTICAL

② DUCT ELBOWS

STRAIGHT TEE   SPLIT TEE

③ DUCT TEES

④ DOUBLE BRANCH

⑤ SINGLE BRANCH WITH
MAIN DUCT REDUCTION

REGISTER
STACK
BOOT
LEADER

⑥ BOOT FITTING FROM
BRANCH TO STACK

VARIOUS TYPES OF DUCT CONNECTIONS

Figure 8.13A   Various types of duct connections.

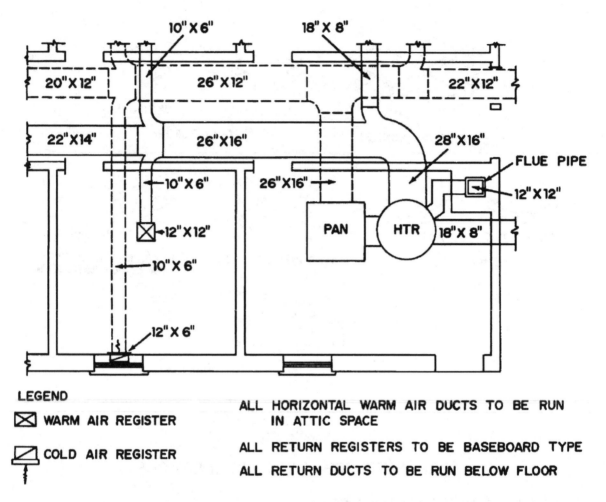

TYPICAL WARM-AIR HEATING SYSTEM PLAN

Figure 8.13B   A typical warm-air heating plan.

| FILLET | PLUG OR SLOT | ARC-SPOT OR ARC-SEAM | GROOVE | | | | | | | BACK OR BACKING | MELT-THRU | SUR-FACING | FLANGE | |
			SQUARE	V	BEVEL	U	J	FLARE-V	FLARE-BEVEL				EDGE	CORNER
◺	⏝	▆	‖	⋁	⋁	⋃	⋃	⋎	⋏	⌒	⬛	⏝	⫫	‖

BASIC ARC AND GAS WELD SYMBOLS

| TYPE OF WELD | | | |
SPOT	PROJECTION	SEAM	FLASH OR UPSET
✳	⟋	✕✕✕		

BASIC RESISTANCE WELD SYMBOLS

| WELD ALL AROUND | FIELD WELD | CONTOUR | |
		FLUSH	CONVEX
○	●	—	⌒

SUPPLEMENTARY SYMBOLS

BASIC AND SUPPLEMENTARY WELD SYMBOLS

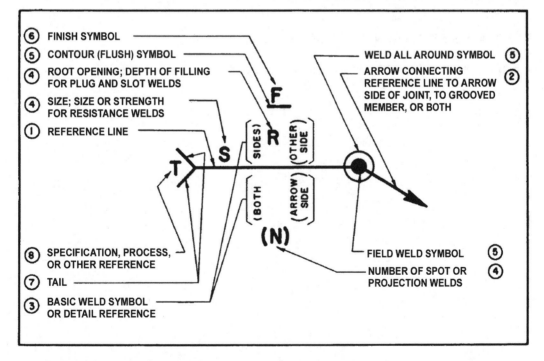

STANDARD LOCATION OF ELEMENTS OF A WELDING SYMBOL

Figure 8.14   Diagram showing basic and supplementary weld symbols and the standard location of elements of a welding symbol.

SYMBOLS – MISCELLANEOUS COMMUNICATONS SYSTEMS

WALL   DESCRIPTION

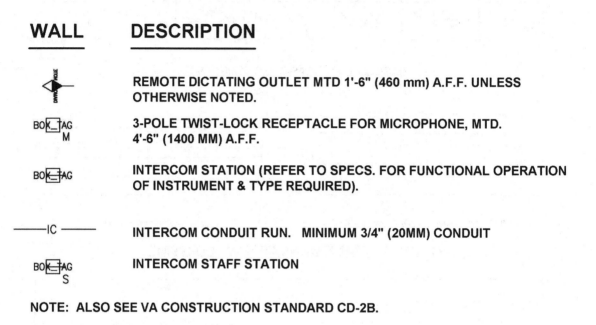

REMOTE DICTATING OUTLET MTD 1'-6" (460 mm) A.F.F. UNLESS OTHERWISE NOTED.

3-POLE TWIST-LOCK RECEPTACLE FOR MICROPHONE, MTD. 4'-6" (1400 MM) A.F.F.

INTERCOM STATION (REFER TO SPECS. FOR FUNCTIONAL OPERATION OF INSTRUMENT & TYPE REQUIRED).

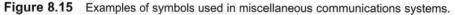

INTERCOM CONDUIT RUN.   MINIMUM 3/4" (20MM) CONDUIT

INTERCOM STAFF STATION

NOTE: ALSO SEE VA CONSTRUCTION STANDARD CD-2B.

Figure 8.15   Examples of symbols used in miscellaneous communications systems.

SYMBOLS - SITE WORK

DESCRIPTION

—— DB —— DIRECT BURIAL CABLE - SIZE AS INDICATED.

—— T —— TELECOMMUNICATIONS DUCT

—— P —— POWER DUCT

EXTERIOR FLOOD LIGHT

DUAL POWER & TELECOMMUNICATIONS MANHOLE

T500 PADMOUNT TRANSFORMER - NUMERAL INDICATES KVA SIZE

SWGR S SWITCHGEAR - SUBSCRIPT DENOTES SUBSTATION

HVS HIGH-VOLTAGE SWITCH ON CONCRETE PAD

LVS LOW-VOLTAGE SWITCH ON CONCRETE PAD

SWBD SWITCHBOARD

SYMBOLS - REWIRING AND DEMOLITION

DESCRIPTION

EXISTING FAN - REMOVE FAN AND BLANK OUTLET

—— E —— EXISTING CONDUIT IN PLACE - REMOVE EXISTING WIRES AND INSTALL NEW WIRES AS INDICATED

—— EX —— EXISTING CONDUIT AND WIRE TO REMAIN

—— X —— EXISTING CONDUIT AND WIRE TO BE ABANDONED OR REMOVED

——□—E—— CONNECTION OF NEW CONDUIT TO EXISTING CONDUIT W/APPROVED TYPE COUPLINGS - INSTALL NEW WIRES

——✕—— CONDUIT SEAL FOR EXPLOSIONPROOF INSTALLATION

B PLACED ADJACENT TO DEVICE SYMBOL DENOTES EQUIPMENT SHALL BE REMOVED AND OUTLET BLANKED

X PLACE ADJACENT TO DEVICE SYMBOL DENOTES EQUIPMENT SHALL BE REMOVED

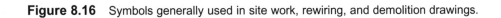

Figure 8.16 Symbols generally used in site work, rewiring, and demolition drawings.

9

Understanding Schedules

9.1 OVERVIEW

A schedule as applied to construction working drawings is an organized method of presenting general notes or lists of materials, building components (doors, windows, etc.), equipment, and so forth in a drawing in tabulated form. The main purpose of incorporating schedules into a set of construction documents is to provide clarity, location, sizing, materials, and information on the designation of doors, windows, roof finishes, equipment, plumbing, and electrical fixtures. Properly done, schedules help keep drawings from becoming cluttered with too much printed information or notes and have proven to be great time-saving devices for the person preparing the drawing as well as the architect, engineer, contractor, and workers on the site. This chapter is intended to assist the reader in interpreting tabulated information on blueprint drawings. It should be understood that schedules and specifications give specific details about actual items, while drawings generally show the size and location of the item.

Schedules are generally organized in a drawing set such that they are near the discipline to which they are related. There are several different approaches to setting up a schedule; it may include all or some of the following information about the product:

- Vendor's name
- Product name
- Model number
- Size
- Quantity
- Rough opening size
- Material
- Color

Many different items or features may be described in schedules: some examples are, doors, windows, lintels, columns, beams, electrical equipment or fixtures, plumbing and mechanical equipment, room finish information, and appliances. These schedules are essentially tables (i.e., a box of columns and rows) that list information about specific items. Schedules allow you to quickly refer to a specific item. On large commercial projects schedules may require several sheets. The exact method of representation depends upon individual company standards. While schedules are usually presented in tabulated form, they are often accompanied by pictorial schedules for additional clarity. This feature is discussed below.

There are many software systems today that allow you to make schedules easily and quickly. For example, in Architectural Desktop schedules are constructed of tabulated data that are extracted from the individual objects in your drawings. To simplify matters, Autodesk has built into ADT2 a good number of schedules that can be used as is or as a basis for customizing. VectorWorks Architect is another excellent program for generating door and window schedules. Figure 9.1 illustrates some of the procedures used to generate door and window schedules using VectorWorks CAD software.

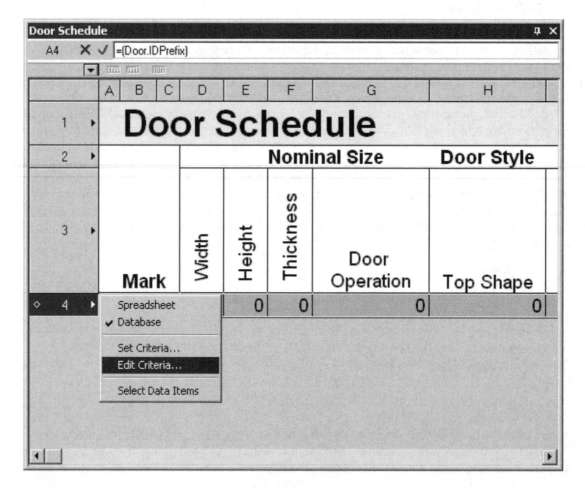

Figure 9.1A   An example of a door schedule generated using VectorWorks CAD software (source: VectorWiki).

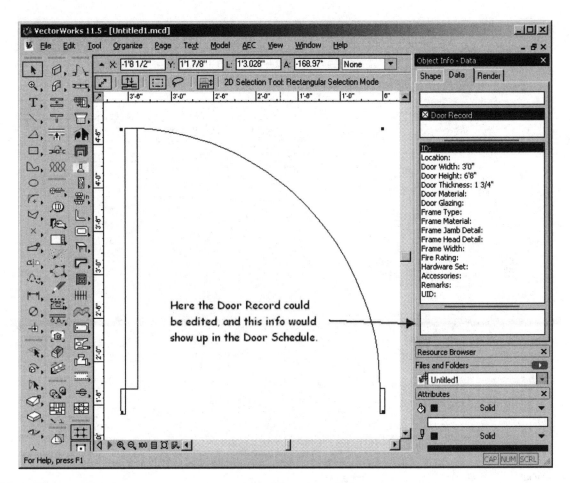

Figure 9.1B   An example of a door schedule generated using VectorWorks CAD software (source: VectorWiki).

9.2   DOOR AND WINDOW SCHEDULES

Door Schedules

Door schedules typically indicate the model or mark, quantity, size, thickness, type, material, function, frame material, fire rating, and remarks. Examples of typical door schedules are shown in Figures 9.2A, B, and C. Sometimes tabulated door and window schedules are accompanied by graphic schedules—in pictorial form with elevations of the door or window types—to facilitate identification (Figures 9.3A and B). Door schedules may include information regarding glazing and louvers if appropriate. Commercial

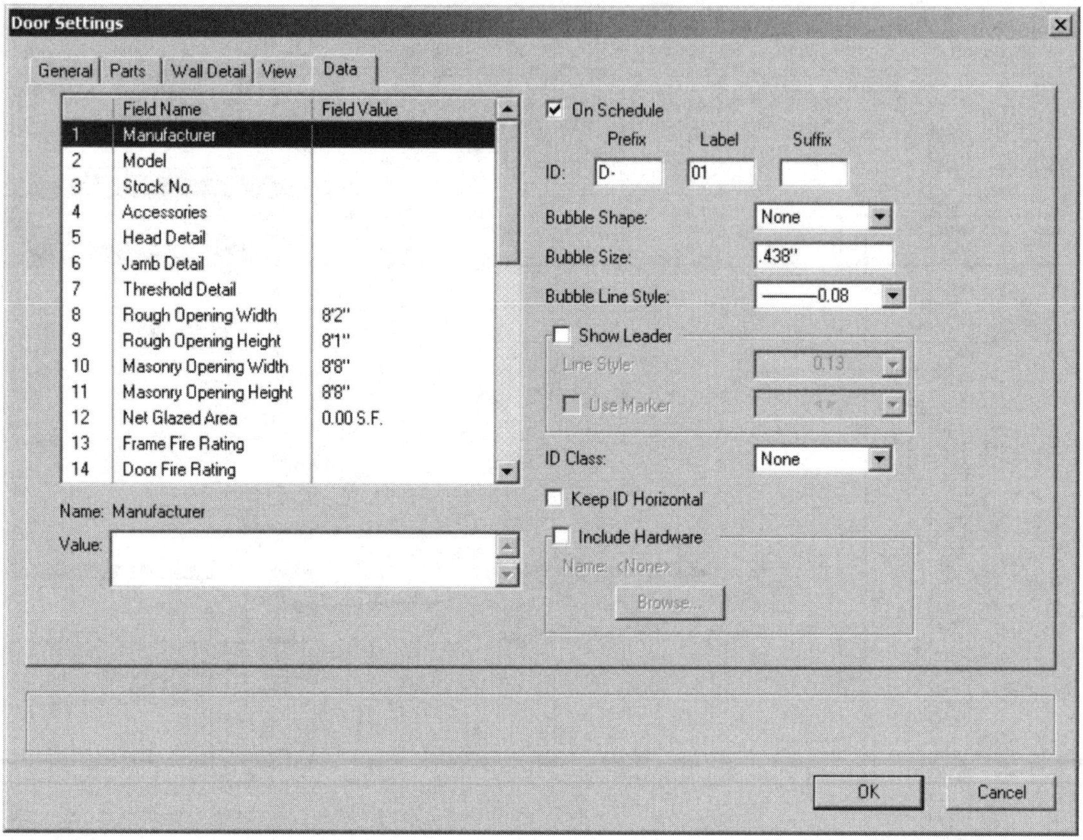

Figure 9.1C An example of a door schedule generated using VectorWorks CAD software (source: VectorWiki).

door schedules may also include hardware and door frame data and key codes. In some cases the designer may decide to show the ironmongery in a separate schedule (Figure 9.4). Door schedules are generally longer and more detailed than window schedules. Doors and door function are usually critical to the success of the building. Doors are in constant use and serve many different functions, including passage, privacy, and security.

Doors may be identified as to size, type, and style, with code numbers placed next to each symbol in a plan view. This code number or mark is then entered on a line in a door schedule, and the principal characteristics of the door are entered in successive columns along the line. The "quantity" column allows a quantity check on doors of the same design as well as the total number of doors required. By using a number with a letter, you will find that the mark serves a double purpose: The number identifies the floor on which the door is located, and the letter identifies the door design. The "remarks or comments" column allows identification by type (panel or flush), style, and material. The schedule is a convenient way of presenting pertinent data without having to refer to the specification.

DOOR AND FRAME SCHEDULE

NO.	DOOR LOCATION	DOOR			TYPE	MATERIAL	GLASS	FRAME		FRAME		LABEL RATING	HARDWARE			REMARKS
		SIZE								DETAIL			PANIC BAR	SELF CLOSING	SET	
		WIDTH	HEIGHT	THICK				MATERIAL	TYPE	JAMB	HEAD	THRESHOLD				

Figure 9.2A Examples of various door-schedule formats. The information on a door schedule is dictated by the complexity and size of the project.

DOOR AND FRAME SCHEDULE

MARK	DOOR			MATERIAL	GLAZING	FIRE RATING LABEL	HARDWARE GROUP	NOTES / REMARKS
	SIZE							
	WD	HGT	THK					

Figure 9.2B Examples of various door-schedule formats. The information on a door schedule is dictated by the complexity and size of the project.

DOOR SCHEDULE

MARK	SIZE	TYPE	MAT L	FINISH	LATCH DEVICE	FRAME	THOLD	CLOSER	REMARKS
101	PR-3-0 X 6-8 X 1 3/4	A	WD/GL	PTD	KEY LOCK	WD	ALUM	CLOSER	TRANSOM, SCREEN DOORS
102	PR-3-0 X 6-8 X 1 3/4	A	WD/GL	PTD	KEY LOCK	WD	ALUM	CLOSER	TRANSOM, SCREEN DOORS
103	3-0 X 6-8 X 1 3/4	C	WD/GL	PTD	PASSAGE	RATED HM	--	CLOSER	60 MIN FIRE RATING
104	3-0 X 6-8 X 1 3/4	C	HM	PTD	KEY LOCK	HM	ALUM	CLOSER	
105	PR-3-0 X 6-8 X 1 3/4	D	WD	PTD	KEY LOCK	HM	ALUM		CHAIN STOP
106	3-0 X 6-8 X 1 3/4	E	WD	PTD	KEY LOCK	HM	ALUM	CLOSER	SCREEN DOOR
107	3-0 X 6-8 X 1 3/4	E	WD	PTD	KEY LOCK	--	--		
108	3-0 X 6-8 X 1 3/4	E	WD	STAIN	KEY LOCK	--	--		
109	3-0 X 6-8 X 1 3/4	E	WD	STAIN	KEY LOCK	--	--		
110	3-0 X 6-8 X 1 3/4	E	WD	STAIN	KEY LOCK	--	--		
111	1-8 X 6-8 X 1 3/4	E	WD	STAIN	KEY LOCK	--	--		
112	3-0 X 6-8 X 1 3/4	E	WD	STAIN	KEY LOCK	--	--		
113	PR-2-6 X 6-8 X 1 1/8	F	WD/SCRN	PTD	--	--	--	CLOSER	SCREEN DOORS
114	PR-3-0 X 6-8 X 1 3/4	B	WD/GL	PTD	KEY LOCK	WD	ALUM		HOLDOPEN AT 180 DEG.
115	PR-3-0 X 6-8 X 1 3/4	B	WD/GL	PTD	KEY LOCK	WD	ALUM		HOLD OPEN AT 180 DEG.
116	PR-2-6 X 6-8 X 1 3/4	F	WD/SCRN	PTD	KEY LOCK	--	--	CLOSER	
117	3-0 X 6-8 X 1 3/4	G	HM	PTD	PANIC	RATED HM	ALUM	CLOSER	60 MIN FIRE RATING
118	3-0 X 6-8 X 1 3/4	E	WD	PTD	KEY LOCK	--	ALUM	CLOSER	SCREEN DOORS
119	3-0 X 6-8 X 1 3/4	E	WD	PTD	KEY LOCK	--	--	CLOSER	
120	3-0 X 6-8 X 1 3/4	E	WD	PTD	KEY LOCK	--	--	CLOSER	
121	2-6 X 6-8 X 1 3/4	E	WD	PTD	KEY LOCK	--	ALUM	CLOSER	SCREEN DOORS

Figure 9.2C Examples of various door-schedule formats. The information on a door schedule is dictated by the complexity and size of the project.

The "mark" heading in door and window schedules usually refers to the window or door location on the blueprint. In other words, on the floor plan of each level of the house, windows and doors are drawn within the wall layout. The architect assigns a number to each window and door beginning with 1.

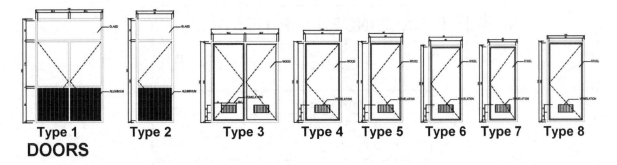

DOORS

Figure 9.3A An example of a pictorial schedule that often accompanies tabulated door schedules.

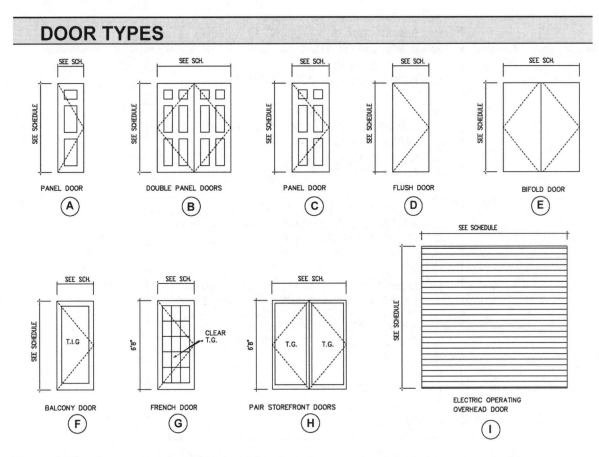

Figure 9.3B An example of a pictorial schedule that often accompanies tabulated door schedules.

IRONMONGERY SCHEDULE

DOOR TYPE	LOCATION	IRONMONGERY
D1	1—BEDROOM FLAT	A
D2	1—BEDROOM/STUDIO	B
D3	STAIRCASE	
D4	TOILET & BATH	C
D5	ACCESS PANEL DOOR	E

DOOR IRONMONGERY SETS

SET A:
1. S.S HINGES 75 X 100. 1 1/2 PAIR
2. V6500 SAFETY LOCK.
3. DOOR CLOSER 1 NO.
4. FLOOR MOUNTED DOOR STOP 1 NO.

SET B:
1. S.S HINGES 75 X 100. 1 1/2 PAIR
2. V6364 LOCKING LATCHBOLT.
3. DOOR CLOSER 1 NO.
4. FLOOR MOUNTED DOOR STOP 1 NO.

SET C:
1. HINGES 75 X 100 D.S.S.W. 1 1/2 PAIR
2. V6300 PRIVACY LOCK SET ON BACK PLATE.
3. FLOOR MOUNTED DOOR STOP 1 NO.

SET D:
1. SLIDING MECHANISM FROM 'HENDERSON'
   OR EQUAL, WITH 2 NOS. COUNTER—SUNK
   SLIDING PLATES (WITH BACKPLATES) AND
   APPROPRIATE OVAL CYLINDER SASH LOCK
2. FLOOR MOUNTED DOOR STOP 2 NOS.

SET E:
1. CUPBOARD LOCKSET
2. HINGES
2. HANDLES.

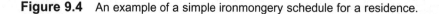

Figure 9.4   An example of a simple ironmongery schedule for a residence.

Window Schedules

A window schedule is similar to a door schedule in that it provides an organized presentation of the pertinent window characteristics. Window schedules include quantity, type, model, manufacturer, size, rough opening, materials, glazing, and finish. There is also usually a column for remarks/notes (Figure 9.5). Also, like door schedules, tabulated schedules cannot always clearly define a specific window. In this case you can add to a schedule a callout with a pictorial drawing of a window adjacent to the window schedule (Figure 9.6).

WINDOW SCHEDULE

SYM.	QTY.	WIDTH	HEIGHT	TYPE	FRAME	SCREEN	GLAZED AREA	VENT. AREA	NOTES
Ⓐ	1	5'-8"	7'-0"	AWNING	WOOD	NO	39.6 sq.ft.	9.9 sq.ft.	
Ⓑ	1	5'-6"	7'-0"	FIXED	WOOD	NO	38.5		1/4" TEMP. GLASS
Ⓒ	3	4'-0"	3'-0"	AWNING	WOOD	YES	12.0	12.0	
Ⓓ	1	5'-6"	7'-0"	FIXED	WOOD	YES	38.5		
Ⓔ	2	1'-8"	3'-3"	CASEMENT	WOOD	YES	5.4	5.4	
Ⓕ	4	1'-4"	5'-8"	FIXED	WOOD	NO	30.2		
Ⓖ	1	4'-0"	3'-6"	GARDEN	WOOD	YES	14.0	5.0	SIDE VENTS
Ⓗ	1	2'-0"	4'-0"	CASEMENT	WOOD	YES	16.0	16.0	

Figure 9.5A A type of window schedule.

ANDERSON WINDOW SCHEDULE

MARK	CATALOG NO.	QTY.	ROUGH DIMENSIONS		FIRST FLOOR		SECOND FLOOR	
			WIDTH	HEIGHT	SILL ROUGH HEIGHT	HEADER ROUGH HT.	SILL ROUGH HEIGHT	HEADER ROUGH HT.
A	TW 2042-2	8	4'-3 15/16"	4'-5 1/4"	2'-6"	6'-11 1/4"	2'-10 3/4"	7'-4"
							3'-8 1/2"	8'-1 3/4"
B	TW 3042	5	3'-2 1/8"	4-5 1/4'	4'-7"	9'-0 1/4"	4'-3 1/2"	8'-8 3/4"
C	TW 2042	5	2'-2 1/8"	4'-5 1/4"	4'-0"	8'-5 1/4"	8'-1 3/4"	12'-7"
							3'-8 1/2"	8'-1 3/8"
D	TW 3052	2	3'-2 1/8"	5'-5 1/4"	3'-4 1/2"	8'-9 3/4"		
	ROUND TOP	1						
E	TW 2856	4	2'-10 1/8"	5'-9 1/4"			9'-0"	14'-9 1/4"

Figure 9.5B A type of window schedule.

WINDOW SCHEDULE

Type	Material	Size	Number	Note
1	ALUMINIUM	170X150 CM	30	
2	ALUMINIUM	250X120 CM	4	
3	ALUMINIUM	75X50 CM	8	
4	ALUMINIUM	170X100 CM	5	
5	ALUMINIUM	170X70 CM	2	
6	ALUMINIUM	250X180 CM	1	

Figure 9.5C   A type of window schedule.

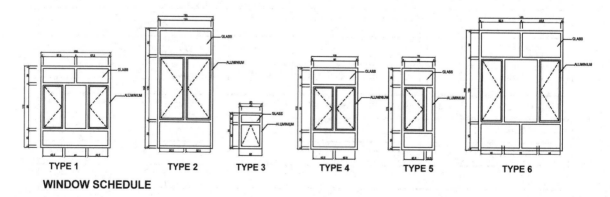

Figure 9.6   A pictorial representation of a window schedule.

9.3   FINISH SCHEDULE

A finish schedule specifies the interior finish material for each room, space, and floor in the building. The finish schedule provides information for the walls, floors, ceilings, baseboards, doors, and window trim. Finish schedules can vary in both format and complexity. A single-family home will generally consist of a limited number of finishes. Commercial buildings, on the other hand, may consist of a larger number of wall and floor finishes. The larger finish schedules are normally in a matrix format and are typically divided into categories such as by space use or by floor. The Y axis lists the spaces, and the X axis is sectioned into floors, base, walls, and ceilings. The room finish schedule should have an entry for every room and hallway in the house or room addition. Finish schedules will also list ceiling heights and marks for each space. The marks can constitute room numbers. Typical examples of finish schedules are shown in Figure 9.7.

ROOM FINISH SCHEDULE

Room No.	Room Name	Floor	Walls				Ceiling		Notes
			N	S	E	W	Matl	Ht	

Figure 9.7A   An example of one type of interior-finish schedule.

INTERIOR FINISH SCHEDULE

ROOM	FLOOR					WALLS				CEILING			NOTES / REMARKS
	VINYL	CARPET	TILE	HARDWOOD	CORK	PAINT	PAPER	TEXTURE	TILE	SMOOTH	BROCADE	PAINT	
FOYER					●		●					●	
KITCHEN	●								●	●		●	
DINING				●				●			●		
FAMILY		●					●				●		
LIVING				●				●			●		
M. BED.		●						●				●	
M. BATH			●				●			●		●	
BED 2		●				●						●	
BED 3		●				●						●	
BATH 2			●						●	●		●	
UTILITY			●						●	●		●	

Figure 9.7B An example of one type of interior-finish schedule.

INTERIOR DOOR SCHEDULE

ROOM	FLOORING	WALLS	CEILING	BASEBOARD	CASINGS	REMARKS
BASEMENT	Concrete	Painted Concrete	Unfinished	None	None	Clean floors and walls and seal floor
LIVING ROOM	Hardwood Oak Plank	1/2" Drywall with Chair Rail	Oak beams and drywall per plans	5" high profile poplar stained	3" wide Profile #415 stained	Floors to have Walnut pegs
KITCHEN	Cork with 3 coats of Urethane	1/2" Drywall with Wallpaper Border	Drywall ceiling	5" high profile #276 pine painted	3" wide Profile #415 painted	Trim paint to high gloss
MASTER BEDROOM	3" Wide hardwood border, Carpet infill	1/2" Drywall with wallpaper	Drywall with stucco finish	4" high profile #355 painted	3" wide Profile #415 painted	Carpet infill to be Berber
HALL "A"	Hardwood with Carpet Runner	1/2" Drywall with paint	Smooth drywall	4" high profile #355 painted	3" wide Profile #415 painted	None

Figure 9.7C An example of one type of interior-finish schedule.

9.4   HVAC SCHEDULES

There are many different types of HVAC schedules, which may include air-handling units, fan-coil units, chiller schedules, etc. These are prepared by the mechanical consultant. HVAC systems are typically either self-contained package units or central systems. Package units include roof top systems, air-to-air heat pumps, and air-conditioning units for rooms. Central systems usually consist of a combination of central-supply subsystem and multiple end-use zone systems (fan systems or terminal units). Figures 9.8A and B show examples of different types of HVAC schedules.

AIR HANDLING UNIT SCHEDULE

UNIT NO.	LOCATION	AREA SERVED	SUPPLY FAN NO.	CFM [l/s]		EXTERNAL STATIC PRESSURE (NOTE 1) IN[mm]	SPECIFIED INTERNAL LOSSES (NOTE 2) IN[mm]	UNSPECIFIED INTERNAL LOSSES (NOTE 3) IN[mm]	FAN TOTAL S.P. (NOTE 4) IN[mm]	TYPE SYSTEM
				SUPPLY	O.A.					
1−AH5	2nd FLOOR	SURGERY	1−SF5	20,000 [9439]	20,000 [9439]	2.5 [63.5]	2.0 [50]	1.0 [25]	5.5 [140]	VAV

NOTES:

1. EXTERNAL STATIC PRESSURE REQUIRED AT DUCT CONNECTIONS TO INLET & OUTLET OF AHU. MEASUREMENTS SHALL BE TAKEN WITHIN 3 FT. [1.0 M] OF INLET AND OUTLET AT A POINT OF MAX. ACCURACY.

2. TOTAL OF MAX. PRESSURE DROPS OF COMPONENTS WHICH ARE SPECIFIED SEPARATELY, IE., PREFILTERS, AFTER FILTERS, HEATING & COOLING COILS, DIFFUSER PLATE, AND SOUND ATTENUATORS.

3. INTERNAL LOSS ALLOWANCE SHALL INCLUDE LOSSES DUE TO ENTRANCE & EXIT OF AHU, MIXING BOXES, DIFFUSER SECTION (OTHER THAN DIFFUSER PLATE) INCLUDING LOSSES DUE TO FAILURE TO PROPERLY CONVERT FAN DISCHARGE VELOCITY PRESSURE TO STATIC PRESSURE, FAN INLET CONDITIONS, CASINGS, HUMIDIFIERS, DAMPERS, ETC.

4. TOTAL FAN S.P. = EXTERNAL STATIC PRESSURE + SPECIFIED INTERNAL LOSSES + UNSPECIFIED INTERNAL LOSSES. MANUFACTURER SHALL PROVIDE SUBMITTAL SHOWING ACTUAL LOSSES OF ALL EQUIPMENT PROVIDED. REFER TO FAN SCHEDULE FOR ADDITIONAL FAN SELECTION INFORMATION.

DESIGNER'S NOTE:
FOR AREA SERVED SHOW FUNCTIONAL AREA SUCH AS SURGERY, KITCHEN, LABORATORIES, ETC., WHERE APPLICABLE, OTHERWISE SHOW FLOORS SERVED. FOR TYPE OF SYSTEM INDICATE DUAL DUCT, MEDIUM PRESSURE, VARIABLE VOLUME, LOW PRESSURE REHEAT, MULTIZONE, VENTILATION SUPPLY UNIT, ETC.

Figure 9.8A   A simple example of an air-handling-unit schedule. Notice the significant number of notes that accompany the schedule. These are critical to a proper understanding of the requirements.

SCHEDULE OF AIR COOLED WATER

CHILLER REF.	CAPACITY-TR	AMPI.TEMP-F	CH.WATER FLOW US.GPM	$\Delta T f^o$
CHILLER 1	85+_3%	115	204	10
CHILLER 2	85+_3%	115	204	10

SCHEDULE OF CH. WATER CIRCULATION PUMPS

PUMP REF.	HEAD OF PUMP	CH.WATER FLOW G.P.M	REMARKS
P1 & P2	TO BE CALCULATED BY CONTRACTOR	408	

SCHEDULE OF WALL / WINDOW EXHAUST FAN

FAN REF.	HEAD OF PUMP	QTY	REMARKS
EX.F-1	100	32	WINDOW/WALL EXTR.
EX.F-2	150	37	FAN WITH ELECTRIC SHUTTER
EX.F-1	250	26	

SCHEDULE OF FAN COIL UNITS

FAN REF.	T.H MBH	S.H MBH	C.F.M	QTY	REMARKS
FC-1	9.6	8.4	400	8	
FC-2	13.8	12.4	600	2	
FC-3	17.7	16.4	800	–	LOW STATIC DUCTED TYPE
FC-4	21.4	19.6	1000	–	
FC-5	25	24	1200	–	
FC-6	13.7	12.1	600	6	
FC-7	18.4	16.9	800	3	
FC-8	23.7	21.4	1000	2	
FC-9	29.3	26.1	1200	21	
FC-10	35.0	31.0	1400	20	DIRECT DRIVE DUCTED TYPE
FC-11	40.8	35.6	1600	8	
FC-12	46.5	40.4	1800	2	
FC-13	–	–	–	–	
FC-14	50.7	45.4	2000	1	
FC-15	7.3	6.4	300	1	
FC-16	9.6	8.4	400	2	
FC-17	13.8	12.4	600	1	
FC-18	17.7	16.4	800	3	LOW STATIC EXPOSED
FC-19	21.4	19.6	1000	2	
FC-20	25	24	1200	–	

Figure 9.8B   Examples of various types of schedules relating to HVAC drawings.

9.5   GRILLE/DIFFUSER SCHEDULES

Grille and diffuser schedules generally show the manufacturer and catalog/model number of each grille and diffuser. They also show the dimensions of each as well as the volume of air in cubic feet per minute (CFM) that each will handle, the quantity required, and the location for installation. A column for notes/remarks is typically included to facilitate the installation of the item.

9.6   LIGHTING-FIXTURE AND ELECTRICAL SCHEDULES

Lighting-fixture schedules are generally used to list the fixture types and identify each fixture type on the drawing of a given project by number. The manufacturer and identification number of each type are given along with the number, size, and type of the lamps for each. One can include a "mounting" column to indicate whether the fixture is wall-mounted, surfaced-mounted on the ceiling, or recessed. Alternatively, this information can be included in the "remarks" column. Also included in this column is information relating to the mounting height above the finished floor, in the case of a wall-mounted lighting fixture, or any other pertinent data for the proper installation of the fixtures (Figure 9.9).

Electrical general notes are placed on the first and subsequent sheets of the electrical drawings. They contain items that are common to all electrical items and may also include information regarding coordination of other trades with the electrical trades

Electrical notes are normally contained on each sheet as required and are relevant to that particular sheet only. They are also used to give specific location and direction regarding electrical issues.

9.7   PANEL-BOARD SCHEDULES

Panel-board schedules are generally found on electrical drawings and are used mainly to indicate relevant information on the service-panel boards within the building. A panel-board schedule should provide sufficient data to identify the panel number (as indicated on the drawings) and the type of cabinet (whether surface-mounted or flush). It should also provide relevant data regarding the panel main bus bars and/or circuitbreakers as well as the number and type of circuitbreakers contained in the panel board and the components fed by each. This type of schedule, however, does not furnish detailed information for the individual circuits (e.g., wire sizes or number of outlets on the circuit); this information needs to be shown elsewhere on the drawing, such as in the plan view or power-riser diagrams.

9.8   MISCELLANEOUS SCHEDULES AND NOTES

Other types of schedules used depend on office procedure and the type of project in hand. Material schedules are used usually to list the approximate quantities of materials needed to complete a project. These schedules are often used for estimating small construction projects such as residential buildings and should not be taken to reflect the exact amount of materials needed. Material schedules are mainly used for guidance.

LIGHT FIXTURE SCHEDULE

FIXTURE	MANUFACTURER	TYPE	COLOR	BULB	MOUNTING	NOTES / REMARKS

Figure 9.9   An example of a a typical lighting-fixture schedule.

Appliance and plumbing-fixture schedules are similar to other types of schedules. For example, the appliance schedule would list each appliance from top to bottom in the first column. Other columns would list manufacturer and model number, color, special features, options, etc. The plumbing-fixture schedule would list the room, fixtures for that room, manufacturer and model number, color, finish, handle options, etc (Figure 9.10). If plumbing-fixture schedules or appliance schedules are not used in a project, the fixture types, manufacturers, catalog numbers, and other information needed must be included in the project specifications. Figure 9.11 is an example of an equipment schedule for a commercial laundry, and Figures 9.12 shows examples of structural/civil-engineering schedules for roof beams, footings, and tie beams.

Usually the specifications will augment information found in the schedules. Examples of information usually found in the specifications include the window manufacturer, the type and manufacturer of the door hardware, and the type and manufacturer of paint for the trim.

9.9   NOTES

Notes are a pivotal aspect of construction drawings, as they often contain critical information regarding the project. Notes should be clear, concise, and easily understood. They do not typically contain technical information but rather clarify and explain conditions or requirements of the project.

There are two basic types of notes:

1. General notes are usually placed at the beginning of the drawings relating to a specific trade or discipline. They include all notes on the drawing not accompanied by a leader and an arrowhead. They are used essentially to explain and specify certain conditions relative to that discipline or the project as a whole.

2. Key notes are contained on a particular page or sheet as needed and relate only to that sheet.

PLUMBING FIXTURE SCHEDULE						
FIXTURE	MANUFACTURER	MODEL NO.	COLOR	SIZE	LOCATION	NOTES / REMARKS

Figure 9.10   An example of a plumbing-fixture schedule.

Project general notes are typically placed near the beginning of the drawing set, usually at least 3 inches below the "revision" block in the right-hand side of the first sheet. The purpose of these notes is to give additional information that clarifies a detail or explains how a certain phase of construction is to be performed. They usually include site issues such as occupancy, security, parking, access, and other general site-related issues (including permits and posting).

Architectural general notes are typically placed on the first and subsequent sheets of the architectural drawings. They contain items that are common to all architectural elements. They may also contain information regarding coordination with other trades. Architectural key notes are contained on each sheet as required and are relevant to that particular sheet only. They provide additional description of architectural details and are typically used with leader lines to show exact detail locations.

All notes, along with the specifications, should be carefully read during the planning of the project.

SCHEDULE OF LAUNDRY EQUIPMENT

Ser	Model	Qty	Item	H (cm)	D (cm)	W (cm)
L01	HS/3110STAT.PM	2 No.'s	Washer Extractor (capacity 110 Kg each)	205	190	177
L02	HS/3055STAT.PM	1 No.	Washer Extractor (capacity 57 Kg)	190	148	155
L03	AD115	4 No.'s	Dryer (capacity 53 Kg each)	206	121	117
L04	PS8030/2M	1 No.	Flat Work Ironer	165	351	440
L05	S-5	1 No.	Flat Work Folder	171	283	421
L06	Y140	1 No.	Temporary Identification Machine	31	35	60
L07	LS/320PM	2 No.'s	Washer Extractor (capacity 21 Kg each)	144	116	86
L08	AD50	2 No.'s	Dryer (capacity 23 Kg each)	183	87	123
L09	BC45	2 No.'s	Utility Laundry Press	125	106	142
L10	ALM3	2 No.'s	Mushrom Press	120	106	91
L11	ALL7	1 No.	Double Legger Utility Laundry Press	120	42	142
L12	AMB10/38	2 No.'s	Steam Finishing Board	96	48	99
L13	ALS12S	2 No.'s	Sleever Press	135	81	78
L14	ALCY/12	2 No.'s	Cuff and Collar Press	125	106	68
L15	DBC	1 No.	Double Buck Shirt Pressing System	167	228	116
L16	Y140	1 No.	Temporary Identification Machine	31	35	60
L17	AFA10	1 No.'s	Spotting Table	111	50	122
L18	345Silver	1 No.	Dry Clean (capacity 20 Kg)	215	142	183
L19	808T2	1 No.	Dry Clean (capacity 10 Kg)	198	150	105
L20	BC45	2 No.'s	Utility Dry Clean Press	125	106	142
L21	AMB10/38	2 No.'s	Steam Finishing Board	96	48	99
L22	AAA21	1 No.	Pants Topper	160	39	55
L23	J/3	1 No.	Triple Head Buff Iron	104	48	116
L24	AQA13151	2 No.'s	Steam / Air Form Finisher	175	155	64
L25	ALL7	1 No.	Double Legger Utility Dry Clean Press	120	42	142
L26	M.S.T.	6 No.'s	Mobile Shelf Trolley	163	56	150
L27	B29-AK2010	4 No.'s	Bishop Shirt Bin Sorter Transporter	25	12.5	40
L28	Work Table	10 No.'s	S.S. Work Table	70	80	185
L29	MATIC02	12 No.'s	Wooden Work Table.	70	80	185
L30	Singer 876	1 No.	Sewing Machine	40	30	60
L31	H603M	2 No.'s	Garment Wrapping Machine	45	25	76
L32	E42P	2 No.'s	Folded Shirts Wrapping Machine	30	40	43
L33	40-700-14	12 No.'s	Dandux 16 Bushel Wet Linen Trolley	90	70	100
L34	51-600-16	12 No.'s	Dandux 14 Bushel Dry Linen Trolley	90	70	100
L35	40-955	2 No.'s	Flat Work Trolley	Not Shown		
L36	M.H.G.T.	20 No.'s	Mobile Hanging Garment Trolley	190	40	200
L37	MUSTEE-M27-F	2 No.'s	Laundry Sink	121	60	101
L38	Blk14	1 No.	Air Compressor	Not Shown		
L39	FH144-11	1 No.	Detecto Scale	173	85	63

Figure 9.11   A laundry-equipment schedule for a small commercial laundry.

SCHEDULE OF ROOF BEAMS

BEAM MARK	DIMENSIONS		BOTTOM STEEL		TOP STEEL						FACE BAR B/F	STIRRUPS		REMARKS
	WIDTH (mm.)	DEPTH (mm.)	BAR "a"	BAR "b"	EXT. SUPPORT * BAR "c"	BAR "d"	BAR "e"	INT. SUPPORT * BAR "f"	BAR "g"			@ L/4	REST OF SPAN	
B1	200	1000	2T20	2T20	2T16	---	---	---	---		6T10	T8@200	T8@200	SIMPLE
B2a	200	500	2T20	2T20	2T16	---	---	2T20	2T20		2T10	T10@150	T10@200	BAR "c" CONTINUOUS
B2b	200	500	2T16	---	2T16	---	---	2T20	2T20		2T10	T10@200	T10@200	BAR "c" CONTINUOUS
B3	200	500	2T20	---	2T16	---	---	2T20	2T20		2T10	T10@200	T10@200	BAR "c" CONTINUOUS
B4	200	500	2T16	---	2T16	---	---	2T16	---		2T10	T8@200	T8@200	BAR "c" CONTINUOUS
B5a	200	500	2T20	2T20	2T16	---	---	2T16	---		2T10	T8@200	T8@200	BAR "c" CONTINUOUS
B5b	200	500	2T16	---	2T16	---	---	---	---		2T10	T8@200	T8@200	CANTILEVER
B6a	200	500	2T20	2T20	2T16	---	---	2T16	2T16		2T10	T10@200	T10@200	BAR "c" CONTINUOUS
B6b	200	500	2T16	---	2T20	2T16	---	---	---		2T10	T10@200	T10@200	CANTILEVER

* NOTE : FOR INTERIOR SPANS, EXTERIOR SUPPORT MEANS SUPPORT AT LEFT AND INTERIOR SUPPORT MEANS SUPPORT AT RIGHT, STIRRUPS TO BEGIN AT 50mm FROM FACE OF SUPPORT UNLESS OTHERWISE SHOWN.

Figure 9.12A An example of a schedule used in civil- and structural-engineering drawings.

SCHEDULE OF FOOTINGS

FOOTING MARK	DIMENSIONS (MM)			BOTTOM BARS		TOP BARS	
	WIDTH	LENGTH	THICKNESS	LONG BARS	SHORT BARS	LONG BARS	SHORT BARS
F1	1300	1600	350	T16@200	T16@200	---	---
F2	1100	1500	350	T16@200	T16@200	---	---
F3	1000	1300	350	T16@200	T16@200	---	---
F4	900	1100	350	T16@200	T16@200	---	---

SCHEDULE OF TIE BEAMS

MARK	DIMENSIONS (MM)		BOTTOM BARS	TOP BARS	SIDE BARS	STIRRUPS	NOTES
	WIDTH	DEPTH					
TB1	200	600	2T16	2T16	2T10	T8@200	---
TB2	200	600	4T16	2T16	2T10	T8@200	BOTTOM BARS 2 LAYERS

Figure 9.12B An example of a schedule used in civil- and structural-engineering drawings.

10

Interpreting Specifications

10.1 GENERAL OVERVIEW

Standard building specifications are written documents that go with the construction drawings and describe the materials as well as the installation methods. They also prescribe the quality standards of construction expected on the project.

In the United States, the Construction Specifications Institute (CSI) has established a widely recognized format of organization for technical specifications. CSI is a nationwide organization composed of architects, engineers, manufacturers' representatives, contractors, and other interested parties who have worked together to develop this system of identification. Prior to 2004 the format consisted of specifications for 16 divisions. These specification standards are noted in the MasterFormat, which in 2004 was expanded to 50 divisions, as will be described later in this chapter.

Specifications are legal documents and should therefore be complete, accurate, and unambiguous. Specification writing has two main roles: defining the scope of work and acting as a set of instructions. At the core of specification writing is defining the scope of work. Ensuring that the required level of quality of the product and services is clearly communicated to bidders and that the completed project conforms to this specified quality is extremely important, although sometimes misunderstood. Most projects now incorporate the specifications within a project manual (a concept first developed by the AIA in 1964) that is issued along with the drawings, bidding requirements, and other contract conditions as part of the contract-documents package.

Construction drawings are supplemented by written project specifications. Project specifications give detailed information regarding materials and methods of work for a particular construction project. They cover various factors relating to the project, such as general conditions, scope of work, quality of materials, standards of workmanship, and protection of finished work. The drawings, together with the project specifications, define the project in detail and show exactly how it is to be constructed. Usually, any set of drawings for an important project is accompanied by a set of project specifications. The drawings and project specifications are inseparable. The drawings indicate what the project specifications do not cover; and the project specifications indicate what the drawings do not portray or clarify further details that are not covered or amplified by the drawings and notes on the drawings. Whenever there is conflicting information on the drawings and project specifications, the project specifications take precedence over the drawings. The general requirements are usually the first specifications listed for the structure, stating the type of foundation, character of load-bearing members (wood frame, steel frame, or concrete), type or types of doors and windows, types of mechanical and electrical installations, and the principal function of the building. Next follows the specific conditions that must be carried out by the constructors.

The impact of new technologies on the way we do business is considerable, and spec writing has not been immune. Specification production and reproduction have advanced by leaps and bounds in just a short time period due to these new technologies. Master systems are now commercially available in electronic form using a number of word processors. The specifier simply loads the master system into the computer and gets instant access to it, complete with drawing checklist and explanation sheets. Upon editing the relevant sections, a printout can be made with an audit trail that indicates what has been deleted and what decisions must be made. Most offices in the United States use an 8 1/2 x 11-inch format, while in Europe an A4 (8 1/4 x 11 3/4-inch) format is typically used.

10.2 WHY SPECIFICATIONS

Drawings alone cannot define the qualitative issues of a scheme, which is why specifications are necessary. Specifications are the written portion of the contract documents that are used to execute the project. Design decisions are continuously made as drawings proceed from schematic sketches to construction documents. Drawings depict the general configuration and layout of the design, including its size, shape, and dimensions. It tells the contractor the quantities of materials needed, their placement, and their general relationship to each other. Technical specifications are a form of materials list, requiring similar decision making that reflects the design intent and describes in detail the quality and character of materials, the standards to which the materials and their installation shall conform, and other issues that are more appropriately described in written rather than graphic form. And no matter how beautiful the designer's concept, the project cannot be correctly implemented without clear, concise, accurate, and easily understood contract documents. Specs are a critical component of the contract documents.

The construction drawings contain as much information about a structure as can be presented graphically. A lot of information can be presented this way, but there is more information that the construction craftsman must have that is not adaptable to the graphic form of presentation. Information of this kind includes quality criteria for materials (for example, maximum amounts of aggregate per sack of cement), specified standards of workmanship, prescribed construction methods, and so on. When there is a discrepancy between the drawings and the specifications, always use the specifications as the final authority. This kind of information is presented in a list of written specifications, familiarly known as the specs. A list of specifications usually begins with a section on general conditions. This section starts with a general description of the building, including type of foundation, types of windows, character of framing, utilities to be installed, and so on. A list of definitions of terms used in the specs comes next, followed by certain routine declarations of responsibility and certain conditions to be maintained on the job.

Even well-drawn construction drawings cannot adequately reveal all the aspects of a construction project. There are many features that cannot be shown graphically. For instance, how would a designer show on a drawing the quality of workmanship required for the installation of electrical equipment or who is responsible for supplying the materials, except by extensive hand-lettered notes? The standard procedure is to supplement construction drawings with written descriptions. These detailed written instructions, commonly called specifications (specs), define and limit the materials and fabrication according to the intent of the engineer or the designer. In fact, when there is a gap between the building as visualized by the designer and the contractor's interpretation of the documents, specifications—not drawings—are the tool to close that gap. The specifications are an important part of the project because they eliminate possible misinterpretation and ensure positive control of the construction. There are several different types of specifications.

Specification Material Sources

Because of time and cost restraints, few individuals (or small firms) would today venture to write a completely new set of specifications for each job. Specifiers would normally rely on the many sources of reference material that are currently available and from which they could compile a set for each new project. Moreover, because of liability issues, specifiers often feel more comfortable relying on specifications that have repeatedly proved satisfactory in the past. When specs have to be modified to fit the conditions of a given job or new specs incorporated, text is generally taken from one of the master spec systems. These contain guideline specifications for many materials, allowing the specifier to edit unnecessary text rather than generate new information each time.

Another advantage of using master systems is that they use correct specification language and format for ease of specification preparation. Listed below are some of the major sources from which specification material is available, much of which can be retrieved via the Internet:

- Master specifications (Masterspec®, SPECSystem™, MasterFormat™, SpecText®, BSDSpeclink®, ezSPECS On-Line™, CAP Studio for the furniture industry, and many others)
- City and national codes and ordinances
- Manufacturers' industry associations (Architectural Woodworking Institute, American Plywood Association, Door and Hardware Institute, Tile Council of America).
- Manufacturers' catalogs (Sweet's Catalog File, Man-U-Spec, Spec-data)
- Manufacturers' on-line catalogs via the Internet
- National standards organizations such as the American National Standards Institute, National Institute of Building Sciences, National Fire Protection Association, National Institute of Standards and Technology, and the Association for Contract Textiles
- Testing societies (American Society for Testing and Materials, Underwriters Laboratories)
- Federal specifications (Specs-In-Tact, G.S.A., N.A.S.A., N.A.F.V.A.C.)
- Magazines and publications (Construction Specifier, Architecture, Architectural Record)
- Books on specifications (see bibliography)
- Individual files of previously written specifications

During recent years, numerous firms that provide online specification-writing services have emerged. These services are discussed later in the chapter.

10.3 TYPES OF SPECIFICATIONS

One of the first things that a specifier has to decide upon when preparing a specification document is the format or method to be used to communicate the desired intent to the contractor. There are essentially two broad categories of specifications, closed or open, and most items can be specified by either method. Within these two broad categories, there are basically four generic types of specifications: propriety specifications, description specifications, performance specifications, and reference-standards specifications. The type chosen depends on several factors (Figure 10.1). These are discussed below.

Closed Specification

Closed (also called prescriptive or restrictive) specification is one that limits acceptable products to one or a few brand-identified types or models and prohibits substitutions. This type of specification is sometimes used where specifiers feel more comfortable resorting to a specific proprietary product with which they are familiar and which will meet the specific criteria of the project. However, it should be noted that this procedure (particularly when only one product is named) is not competitive and rarely attracts the most favorable price for the owner. Also, while a closed specification is common in private construction work, most public projects are required by law to be bid under open specifications.

The closed proprietary specification method is the easiest form to write but the most restrictive, in that it names a specific manufacturer's product. It generally establishes a narrower definition of acceptable quality than do performance or reference-standard methods, and gives the designer/space planner complete control over what is installed. The specification can also be written as an open proprietary section, in which multiple manufacturers or products are named or alternatives solicited. This would increase the potential competition and encourage a lower installation price from vendors. In some cases, a multiple choice may not be appropriate, as, for example, where a specific brick is required for repairs to an existing brick facade. When the specification does not allow for any substitution of materials, it is known as a base bid proprietary specification.

Open Specification

Also called performance or nonrestrictive, this type of specification gives the contractor some choice in how to achieve the desired results. Proprietary specifications may also be used as open specifications but with the addition of the "or equal" clause, which allows the contractor to consider other products for bid if they are shown to be equal in performance and specifications.

Due to the ambiguity surrounding this clause and the disagreements it often perpetuates, specifiers generally shy away from incorporating it into proprietary specifications.

A second method of open specification that is gaining popularity is descriptive specification.

This type of specification describes in detail the requirements for the material or product and the workmanship required for its fabrication and installation without providing a trade name. This type of specification is often stipulated by some government agencies to allow the maximum competition among product manufacturers. Descriptive specifications are also more difficult to write than proprietary ones because the specifier must include all the product's relevant characteristics in the specification.

A third type of specification that is often used is the reference standard. This standard simply describes a material, product, or process referencing a recognized industry standard or test method as the basis for the specification and is often used to specify generic materials such as portland cement and

PRESCRIPTIVE – RESTRICTIVE				PERFORMANCE - NONRESTRICTIVE	
Proprietary Base Bid	Proprietary or Equal	Descriptive	Reference Standards	Performance	Performance plus Descriptive
Part 2: Materials Manufacturer: Corporation A: Brand Name Style: #245 Cascade Color: 17849 Base Bid: When brand names are used no substitute products will be considered.	**Part 2: Materials** Manufacturers: Corporation A: Brand name Corporation B: Brand name Corporation C: Brand name or Approved Equal Proposed substitutions must be given complete consideration, statement of procedure for determining what is "equal" must be included or Modified or Approved Equal (1) Required definite prebid deadline for submittal of request. (2) Federal, state or city projects require specifier to include salient characteristics of the item to provide a basis for judging equality.	**Part 2: Materials** Carpet: (no brand name) Construction Type: Tufted or other Pile Yarn: Generic Gauge or Pitch Stitches per inch or Wires Pile Height or Wire Height Dyeing Method Yarn size and Ply Width Pile Yarn weight and Total Weight Primary and Secondary Backing Detailed Installation Information	**Part 1: General** Quality Assurance References: ANSI A117.1-1986 "Specifications for Making Buildings and Facilities Accessible to, and Usable by, the Physically Handicapped." American National Standards Institute, 1430 Broadway, New York, NY 10018 (A-210, A-5-2.2.4a., A-8-4.5, A-10-1.1.2). NFPA 255-1984, "Standard Method of Test for Critical Radiant Heat Energy Source." National fire Protection Association, Batterymarch Park, Quincy, MA 02269 (A-12-3-3.2).	**Subsystem: Carpet** Attribute: Fire Safety Requirement: Provide flame-spread resistance. Criteria: This subsystem shall provide a maximum flame-spread rating of 25. Test Type: Calculation Method ASTM-E 84, "Surface Burning Characteristics of Building Materials." American Society for Testing and Materials.	**Part 2: Materials** Carpet Performance and Design Requirements: Design carpet for the following conditions: 1. Maximum static electricity ...2.5 Kv 2. Tuft bind to resist ...12 lbs/in. Carpet Construction Materials to conform to AATCC 134-1979. Electrostatic Propensity of Carpets. 2.5 Kv at 20% RH and 70° F.
These types of "closed" or restrictive specifications require exact products by brand name, model number, and all important characteristics. Quoted test data is supplied by the manufacturer. Drawings also show dimensions and engineering aspects of the particular product.		Descriptive: (1) Avoid conflicts with drawings. (2) Research all products. (3) Compare costs vs. performance. (4) State required submittals, including tests and standards. Reference: (1) Know the standard. Have a copy in designers' office files. (2) Avoid use of only minimum standards. (3) Use full names and dates of the referenced standards. (4) Enforce the requirements.		Performance specifications: (1) Use no manufacturers or trade names. (2) Desired results are stated. (3) Method to achieve end result not included in document. Performance combination: Descriptive and reference standard specifications can produce nonrestrictive specifications. Only essential criteria that will meet intended use are stated. Specifier must ensure that criteria can be met by several manufacturers.	

Figure 10.1 Various types of specifications (source: Specifications for Commercial Interiors by S.C. Raznicoss).

clear glass. Thus, in specifying gypsum wallboard, for example, you can state that all gypsum-wallboard products shall meet the requirements of ASTM C36.

As this document describes in detail the requirements for this product, the specifier is relieved of having to repeat the requirements and can instead refer to the recognized industry standard.

In using a reference standard, the specifier should not only have a copy of that standard, but should also know what is required by the standard, including choices that may be contained therein, that should be enforced by all suppliers. This type of specification is fairly easy to write and is generally short. In addition, the use of reference-standard specifications reduces your liability and the possibility for errors.

The fourth major type of specification used is the performance specification. This type of specification establishes the performance requirements without dictating the methods by which the end results are to be achieved. This gives the greatest leeway to contractors because it allows them to use any material or system that meets the performance criteria specified, provided the results can be verified by measurement, tests, or other types of evaluation.

Performance specifications are not often used by architects and engineers because they are the most difficult to write. The specifier must know all the criteria for a product or system, state the appropriate methods for testing compliance, and write an unambiguous document. In addition, sufficient data must be provided to ensure that the product can be demonstrated. Performance specifications are mostly used in specifying complex systems and where a specifier wants to encourage new ways of achieving a particular result.

Product specifications often use a combination of methods to convey the designer's intent.

For example, a specification for a ceramic tile would use a proprietary specification to name the product or products selected by the specifier, a descriptive specification to specify the size and design, and a reference standard to specify the ASTM standard, grade, and type required.

10.4 ORGANIZING THE PROJECT MANUAL

Traditionally, the organization of the project manual has been a matter of individual preference by the design firm producing it, resulting in a wide diversity of method around the country that became confusing. As design firms and contractors became increasingly nationwide in their operations, a pressing need grew for a consistent arrangement of building-construction specifications. To meet this challenge, the American Institute of Architects (AIA) in 1964 developed the concept of the project manual, which has now gained wide acceptance. Essentially, it contains the technical specifications as well as several other types of documents, which, together with the drawings, constitute the contract documents. A typical table of contents for the project manual might show the following major divisions:

- General project information. All parties responsible for the development of the project should be included on the Project Manual's cover page, which identifies the names and addresses of the owners, architects, civil engineers, mechanical engineers, electrical engineers, and structural engineers.

- Bidding requirements. These apply to contracts awarded through a bidding process and include an invitation to bid (or advertisement), prequalification forms, instructions to bidders, bid form, and information available to bidders.

- Contract forms, which may include the agreement (the contract between owner and contractor), performance bond, labor and materials payment bond, and certificates of insurance.

- Contract conditions (general and supplementary). These include general conditions of the contract such as AIA Form 201 or similar preprinted forms. Supplementary conditions include anything that is not covered in the general conditions, such as addenda (changes made before contract signing) and change orders (changes made after contract signing).

- Technical specifications. These provide technical information concerning the building materials, components, systems, and equipment shown on the drawings with regard to quality, performance characteristics, and stipulated results to be achieved by application of construction methods (Figure 10.2).

Writing and Coordination Guidelines

As mentioned earlier, specifications are legal documents, and their language must be precise. If the written text is ambiguous or inadequate, the specification will not communicate. Moreover, a convention has developed over the years as to the information that should be shown on the drawings and that which should be indicated in the specifications. This is essentially based on a number of broad general principles, which include:

- Drawings should convey information that can be most readily and effectively expressed graphically by means of drawings and diagrams. This would include data such as dimensions, sizes, gauges, proportions, arrangements, locations, and interrelationships.

- Specifications should convey information that is easier to convey through the written word, such as descriptions, standards, procedures, guarantees, and names.

- Drawings are used to express quantity, whereas specifications should describe quality.

- Drawings should denote type (for example, wood), while specifications will clarify the species (for example, oak).

Some exceptions to these understandings can create confusion. For example, building departments of the majority of municipalities will only accept drawings with applications for building permits and will not accept a project manual with specifications. Furthermore, all data demonstrating compliance with the building code must be indicated on the drawings.

This stipulated repetition of identical data on both the specifications and the drawings exposes the documents to errors and inconsistency. Nevertheless, this aside, to achieve better communication, the specifier should:

- Have a good understanding of the most current standards and test methods referred to and the sections that are applicable to the project. Use accepted standards to specify quality of materials or workmanship required, such as "Lightweight concrete masonry units: ASTM C-90-85; Grade N. Type 1."

- Avoid specifications that are impossible for the contractor to carry out. Also refrain from specifying the results and the methods proposed to achieve those results, as the two may conflict. For example, if you specify that a fabric should meet certain ASTM standards and then specify a specific fabric that fails to meet the stated requirements, the specification will be impossible to comply with.

- Do not specify standards that cannot be measured. Using phrases such as "a workmanlike job," for example, should be avoided, as they are subject to wide interpretation.

MASTERSPEC®

TABLE OF CONTENTS   SMALL PROJECT™ Specifications

©2001 American Institute of Architects

Issue Date	Sect. No.	SECTION TITLE	Issue Date	Sect. No.	SECTION TITLE
DIVISION 1 - GENERAL REQUIREMENTS			**DIVISION 6 - WOODS AND PLASTICS**		
5/1/01	00000	SECTION TEMPLATE	5/1/01	06100	ROUGH CARPENTRY
5/1/01	01100	SUMMARY	5/1/01	06105	MISCELLANEOUS CARPENTRY
5/1/01	01200	PRICE AND PAYMENT PROCEDURES	5/1/01	06176	METAL-PLATE-CONNECTED WOOD TRUSSES
5/1/01	01300	ADMINISTRATIVE REQUIREMENTS	5/1/01	06185	STRUCTURAL GLUED-LAMINATED TIMBER
5/1/01	01400	QUALITY REQUIREMENTS	5/1/01	06200	FINISH CARPENTRY
5/1/01	01420	REFERENCES	5/1/01	06401	EXTERIOR ARCHITECTURAL WOODWORK
5/1/01	01500	TEMPORARY FACILITIES AND CONTROLS	5/1/01	06402	INTERIOR ARCHITECTURAL WOODWORK
5/1/01	01600	PRODUCT REQUIREMENTS			
5/1/01	01701	EXECUTION AND CLOSEOUT REQUIREMENTS	**DIVISION 7 - THERMAL AND MOISTURE PROTECTION**		
5/1/01	01732	SELECTIVE DEMOLITION	5/1/01	07115	BITUMINOUS DAMPPROOFING
			5/1/01	07131	SELF-ADHERING SHEET WATERPROOFING
DIVISION 2 - SITE CONSTRUCTION			5/1/01	07210	BUILDING INSULATION
5/1/01	02230	SITE CLEARING	5/1/01	07241	EXTERIOR INSULATION AND FINISH SYSTEMS - CLASS PB
5/1/01	02300	EARTHWORK	5/1/01	07311	ASPHALT SHINGLES
5/1/01	02361	TERMITE CONTROL	5/1/01	07317	WOOD SHINGLES AND SHAKES
5/1/01	02510	WATER DISTRIBUTION	5/1/01	07320	ROOF TILES
5/1/01	02525	WATER SUPPLY WELLS	5/1/01	07411	METAL ROOF PANELS
5/1/01	02530	SANITARY SEWERAGE	5/1/01	07412	METAL WALL PANELS
5/1/01	02540	SEPTIC TANK SYSTEMS	5/1/01	07460	SIDING
5/1/01	02553	NATURAL GAS DISTRIBUTION	5/1/01	07511	BUILT-UP ASPHALT ROOFING
5/1/01	02554	FUEL OIL DISTRIBUTION	5/1/01	07531	EPDM MEMBRANE ROOFING
5/1/01	02620	SUBDRAINAGE	5/1/01	07552	SBS-MODIFIED BITUMINOUS MEMBRANE ROOFING
5/1/01	02630	STORM DRAINAGE	5/1/01	07610	SHEET METAL ROOFING
5/1/01	02741	HOT-MIX ASPHALT PAVING	5/1/01	07620	SHEET METAL FLASHING AND TRIM
5/1/01	02751	CEMENT CONCRETE PAVEMENT	5/1/01	07710	MANUFACTURED ROOF SPECIALTIES
5/1/01	02780	UNIT PAVERS	5/1/01	07720	ROOF ACCESSORIES
5/1/01	02810	IRRIGATION SYSTEMS	5/1/01	07811	SPRAYED FIRE-RESISTIVE MATERIALS
5/1/01	02821	CHAIN-LINK FENCES AND GATES	5/1/01	07841	THROUGH-PENETRATION FIRESTOP SYSTEMS
5/1/01	02832	SEGMENTAL RETAINING WALLS	5/1/01	07920	JOINT SEALANTS
5/1/01	02920	LAWNS AND GRASSES			
5/1/01	02930	EXTERIOR PLANTS	**DIVISION 8 - DOORS AND WINDOWS**		
			5/1/01	08110	STEEL DOORS AND FRAMES
DIVISION 3 - CONCRETE			5/1/01	08163	SLIDING ALUMINUM-FRAMED GLASS DOORS
5/1/01	03300	CAST-IN-PLACE CONCRETE	5/1/01	08211	FLUSH WOOD DOORS
5/1/01	03371	SHOTCRETE	5/1/01	08212	STILE AND RAIL WOOD DOORS
5/1/01	03410	PLANT-PRECAST STRUCTURAL CONCRETE	5/1/01	08263	SLIDING WOOD-FRAMED GLASS DOORS
5/1/01	03450	PLANT-PRECAST ARCHITECTURAL CONCRETE	5/1/01	08311	ACCESS DOORS AND FRAMES
5/1/01	03470	TILT-UP PRECAST CONCRETE	5/1/01	08331	OVERHEAD COILING DOORS
			5/1/01	08351	FOLDING DOORS
DIVISION 4 - MASONRY			5/1/01	08361	SECTIONAL OVERHEAD DOORS
5/1/01	04810	UNIT MASONRY ASSEMBLIES	5/1/01	08410	ALUMINUM ENTRANCES AND STOREFRONTS
5/1/01	04815	GLASS UNIT MASONRY ASSEMBLIES	5/1/01	08520	ALUMINUM WINDOWS
5/1/01	04860	STONE VENEER ASSEMBLIES	5/1/01	08550	WOOD WINDOWS
			5/1/01	08561	VINYL WINDOWS
DIVISION 5 - METALS			5/1/01	08610	ROOF WINDOWS
5/1/01	05120	STRUCTURAL STEEL	5/1/01	08620	UNIT SKYLIGHTS
5/1/01	05210	STEEL JOISTS	5/1/01	08710	DOOR HARDWARE
5/1/01	05310	STEEL DECK	5/1/01	08716	POWER DOOR OPERATORS
5/1/01	05400	COLD-FORMED METAL FRAMING	5/1/01	08800	GLAZING
5/1/01	05500	METAL FABRICATIONS	5/1/01	08960	SLOPED GLAZING SYSTEMS
5/1/01	05520	HANDRAILS AND RAILINGS			

MASTERSPEC SMALL PROJECT TABLE OF CONTENTS – MAY 2001 - Page 1

Figure 10.2   The Masterspec Table of Contents for a small project (source: American Institute of Architects).
(continued)

Issue Date	Sect. No.	SECTION TITLE		Issue Date	Sect. No.	SECTION TITLE

DIVISION 9 - FINISHES

5/1/01	09210	GYPSUM PLASTER
5/1/01	09215	GYPSUM VENEER PLASTER
5/1/01	09220	PORTLAND CEMENT PLASTER
5/1/01	09260	GYPSUM BOARD ASSEMBLIES
5/1/01	09271	GLASS-REINFORCED GYPSUM FABRICATIONS
5/1/01	09310	CERAMIC TILE
5/1/01	09385	DIMENSION STONE TILE
5/1/01	09511	ACOUSTICAL PANEL CEILINGS
5/1/01	09512	ACOUSTICAL TILE CEILINGS
5/1/01	09638	STONE PAVING AND FLOORING
5/1/01	09640	WOOD FLOORING
5/1/01	09651	RESILIENT FLOOR TILE
5/1/01	09652	SHEET VINYL FLOOR COVERINGS
5/1/01	09653	RESILIENT WALL BASE AND ACCESSORIES
5/1/01	09680	CARPET
5/1/01	09681	CARPET TILE
5/1/01	09720	WALL COVERINGS
5/1/01	09751	INTERIOR STONE FACING
5/1/01	09771	FABRIC-WRAPPED PANELS
5/1/01	09910	PAINTING

DIVISION 10 - SPECIALTIES

5/1/01	10155	TOILET COMPARTMENTS
5/1/01	10200	LOUVERS AND VENTS
5/1/01	10265	IMPACT-RESISTANT WALL PROTECTION
5/1/01	10431	SIGNS
5/1/01	10520	FIRE- PROTECTION SPECIALTIES
5/1/01	10651	OPERABLE PANEL PARTITIONS
5/1/01	10750	TELEPHONE SPECIALTIES
5/1/01	10801	TOILET AND BATH ACCESSORIES

DIVISION 11 - EQUIPMENT

5/1/01	11132	PROJECTION SCREENS
5/1/01	11451	RESIDENTIAL APPLIANCES
5/1/01	11460	UNIT KITCHENS

DIVISION 12 - FURNISHINGS

5/1/01	12356	KITCHEN CASEWORK
5/1/01	12484	FLOOR MATS AND FRAMES
5/1/01	12491	HORIZONTAL LOUVER BLINDS
5/1/01	12492	VERTICAL LOUVER BLINDS
5/1/01	12496	WINDOW TREATMENT HARDWARE

DIVISION 13 - SPECIAL CONSTRUCTION

5/1/01	13038	SAUNAS
5/1/01	13100	LIGHTNING PROTECTION
5/1/01	13125	METAL BUILDING SYSTEMS
5/1/01	13851	FIRE ALARM
5/1/01	13930	WET-PIPE FIRE SUPPRESSION SPRINKLERS

DIVISION 14 - CONVEYING SYSTEMS

| 5/1/01 | 14240 | HYDRAULIC ELEVATORS |
| 5/1/01 | 14420 | WHEELCHAIR LIFTS |

DIVISION 15 - MECHANICAL

5/1/01	15050	BASIC MECHANICAL MATERIALS AND METHODS
5/1/01	15080	MECHANICAL INSULATION
5/1/01	15110	VALVES
5/1/01	15130	PUMPS
5/1/01	15140	DOMESTIC WATER PIPING
5/1/01	15150	SANITARY WASTE AND VENT PIPING
5/1/01	15160	STORM DRAINAGE
5/1/01	15181	HYDRONIC PIPING
5/1/01	15183	REFRIGERANT PIPING
5/1/01	15194	FUEL GAS PIPING
5/1/01	15410	PLUMBING FIXTURES
5/1/01	15430	PLUMBING SPECIALTIES
5/1/01	15480	DOMESTIC WATER HEATERS
5/1/01	15512	CAST-IRON BOILERS
5/1/01	15513	CONDENSING BOILERS
5/1/01	15519	ELECTRIC BOILERS
5/1/01	15530	FURNACES
5/1/01	15554	FLUES AND VENTS
5/1/01	15628	RECIPROCATING/SCROLL WATER CHILLERS
5/1/01	15671	CONDENSING UNITS
5/1/01	15731	PACKAGED TERMINAL AIR CONDITIONERS
5/1/01	15732	ROOFTOP AIR CONDITIONERS
5/1/01	15745	WATER-SOURCE HEAT PUMPS
5/1/01	15763	FAN-COIL UNITS
5/1/01	15764	RADIATORS
5/1/01	15766	UNIT HEATERS
5/1/01	15772	RADIANT HEATING PIPING
5/1/01	15810	DUCTS AND ACCESSORIES
5/1/01	15838	POWER VENTILATIORS
5/1/01	15855	DIFFUSERS, REGISTERS, AND GRILLES
5/1/01	15900	HVAC INSTRUMENTATION AND CONTROLS

DIVISION 16 - ELECTRICAL

5/1/01	16050	BASIC ELECTRICAL MATERIALS AND METHODS
5/1/01	16122	UNDERCARPET CABLES
5/1/01	16140	WIRING DEVICES
5/1/01	16410	ENCLOSED SWITCHES AND CIRCUIT BREAKERS
5/1/01	16420	ENCLOSED CONTROLLERS
5/1/01	16442	PANELBOARDS
5/1/01	16461	DRY-TYPE TRANSFORMERS (600 V AND LESS)
5/1/01	16500	LIGHTING
5/1/01	16750	VOICE AND DATA COMMUNICATION CABLING

MASTERSPEC SMALL PROJECT TABLE OF CONTENTS - FEBRUARY 2001 - Page 2

Figure 10.2 The Masterspec Table of Contents for a small project (source: American Institute of Architects).

- The clarity of specifications depends on the use of simple, direct statements, concise use of terms, and attention to grammar and punctuation. Avoid the use of words or phrases such as etc. and/or, any, and either, which are ambiguous and imply a choice that may not be intended.

- Avoid exculpatory clauses such as,"the general contractor shall be totally responsible for all...," which try to shift responsibility. Be fair in designating responsibility.

- Keep specifications as short as possible, omitting words like "all," "the," "an," and "a." Describing only one major idea per paragraph makes reading easier while improving comprehension. It also facilitates editing and modifying the specifications at a later date.

- Capitalize the following: major parties to the contract, such as Contractor, Client, Owner,Architect; the contract documents, such as Specifications, Working Drawings, Contract, Clause, Section, Supplementary Conditions; specific rooms within the building, such as Living Room, Kitchen, Office; grade of materials, such as No.1 Douglas Fir and FAS White Oak; and, of course, all proper names. The specifier should never underline anything in a specification, as this implies that the remaining material can be ignored.

- Use "shall" and "will" correctly. "Shall" is used to designate a command: "The Contractor shall...." whereas "will" implies a choice: "The Owner or Space Planner will....."

The coordination of the specifications with the construction drawings is essential, as they complement each other. They should not contain conflicting requirements, omissions, duplications, or errors. To minimize the possibility of errors, the specifier should:

- Ensure that the specifications contain requirements for all the materials and construction depicted on the drawings.

- Use the same terminology in both documents (i.e., drawings and specifications). If metal studs are used in the specifications, the same term should be indicated on the drawings.

- Check that dimensions and thicknesses are shown only on one document and not duplicated. Typically, sizes are indicated on the drawings, and the standards for the materials and components that those sizes reference are written into the specifications (unless the project is a very small one without a project manual).

- Make sure that notes on drawings do not describe installation methods or material qualities, as these normally belong in the specifications.

10.5 SPECIFICATION FORMATION AND ORGANIZATION

The 16-division MasterFormat™ was originally created in 1963 and is a product of The Construction Specifications Institute and Construction Specifications Canada. It is a widely used format both in the United States and Canada for specifications of nonresidential building projects. MasterFormat is the standard for titling and arranging construction project manuals containing bidding requirements, contracting requirements, and specifications. The Construction Specification Institute (CSI) has been working since its inception on trying to standardize the specification numbering system and the format of the sections, which was modified in the MasterFormat version of 1995. In recent years the CSI actively sought to add new divisions to address the rapidly evolving and growing computer and communications technology. A modified MasterFormat was introduced in 2004 that increased its division numbers from

16 to 50, of which 13 divisions were left blank to provide room for future revisions as construction products and technology evolve (Figure 10.3). The consensus at CSI is that adding divisions is better than trying to fit everything into the previous format of 16 divisions.

This move to modify and enhance the MasterFormat is driven in part by changes in the construction marketplace. Construction technology has advanced rapidly since 1995. For example, there have been major developments in the scope and complexity of computer and communications systems for buildings and security systems.

The Construction Specification Institute describes "MasterFormat" as a master list of numbers and titles for organizing information about construction requirements, products, and activities into a standard sequence. Construction projects use many different delivery methods, products, and installation techniques. Successful completion of projects requires effective communication among the people involved on a project. Information retrieval is nearly impossible without a standard filing system familiar to each user. MasterFormat facilitates standard filing and retrieval schemes throughout the construction industry. MasterFormat is a uniform system for organizing information in project manuals, for organizing cost data, for filing product information and other technical data, for identifying drawing objects, and for presenting construction market data.

The MasterFormat standard is the most widely used standard for organizing specifications and other written information for commercial and institutional building projects in the U.S. and Canada. It provides a master list of divisions, and section numbers and titles within each division, to follow in organizing information about a facility's construction requirements and associated activities. Standardizing the presentation of such information improves communication among all parties involved in construction projects. Each division is further defined in MasterFormat™ by level two and three numbers and titles and suggested level four titles. Level two numbers and titles identify clusters of products and activities with an identifying characteristic in common. An explanation of the titles used in MasterFormat is provided, giving a general description of the coverage for each title. A keyword index of requirements, products, and activities is also provided to help users find appropriate numbers and titles for construction subjects.

The current MasterFormat consists essentially of dividing the specifications into 50 divisions. MasterFormat 2004 Edition divisions are:

Specification Section Format

Each specification section covers a particular trade or subtrade (e.g., drywall, carpet, ceiling tiles). Furthermore, each section is divided into three basic parts, each of which contains the specifications about a particular aspect of each trade or subtrade.

Part 1: General

This part of the specification outlines the general requirements for the section and describes the scope of work of the project as well as providing the bidder or contractor with the administrative requirements for the section. In general, it sets the quality control, requirements for delivery and job conditions, notes the related trades with which this section needs to be coordinated, and specifies the submittals required for review prior to ordering, fabricating, or installing material for that section. It consists generally of the following:

- Description and scope: This article should include the scope of the work and the interrelationships between work in this section and the other sections. In addition, it should include definitions and options.

PROCUREMENT AND CONTRACTING REQUIREMENTS GROUP:
- Division 00 — Procurement and Contracting Requirements

SPECIFICATIONS GROUP

General Requirements Subgroup
- Division 01 — General Requirements

Facility Construction Subgroup
- Division 02 — Existing Conditions
- Division 03 — <u>Concrete</u>
- Division 04 — <u>Masonry</u>
- Division 05 — Metals
- Division 06 — <u>Wood</u>, Plastics, and Composites
- Division 07 — <u>Thermal and Moisture Protection</u>
- Division 08 — Openings
- Division 09 — Finishes
- Division 10 — Specialties
- Division 11 — Equipment
- Division 12 — <u>Furnishings</u>
- Division 13 — Special Construction
- Division 14 — Conveying Equipment
- Division 15 — RESERVED FOR FUTURE EXPANSION
- Division 16 — RESERVED FOR FUTURE EXPANSION

Facility Services Subgroup:
- Division 20 — RESERVED FOR FUTURE EXPANSION
- Division 21 — <u>Fire Suppression</u>
- Division 22 — <u>Plumbing</u>
- Division 23 — <u>Heating, Ventilating, and Air Conditioning</u>
- Division 24 — RESERVED FOR FUTURE EXPANSION
- Division 25 — Integrated Automation
- Division 26 — Electrical
- Division 27 — <u>Communications</u>
- Division 28 — <u>Electronic Safety and Security</u>
- Division 29 — RESERVED FOR FUTURE EXPANSION

Site and Infrastructure Subgroup:
- Division 30 — RESERVED FOR FUTURE EXPANSION
- Division 31 — Earthwork
- Division 32 — Exterior Improvements
- Division 33 — Utilities
- Division 34 — Transportation
- Division 35 — Waterways and Marine Construction
- Division 36 — RESERVED FOR FUTURE EXPANSION
- Division 37 — RESERVED FOR FUTURE EXPANSION
- Division 38 — RESERVED FOR FUTURE EXPANSION

Figure 10.3   The newly revised MasterFormat System (source: Construction Specifications Institute, Inc.).
(continued)

- Division 39 — RESERVED FOR FUTURE EXPANSION

Process Equipment Subgroup:
- Division 40 — Process Integration
- Division 41 — Material Processing and Handling Equipment
- Division 42 — Process Heating, Cooling, and Drying Equipment
- Division 43 — Process Gas and Liquid Handling, Purification, and Storage Equipment
- Division 44 — Pollution Control Equipment
- Division 45 — Industry-Specific Manufacturing Equipment
- Division 46 — RESERVED FOR FUTURE EXPANSION
- Division 47 — RESERVED FOR FUTURE EXPANSION
- Division 48 — Electrical Power Generation
- Division 49 — RESERVED FOR FUTURE EXPANSION

Figure 10.3   The newly revised MasterFormat System (source: Construction Specifications Institute, Inc.).

- Quality assurances: This article should include requirements for qualification of consultants, contractors, and subcontractors. Also included here are the standards and test requirements, and any full-size "mock-up" models of items for testing.
- Submittals: Instructions for submittal of product samples and other relevant information, including warranties, certificates, product data, and installation instructions.
- Product handling, delivery, and storage: This includes instructions for aspects like packing, location for delivery, temperature control, and protection for the product after delivery.
- Project and site conditions: This stipulates the requirements and conditions that must be in place prior to installation, such as temperature control and the use of necessary utilities. For example, all wall tiling should be completed prior to cabinet installation.
- Alternatives: Whether alternatives are acceptable is detailed in the General Requirements.
- Sequencing and scheduling: This is used where timing is critical and where tasks and/or scheduling need to follow a specific sequence.
- Warranties: This section typically includes warranties that exceed one year. Terms and conditions of the warranty should be spelled out, and the owner should be provided with copies.

Part 2: Products

This section defines and details the materials and products being specified, including fabrication or manufacturing of the product, the standards to which the materials or products must conform to so as to fulfill the specifications (Figure 10.4), and similar concerns. The itemized subsections would therefore include:

- Manufacturers: This section is used when writing a proprietary specification and lists approved manufacturers. The section should be coordinated with the product options and substitutions section.

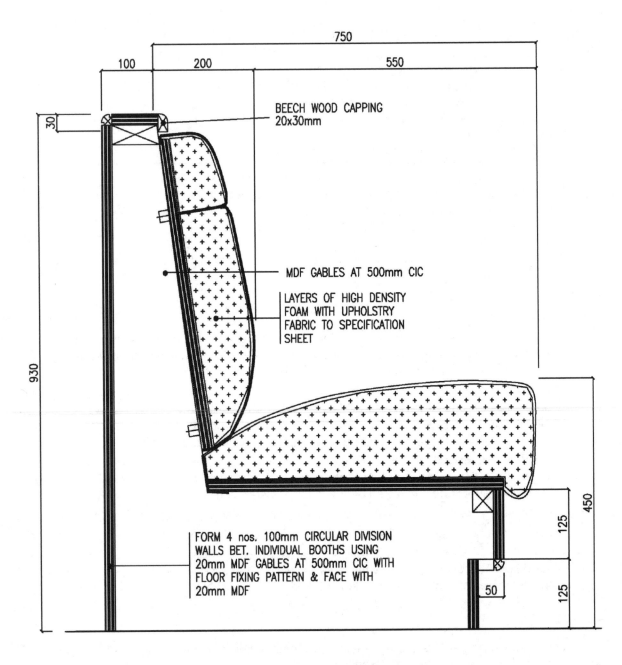

750

100   200   550

30

930

BEECH WOOD CAPPING
20x30mm

MDF GABLES AT 500mm CIC

LAYERS OF HIGH DENSITY
FOAM WITH UPHOLSTRY
FABRIC TO SPECIFICATION
SHEET

FORM 4 nos. 100mm CIRCULAR DIVISION
WALLS BET. INDIVIDUAL BOOTHS USING
20mm MDF GABLES AT 500mm CIC WITH
FLOOR FIXING PATTERN & FACE WITH
20mm MDF

450

125

125

50

SECTION
SCALE: 1" = 10'-0'

Figure 10.4   Section of a custom-upholstered restaurant banquette seating detail (source: Kubba Design).

- Materials, furnishings, and equipment: A list should be provided of materials to be used. If writing descriptive or performance specifications, detail the performance criteria for materials, furnishings, and equipment.
- Mixes: This section specifies the proportions of materials to be used when mixing a particular product.
- Fabrication: In this section, fabrication and construction details should be given.

Part 3: Execution

This part of the specification describes the quality of work-the standards and requirements specified in the installation of the products and materials. It also describes the conditions under which the products are to be installed, the protection required, and the closeout and cleanup procedures. The subheadings in this section include:

- Inspection: The section outlines what the contractor is required to do—for example, to the subsurface—prior to installation. Wording may include such phrases as "the moisture content of the concrete should meet manufacturer's specifications prior to installation of the flooring material."
- Preparation: This stipulates the improvements to be made prior to installation.
- Installation and performance: The specific requirements for each finish should be specified, as well as the quality of work to be achieved.
- Field quality control: This specifies the tests and inspection procedures to be used to determine the quality of the finished work.
- Protection: Where special protection is necessary for a particular installation, such as marble flooring, this section must be included.
- Adjust and clean: This outlines in detail the cleaning and adjustments requirements.
- Schedules: This is used only if deemed necessary.

10.6 AUTOMATED SPECIFICATION-WRITING SYSTEMS

Over the past decade, several firms have developed various versions of automated specification-writing systems, and many now offer these services on-line to architects, interior designers, engineers, and others. One such firm is Building Systems Design's (BSD) SpecLink, which is an electronic specification system that uses master-guide specifications in CSI three-part format and has a database of over 780 master-specification sections and over 120,000 data links that automatically include related requirements and exclude incompatible options as you select specification text (Figure 10.5). BSD also developed the Perspective early design-performance specifications organized by CSI UniFormat.

Interspec LLC is another firm that uses a proprietary technology that connects a large database of building specifications to an electronic architectural drawing of the project. The customer can also access the specs through the Internet. Moreover, the customer can make alterations as the specs are being written. Interspec also has a do-it-yourself program for designers with small projects. Using the e-Specs service will enable companies to increase their productivity while simultaneously reducing their costs. By linking the architect's CAD drawings to the master-guide specifications, the need to mail or deliver large blueprint drawings to the spec writer is eliminated. With these automated systems, the designer can input all necessary information at the earliest stages of the project, before any drawings are available, and instantly obtain an outline or preliminary specification.

SpecsIntact System is another automated software system available for preparing standardized facility construction specifications. SpecsIntact was designed by NASA to help architects, engineers, specification writers and other professionals doing business with the three government agencies using it, i.e. the National Aeronautics and Space Administration (NASA), the U.S. Naval Facilities Engineering Command (NAVFAC), and the U.S. Army Corps of Engineers (USACE).

There are many other systems on the market such as e-Specs Online, which is a proprietary browser-based specification-management system. These new systems are transforming the way architects and interior designers prepare specifications for commercial and residential buildings. They can provide greater accuracy, in less time, at a lower cost. These systems also eliminate or minimize costly construction changes due to omissions, discrepancies, or improper quality controls. A firm's proprietary interactive online editing system can be integrated into the specification-development process over the Internet with secure password access. A completed specification manual can be delivered on-line for client downloading or printed and bound, as well as saved on CD-ROM. The bottom line is whether outsourcing is the most effective way to go for a particular design firm.

10.7 THE PROBLEM OF LIABILITY

Architects and engineers, like other professionals, are expected to exercise reasonable care and skill in carrying out their work. Although this does not imply 100 percent perfection at all times, the level of performance should be consistent with that ordinarily provided by other qualified practitioners under similar circumstances. Law relating to professional responsibility and liability has become very active in recent years, and the zone of risk and exposure has expanded dramatically in professional practice. Indeed, under current law, whenever a designer enters into a contractual agreement and specifies a subsystem of a commercial or institutional space, he or she becomes responsible for the performance of that system.

Among the more significant areas of exposure are the liability of the architect engineer to third parties unconnected with the contract for claims of negligence or errors in design that lead to alleged injury of persons using the building. The legal bases for the majority of current liability suits include professional negligence, implied warranty or misrepresentation, implied fitness warranty, breach of contract, joint and several liability, and liability without fault for design defects. Often, these legal bases overlap. Thus, a designer who fails to reject defective work by a contractor or supplier may be considered to be professionally negligent and in breach of contract.

Designers can protect themselves from possible liability suits by working within their area of expertise, using concise contracts and specifications, complying with codes and regulations, using reputable contractors, maintaining accurate records, and securing legal counsel and liability insurance.

Another area of exposure is building product performance—that is, holding the architect responsible for damages caused by faulty materials and components and sometimes for the cost of their replacement. This tends to place a heavy emphasis on the selection and specification of building products with long records of satisfactory performance, thus inhibiting the introduction of new materials and methods.

Product liability is mainly concerned with negligence. And while it greatly affects manufacturers, retailers, wholesalers, and distributors, designers and specifiers are increasingly becoming involved in product-liability suits. Designers can minimize product-liability actions by specifying products manufactured for the intended use.

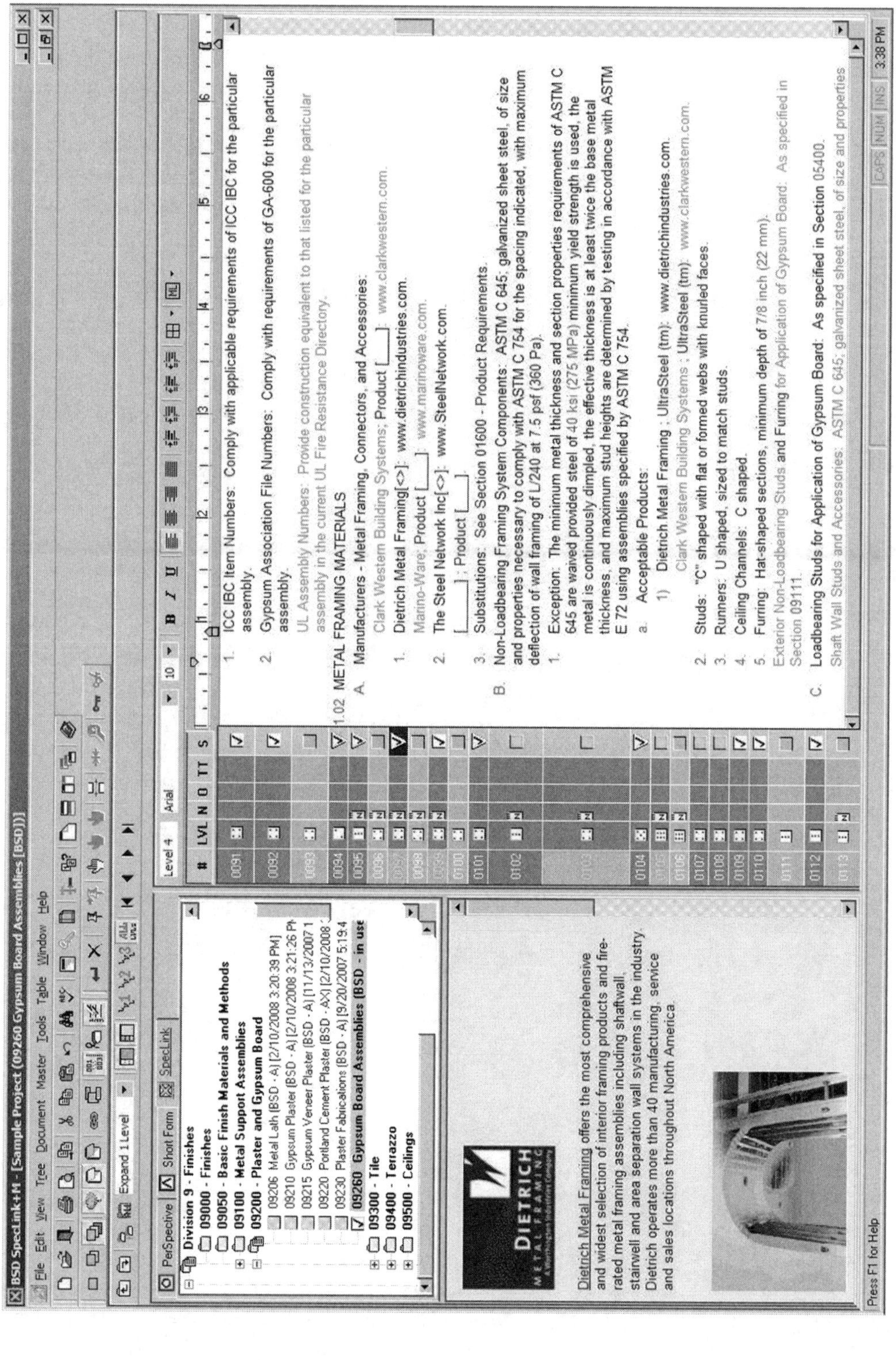

Figure 10.5 BSD SpecLink summary catalog listing and computer screen printout. SpecLink is one of the many electronic specification services that have emerged in recent years (source: Building Systems Design, Inc.).

11

Building Codes and Barrier-Free Design

11.1 GENERAL

The purpose of building codes is to govern the construction of public buildings, commercial buildings, and places of residence and to regulate construction and thereby provide occupants with a safe and healthy environment. Building codes are an imperative part of the design and construction process. They define minimum standards for safety and comfort that must be met in new construction and major renovations. Prior to having obtained a building permit to construct a commercial property, the developer is required to produce design plans that conform to the building codes in effect at the time.

Existing properties are not normally required to conform to newer code requirements unless major renovations are performed. When older properties are to be updated, local regulations dictate the conditions when compliance with newer codes is required. Typically, when interior renovation includes reconstruction of 25 to 50 percent of a floor, local regulations require compliance with existing life-safety code requirements. It is therefore important to determine the functional obsolescence of all major life-safety elements. This is particularly relevant to office buildings and hotels, where interior renovations and reconfigurations are periodically performed.

Building codes are essentially local laws, and each municipality (county or district in sparsely populated areas) enforces its own set of regulations (Figure 11.1). A strong and sustained movement has been underway for many years to unify the various local codes around the nation in response to the building industry's continuous requests for a single unified building regulatory system. The majority of states have already moved in that direction, and the three main code organizations have come together to form the International Code Council (ICC), whose cardinal mission is to unify the code system into a single set of comprehensive building codes that can be used anywhere in the United States.

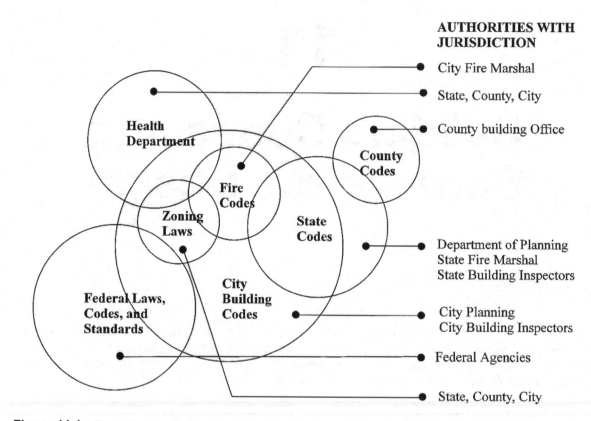

Figure 11.1   Illustration showing overlapping code structure and complexity of current regulations (source: Specifications for Commercial Interiors by S.C. Reznikoff).

11.2 BUILDING CODES TODAY

One of the most confusing aspects of American codes and standards is that, unlike Europe, Canada, and many other parts of the world, there is a complete absence of uniformity between federal agencies, states, counties, and municipalities, although in recent years there have been major efforts to unify codes on the national level.

When dealing with counties and municipalities we are confronted with other issues. For example, cities such as Houston have large oil refineries that create certain hazards, and cities such as Chicago and New York require special codes and standards that relate to high-rise buildings and population density. The state of California has also decided against using the IBC codes and elected to continue using the 1997 Uniform Building Code T as the basis for the 2001 edition of the California Standards Code.

On the other hand, a town that lies in the path of hurricanes may require special storm-protection standards. It is no surprise therefore that some codes have evolved through modifications necessitated by particular geographic and population needs.

The terrorist attacks of September 11 and Oklahoma City continue to impact code development, and a change to the International Building Code (IBC) related to the World Trade Center collapse was recently approved. The IBC now requires that buildings 420 feet and higher have a minimum 3-hour structural fire-resistance rating. The previous requirement was limited to 2 hours. This change provides increased fire resistance for the structural system, leading to enhanced tenability of the structure, and gives firefighters additional protection while fighting a fire.

The IBC establishes minimum standards for the design and construction of building systems. It addresses issues such as use and occupancy, entry and exit during emergencies, engineering practices, and construction technology. Figure 11.2 is a general checklist to indicate whether a project is code-compliant or not.

CODE COMPLIANCE CHECKLIST

1 **DETERMINE WHICH CODES ARE REQUIRED**
- Building Code and Other Code Publications
- Standards and Tests
- Government Regulations
- Local Codes and Ordinances

2 **OCCUPANCY REQUIREMENTS**
- Determine Types of Occupancy Classification(s)
- Calculate Occupancy Load(s)
- Review Specific Occupancy Requirements
- Compare Code and Accessibility Requirements

3 **MINIMUM TYPE OF CONSTRUCTION**
- Determine Construction Type
- Determine Ratings of Structural Elements
- Calculate Maximum Floor Area (as required)
- Calculate Building Height (as required)
- Check All Enforced Standards

4 **MEANS OF EGRESS REQUIREMENTS**
- Determine Quantity & Type of Each Means of Egress
- Calculate Travel Distance
- Calculate Minimum Widths
- Determine Required Signage
- Compare Code and Accessibility Requirements
- Check All Enforce Standards

5 **FIRE RESISTANT REQUIREMENTS**
- Determine Fire and Smoke Barriers
- Determine Through Penetration Opening Protective
- Review Types of Fire Tests and Ratings Required
- Compare Code and Accessibility Requirements
- Check All Enforced Standards

6 **FIRE PROTECTION REQUIREMENTS**
- Determine Fire and Smoke Detection Systems
- Determine Fire Suppression Systems
- Review Possible Sprinkler Tradeoffs (as required)

7 **REVIEW PLUMBING REQUJIREMENTS**
- Determine Types of Fixtures Required
- Calculate Number of Each Fixture Required
- Compare Code and Accessibility Requirements
- Coordinate with Engineer (as required)

8 **REVIEW MECHANICAL REQUIREMENTS**
- Determine Access and Clearance Requirements
- Figure Zoning and Thermostat Locations
- Determine Type of Air Distribution System
- Check for Accessibility Compliance
- Coordinate with Engineer (as required)

9 **REVIEW ELECTRICAL REQUIREMENTS**
- Determine Location of Outlets, Switches, and Fixtures
- Determine Emergency Power & Lighting Requirements
- Determine Types of Communication Requirements
- Check for Accessibility Compliance
- Coordinate with Engineer (as required)

10 **FINISH AND FURNITURE REQUIREMENTS**
- Review Tests and Types of Ratings Required
- Determine Special Finish Requirements
- Determine Special Furniture Requirements
- Compare Code and Accessibility Requirements
- Check All Enforced Standards

NOTE: Consult the jurisdiction having authority at any step in question.

Figure 11.2 A general checklist used to determine general code compliance (source: The Codes Guidebook for Interiors by S.K. Harmon and K.G. Kennon).

11.3 MODEL CODE ORGANIZATIONS

International Code Council

The International Code Council (ICC) was established in 1994 as a nonprofit organization dedicated to developing a single set of comprehensive and coordinated national model construction codes. The founders of the ICC are Building Officials and Code Administrators International, Inc. (BOCA), International Conference of Building Officials (ICBO), and Southern Building Code Congress International, Inc. (SBCCI). Although regional code development has been effective and responsive to our country's needs, the demand arose for a single set of codes. The nation's three model code groups responded by creating the International Code Council, which combines the strengths of the regional codes without regional limitations. The ICC developed the series of International Codes (I-Codes). I-Codes respond to the needs of the construction industry and public safety. A single set of codes has strong support from government, code-enforcement officials, fire officials, architects, engineers, builders, developers, and building owners and managers.

ICC made great strides when in 2000 it published the International Building Code (IBC)—a single family of codes that is being adopted across the nation. It is hoped that, as the IBC gains popularity, the existing regional and local model codes will be phased out.

Today, an overwhelming majority (97 percent) of cities, counties, and states that adopt building and safety codes are using documents published by the International Code Council and its members. The IBC, like its predecessors, will be updated every three years and will gradually replace the existing model codes. BOCA, ICBO, and SBCCI have agreed to merge their respective organizations into one model code group. This will allow a single approach to the proper interpretation, training, and other services for the International Codes.

The ICC published the first complete set of International Codes (I-Codes) in 2000, followed by the 2003 and 2006 editions. In 2007, one or more of the I-Codes were in use within 47 states as well as the District of Columbia, Puerto, Rico and the United States Department of the Navy, either enforced statewide or at the local level. ICC has developed and made available numerous publications pertaining to: building, energy conservation, fire, fuel gas, mechanical, plumbing, residential, property maintenance, private sewage disposal, zoning, and electricity as well as code administrative provisions and the ICC Performance Code for Buildings and Facilities. All of the above codes are comprehensive and coordinated with one another.

The Building Officials and Code Administrators (BOCA) model code was incorporated in 1938 and is the oldest professional association of construction code officials in America. BOCA was specifically set up as a forum for the interchange of information and expertise concerning building safety and construction regulation. BOCA is now incorporated in the International Code Council (ICC).

The Uniform Building Code (UBC) is published by the International Conference of Building Officials (ICBO). The Uniform Building Code is used mainly on the west coast of the United States.

The Standard Building Code (SBCCI) is used in much of the southeastern United States. SBCCI joined with BOCA and ICBO in 1994.

The Council of American Building Officials (CABO) One- and Two-Family Dwelling Code is a compilation of BOCA, SBCCI and NFPA. The latest edition (2000) of the code has been renamed the International Residential Code (IRC). CABO also established the Building Officials Certification Program to enhance professionalism in the field of building-code enforcement.

11.4 STANDARD WRITING ORGANIZATIONS AND INSTITUTIONS

There are numerous organizations involved in writing and maintaining standards. The vast majority are developed by trade associations, government agencies, or standards-writing organizations. Likewise, there is a long-standing relationship between construction codes and standards that address design, installation, testing, and materials related to the building industry. The pivotal role standards play in the building regulatory process is that they represent an extension of the code requirements and are therefore equally enforceable.

However, standards only have legal standing when stipulated by a particular code that is accepted by a jurisdiction. Building standards function as a valuable design guideline to architects while establishing a framework of acceptable practices from which many codes are later taken. When a standard is stipulated, an acronym formed from the standard organization and a standard number is called out. The most important and relevant of these organizations for building owners and consultants are:

- The American National Standards Institute (ANSI) approves standards as American National Standards and provides information and access to the world's standards. It is also the official U.S. representative to the world's leading standards bodies, including, the International Organization for Standardization (ISO). It provides and administers the only recognized system in the United States for establishing standards.

- American Society of Heating, Refrigerating and Air-Conditioning Engineers (ASHRAE) is an international organization whose sole purpose is to advance the arts and sciences of heating, ventilation, air conditioning and refrigeration for the public's benefit. ASHRAE's stated mission is to write, "standards and guidelines in its fields of expertise to guide industry in the delivery of goods and services to the public." ASHRAE standards and guidelines include standard methods of testing for rating purposes, outline and specify preferred procedures for designing and installing equipment and provide other information to guide the industry. In addition, ASHRAE "sets design standards for occupant comfort, building commissioning, and specification of building automation control networks."

- ASTM International (previously known as American Society of Testing and Materials) is one of the largest voluntary standards development organizations in the world providing a global forum for the development and publication of voluntary consensus standards for materials, products, systems and services having internationally recognized quality and applicability.

- National Standards Systems Network (NSSN) has as its primary mission to promulgate standards information to a broad constituency, and serves as a one-stop information repository. Its stated goal is to become a leader in providing technical data and information regarding major developments in a world wide standardization arena.

- National Fire Protection Association (NFPA) is a worldwide leader in providing fire, electrical, and life-safety information to the public.

- Underwriters Laboratory (UL) maintains and operates laboratories around the world for the testing and examination of devices, systems, and materials to determine their properties and their relation to life, fire, casualty hazards, and crime prevention.

Various federal agencies and departments collaborate with trade associations, private corporations, and the general public to develop federal laws for building construction. Federal agencies also use rules and regulations to implement laws passed by Congress. These regulations are often national laws that supersede or supplement local building codes. Each federal agency has its own set of rules and regulations that are published in the Code of Federal Regulations (CFR). Readers should be familiar with the following governmental agencies that produce building regulations that may impact projects under review:

- Access Board (previously named the Architectural and Transportation Barriers Compliance Board)
- The Department of Energy (DOE)
- The Environmental Protection Agency (EPA)
- The Federal Emergency Management Agency (FEMA)
- The General Services Administration (GSA)
- The Department of Housing and Urban Development (HUD)
- The National Institute of Standards and Technology (NIST)
- Occupational Safety and Health Administration (OSHA)

There are many national organizations that support the organizations that produce codes and standards and are essential to their development, although they are not themselves directly responsible for their production. Two such organizations are the National Conference of States on Building Codes and Standards (NCSBCS), and the National Institute of Building Sciences.

11.5 BUILDING-CODE ELEMENTS AND APPLICATIONS

Most code requirements for fire and smoke protection are based on occupancy classifications. "Occupancy" refers to the type of use a building or interior space is put to, such as a residence, an office, a school, or a restaurant. "Occupant load" is a term used to specify the number of people that a building code assumes will occupy a given structure or portion of it. Occupant-load calculations are based on the assumption that certain categories of occupancy have greater densities of people than others and that exiting provisions should adequately reflect this. Load factors are depicted in either gross square feet or net square feet.

The formula used to determine the occupancy load is:

$$\text{occupancy load} = \text{floor area (sq. ft.)} \div \text{occupancy factor.}$$

Thus, the square footage of the interior space that is assigned to a particular use is divided by the occupant load factor for the occupancy use as given in the code. Occupant load factors help determine the required occupant loads of a space or building and range from a low of 3 square feet per person for a waiting space to a high of 500 square feet per person for storage areas. In ascertaining the occupant load, it is presumed that all parts of the building will be occupied at the same time. If a building or building area provides more than one use—i.e., has mixed occupancies—the occupant load is determined by the use that reflects the highest concentration of people.

Types of Occupancy

Occupancy, as stated, refers to the type of use of the building or interior space, such as a residence, office, store, or school. An occupancy classification must be assigned to any building or space, and determining the occupancy classification is an essential part of the code process. The concept behind occupancy classification is that certain building uses are more hazardous than others. For example, a large theater with hundreds of people is more dangerous than a single-family residence.

Code publications divide their occupancies into different categories, based on the activities occurring in the space, the associated level of hazards present, and the anticipated number of people occupying the space at any given time. Buildings that house more than one use, will result in more than one occupancy group being designated for the building. For example if an office building has underground parking, each occupancy must be considered separately because of the different kinds and degrees of hazards.

Ten of the most common occupancy classifications used by model codes are:

1. Assembly: With certain exceptions, it consists of structures, buildings or portions of that are used for the gatherings for purposes such as civic, social, religious, recreational functions, the consumption of food or drink, or waiting areas.

2. Business: Consists of office, professional or service-type transactions.

3. Educational: Structures that are used for educational purposes by 6 or more persons over 2 ½ years of age through the 12th Grade.

4. Factory and Industrial: Structures used for assembling, dissembling, fabricating, finishing, manufacturing, packaging, repair or processing operations not classified as Group H or Group S.

5. Hazardous: With noted exceptions these are structures that are used for manufacturing, processing, generation or storage of materials in quantities that are a physical or health hazard.

6. Institutional: Used for the care or supervision of people having physical limitations due to health or age are harbored for medical treatment or other care or treatment or in which occupants are detained for penal or correctional purposes or where the liberty of the occupants are restricted.

7. Mercantile: Buildings that are utilized for the display and sale of merchandise involving stocks of goods accessible to the public

8. Residential: Used for sleeping purposes not classified as Group I or per the IRC (International Residential Code).

    R-1 - Multi-unit transient

    R-2 - Multi-unit non transient

    R-3 - Not classified elsewhere

    R-4 - Residential care

9. Storage Group: Used for storage not classified as hazardous occupancy and consists of:

    S-1 - Moderate hazard storage

    S-2 - Low hazard storage

10. Utility and Miscellaneous: Consists of accessory and miscellaneous structures not classified in other occupancies.

Classification by Construction Type

Construction type indicates the fire resistance of certain building elements such as fire and party walls, stair and elevator enclosures, exterior and interior bearing and nonbearing walls, columns, shaft enclosures, smoke barriers, floors, ceilings, and roofs. Fire ratings are based on the number of hours a building element will resist fire before it is adversely affected by the flame, heat, or hot gases.

All buildings are classified into one of five or six types of construction. Type I buildings are the most fire-resistive and typically contain structural members that are noncombustible. Type I buildings also have the highest fire rating, usually 2 to 4 hours. Type V buildings (Type VI in the SBCCI codes) have the lowest fire rating and are typically of wood-frame construction.

Adjuncts to Building Codes

Building codes typically have additional companion codes and standards that govern other aspects of construction, which, with the exception of the electrical code, are usually published by the same group that publishes the model building codes.

Model codes frequently use industry standards developed by trade associations, government agencies, and standards-writing agencies such as the American Society for Testing and Materials (ASTM), the American National Standards Institute (ANSI), and the National Fire Protection Association (NFPA). Building codes reference these standards by name, number, and date of latest revision and become law when a code is accepted by a jurisdiction. In addition, there may be local jurisdictions that maintain energy-conservation codes, health and hospital codes, fabric flammability regulations, and codes that regulate construction and finishes.

Test Ratings and Fire-Resistant Materials and Finishes

It is estimated that roughly 75 percent of all codes deal with fire and life-safety issues, and the primary aim of fire codes is to confine a fire to its area of origin, thus limiting its spread and preventing flashover. To facilitate this, all approved materials and construction assemblies referred to in building codes are assigned ratings based on standardized testing procedures. The rating of an assembly is ascertained by evaluating its performance during testing and by examining its fire-resistive properties. There are hundreds of standardized tests for building materials and construction assemblies.

Any approved testing laboratory can undertake the testing of building materials, provided that standardized procedures are followed. The American Society for Testing and Materials (ASTM), the National Fire Protection Association (NFPA), and Underwriters Laboratories (UL), in collaboration with the American National Standards Institute (ANSI), are among the best-known organizations that have developed a large variety of standardized tests and testing procedures.

Upon being subjected to one of the standard tests, a material is given a rating based on its performance during the test. For construction assemblies tested according to ASTM E-119, the rating given is according to time—that is, how long an assembly will contain a fire, retain its structural integrity, or both. The test evaluates a construction assembly's performance in the face of the temperature rise on the protected side of the assembly, the amount of smoke, gas, or flame that penetrates the assembly, and the assembly's structural performance during exposure to fire. The ratings are 1-hour, 2-hour, 3-hour, and 4-hour; 20-, 30-, and 45- minute ratings are also used for doors and other opening assemblies. Assemblies that consultants and field observers must be concerned with include fire walls, fire-separation walls, shaft enclosures (such as stairways, exits, and elevators), floor/ceiling constructions, and doors and rated glazing.

Building codes typically have tables that stipulate the type of construction that meets the different hourly ratings. Thus, when a building code states that a 1-hour-rated partition assembly is required between an exit corridor and an adjoining tenant space, the designer must select and detail a design that incorporate the requirement for 1-hour construction.

Means of Egress

Exiting is one of the most critical requirements of building codes. It comprises three main categories:. exit access, exit, and exit discharge (Figure 11.3).

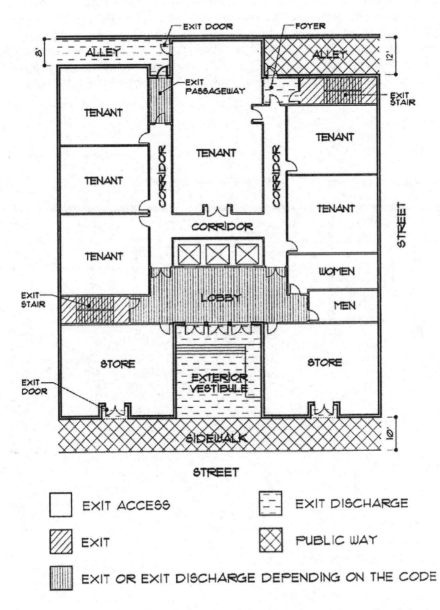

Figure 11.3 Typical example of means of egress from a building (source: The Codes Guidebook for Interiors by S.K. Harmon and K.G. Kennon).

Arrangement of exits is specified by code. They should be located as far apart from one another as possible so that if one is blocked in an emergency, the other(s) can still be reached. The code states that when two or more exits are required, they must be placed a distance apart equal to and not less than one-half the length of the longest diagonal dimension within the building or area to be served, as measured in a straight line between the exits. This is known as the half-diagonal rule and is shown diagrammatically in Figure 11.4).

The codes limit the length of travel distance from within a single space to an exit-access corridor. This is defined as the maximum distance and cannot exceed 200 feet (61 m) in an unsprinklered building and 250 feet (76.25 m) in a sprinklered building (Figure 11.5). There are exceptions to the rule, such as when the last portion of the travel distance is entirely within a 1-hour-rated exit corridor. Basically, codes classify travel distances into two types: The first relates to the length of travel distance from within a single space to the exit-access corridor (also known as the common path of travel), and the second

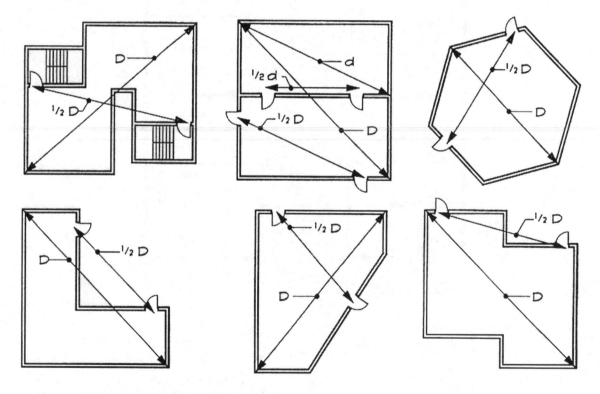

D = DIAGONAL OR MAXIMUM DISTANCE
½D = HALF OF DIAGONAL OR MINIMUM DISTANCE

Figure 11.4   The half-diagonal rule (source: The Codes Guidebook for Interiors by S.K. Harmon and K.G.Kennon).

regulates the length of travel distance from anywhere in a building to the floor or building's exit. Typically, however, if the travel distance within a tenant space exceeds 75 feet (22.9m), an additional exit is required, even if it is not required by the occupant load.

Codes usually allow a room to have a single exit through an adjoining or intervening room, provided that it affords a direct and unobstructed means of travel to an exit corridor or other exit and as long as the total stipulated maximum travel distances are not exceeded. Exiting is not permitted through kitchens, storerooms, rest rooms, closets, or spaces used for similar purposes. Codes normally categorize foyers, lobbies, and reception rooms constructed as required for corridors with a one-hour-rated wall as intervening rooms, thereby allowing them to be used for exit purposes.

Typically, corridor construction must be of 1-hour fire-resistive construction when serving an occupant load of 10 or more in R-1 and I occupancies and when serving an occupant load of 30 or more in other occupancies. The 1-hour-rated corridors must extend through the ceiling to the rated floor or roof above unless the ceiling of the entire story is 1-hour-rated. Where a duct penetrates a fire-rated corridor, a fire damper that closes automatically upon detection of heat or smoke so as to restrict the passage of flame must be provided.

There are different types of stairs including straight run, curved, winder, spiral, scissor, etc. Exit stairs should be wide enough to allow for two people to descend side by side with no sudden decrease in width along the path of travel. Stairs must also adhere to specific code and accessibility requirements and be constructed in a manner and using materials consistent with the construction type of the building. Typically, new stairs are required to have a minimum width of 44 inches, an 11-inch tread depth, and a maximum riser height of 7 inches (Figure 11.6). Handrails and guardrails are likewise regulated.

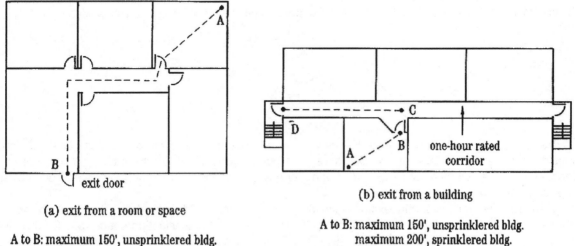

(a) exit from a room or space

A to B: maximum 150', unsprinklered bldg.
maximum 200', sprinklered bldg.

(b) exit from a building

A to B: maximum 150', unsprinklered bldg.
maximum 200', sprinklered bldg.
C to D: maximum 100'

Figure 11.5 Maximum acceptable distances required to exits (source: Interior Design Reference Manual by D.K. Ballast).

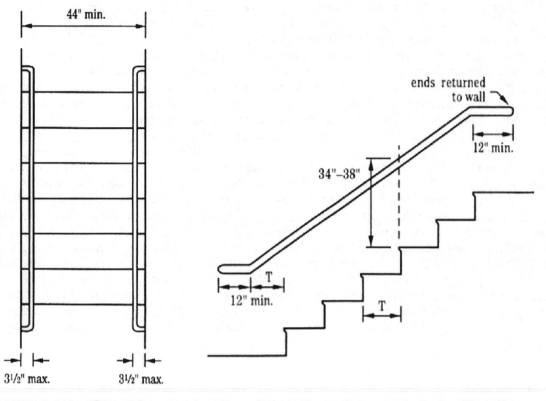

Stair and Handrail Design

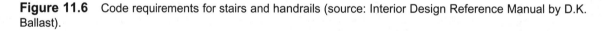

Figure 11.6 Code requirements for stairs and handrails (source: Interior Design Reference Manual by D.K. Ballast).

Escalators and moving walkways, like elevators, are not usually allowed as a means of egress and should not be taken into account as such in egress calculations, although there may be some exceptions, in which case they must be provided with standby power and must comply with emergency operation and signaling-device requirements.

Requirements for residential exiting (individual dwelling units and single-family houses) are not as strict as for commercial occupancies. Codes typically have a subclassification specifically for dwelling units. The International Residential Code (IRC) is specifically designed for one- and two-family houses. The designer must verify which code is applicable to a particular project.

The IRC requires at least one regulated exterior door per residence with minimum dimensions of 30 inches x 80 inches. Bedrooms located on upper floors typically require an emergency means of egress for these areas—which can usually be an operable window as long as it is not more than 44 inches from the floor. Stair and ramp dimensions are also regulated but are not as strict as those for commercial use. One handrail is normally required in residential stairs and ramps.

Plumbing Systems

Previously major code organizations published separate plumbing and mechanical codes. The International Code Council (ICC) published the first International Plumbing Code (IPC) in 1997 and International Mechanical Codes in 1998. Model codes specify in great detail how a plumbing or mechanical system should be designed. Plumbing codes specify the number of sanitary fixtures required based on the type of occupancy.

Sound Ratings

Model building codes sometimes require the use of insulation to control sound transmission in wall and floor assemblies separating dwelling units or guest rooms in residential occupancies from one another and from public spaces. Codes usually specify the minimum sound-transmission class (STC) for walls or impact-insulation class (IIC) for floors. Construction details can then be designed to satisfy these requirements.

11.5 BARRIER-FREE DESIGN AND ADA REQUIREMENTS

The two most recent pieces of legislation dealing with accessible design are the Americans with Disabilities Act (ADA), which is a federal civil law, and the Fair Housing Amendments Act. The latter extends the nondiscrimination protections of the Fair Housing Act to persons with disabilities as well as persons with families. The main purpose of ADA legislation is to make American society more accessible to persons with disabilities. With few exceptions all existing buildings as well as new construction must comply with the Americans with Disabilities Act.

While the employment provisions of the ADA apply to employers of fifteen employees or more, its public-accommodation provisions apply to all sizes of business, regardless of number of employees. State and local governments are covered regardless of size. The ADA consists of five titles:

1. Title I. Employment: Business must provide reasonable accommodations to protect the rights of individuals with disabilities in all aspects of employment.

2. Title II. Public Services: The regulations of Title II apply to public services provided by state and local governments and include public school districts, the National Railroad Passenger Corporation, port authorities, and other government units, whether or not they receive federal funds.

3. Title III. Public Accommodations: This segment of the law applies mainly to commercial facilities and prohibits privately owned and operated businesses from denying goods, programs, or services to persons with disabilities. All new and altered commercial facilities are subject to the accessibility requirements of Title III

4. Title IV. Telecommunications: This section is aimed at federally regulated telecommunications companies and federally funded public-service television offering services to the general public.

5. Title V. Miscellaneous: This section includes a provision prohibiting coercion, threatening, or retaliation against the disabled or those attempting to aid people with disabilities in the assertion of their rights under the ADA.

11.6 RELEVANT ADA COMPONENTS

Of particular importance to architects and engineers are Titles III and IV of the ADA. The Department of Justice (DOJ) and the Department of Transportation (DOT) enforce Titles III and IV of the Act throughout the United States to make American society more accessible to people with disabilities. The information provided here is baseline data that gives an overview of the types of issues to be reviewed and standard handicapped-compliance modifications required at a typical facility. Figure 11.7 shows the typical components that are evaluated for ADA compliance.

Figure 11.7 Typical components to be evaluated for the Americans with Disabilities Act (ADA).

Accessibility Guidelines

The site surrounding the facility should be surveyed during the review for handicapped accessibility and pedestrian and vehicular circulation, and layout should be reviewed for unimpeded access to the entrance of the facility. Likewise, the accessibility of the parking area, including the adequacy of location, dimensions, and identification of parking stalls, should be reviewed and recorded. Walkways should provide adequate access between various site areas and the building. These walkways should not be less than 48 inches wide and should not have a slope in excess of 5 percent (1 foot, 0 inches rise in 20 feet, 0 inches).

Accessibility is achieved by addressing the requirements of the ADA and other applicable codes as well as state and local regulations. There are essentially three accessibility documents that designers and consultants most frequently use and should be familiar with:

- ANSI A117.1 was developed by the American National Standard Institute (ANSI) and is one of the first accessibility guidelines to be used in the United States. The latest edition of ANSI A117.1/ICC is the 1998 edition, which was developed jointly with the International Code Council and the Access Board. This edition has been modified to be more in step with the ADAAG.

- The Americans with Disabilities Act Accessibility Guidelines (ADAAG) was developed by the Architectural and Transportation Barriers Compliance Board (ATBCB or Access Board) as a guideline for ADA legislation. It was based on the 1986 ANSI A117.1, but after the incorporation of additional requirements it became stricter than ANSI. The Access Board is responsible for making revisions to the ADAAG and is currently working with the DOJ on updating the ADAAG.

- The Uniform Federal Accessibility Standards (UFAS) are based on the 1980 ANSI standard and applies mostly to government buildings and organizations that accept federal funding. These buildings are not currently required to conform to ADA regulations.

The ADA stipulates that new construction and alterations to existing facilities comply with the ADAAG. For example, a new tenant space within an existing building is now considered by the ADA to be new construction and must comply with the ADAAG. Rules of compliance for alterations and renovations to existing buildings are sometimes complex, and under Title III of the ADA altered buildings must be made accessible if that is readily achievable. When prevailing conditions prevent barrier removal, a public accommodation has to make its services available through alternative means, such as relocating activities to accessible locations.

Accessible Routes

The ADAAG defines an accessible route as, "a continuous, unobstructed path connecting all accessible elements and spaces in a building or facility." This includes pathways, corridors, doorways, floors, ramps, elevators, and clear floor space at fixtures. Designing safe and barrier-free accessible routes is essential for people with disabilities, and enhancing their movement is critical to their wellbeing.

Adequate corridor width is essential to passage for someone with mobility or vision impairment. The ADAAG puts great emphasis on the provisions for access and egress and clearly delineates the requirements for length, space, lighting, signage, and safety measures. Corridors, for example, should ideally be a minimum of 42 inches (1065 mm) wide and not more than 75 feet (22.9 m) long. They should be well lit with indirect lighting to prevent glare. Wall finishes should incorporate blends of contrasting colors to increase visual acuity. Openings that form part of an accessible route should not be less than 32 inches (815 mm) wide. The minimum passage width for two wheelchairs is 60 inches (1525 mm). If an accessible route is less than 60 inches (1525 mm) wide, passing spaces at least 60 inches by 60 inches must be provided at intervals not to exceed 200 feet (61 m).

The ADAAG stipulates that the minimum clear floor space required to accommodate one stationary wheelchair is 30 inches (762 mm) by 48 inches (1220 mm). For maneuverability, a minimum 60-inch (1525-mm)-diameter circle is required for a wheelchair to make one 180-degree turn. In place of this, a T-shaped space may be provided.

Doors and Doorways

Doors should have a clear opening width of between 32 (815 mm) and 36 inches (915 mm) when the door is opened at 90 degrees (Figure 11.8). The maximum depth of a 32-inch-wide (815-mm) doorway is 24 inches (610 mm). If the depth exceeds this, the width must be increased to 36 inches (915 mm). Threshold heights should not exceed 1/2 inch (12.7 mm) and should not contain any sharp slopes or abrupt changes but should be beveled so no slope of the threshold is greater than 1:2. Door closers should not hamper a door's use by the disabled. No part of an accessible route may have a slope more than 1:20 (1 inch rise for every 20 inches/508 mm distance). If a slope is greater than this, it is classified as a ramp and must meet different requirements including the handrail provision.

Barrier-free codes also require that door hardware meet certain specifications. Lever handles on doors for disabled people are usually cost-effective. All hardware on doors, cabinets, and windows should be easy to grasp and operate with one hand and should not need a tight grip for turning. This includes lever-operated, push-type mechanisms, and U-shaped handles. Standard doorknobs are not allowed.

Plumbing Fixtures and Public Lavatories

If a toilet-stall approach is from the latch side of the stall door, clearance between the door side of the stall and any obstruction may be reduced to a minimum of 42 inches (1065 mm). Many toilet stalls are positioned at the end of a path of travel between the row of stalls and the wall (Figure 11.9). The advantage of using an end toilet for the accessible stall is that the grab bars can be fixed to the wall rather than to a partition, which allows sturdier anchoring to meet minimum strength requirements.

There are several toilet-stall layouts that meet ADA requirements. Toilet rooms as well as toilet stalls must have a minimum 60-inch (1525-mm) clear internal turning space. However, the clear floor space at fixtures and controls may extend up to 19 inches (483 mm) under a wall-mounted sink. The clearance depth varies depending on whether a wall-hung or floor-mounted water closet (60-inch by 56-inch minimum inside dimensions) is used. In most cases, the door must provide a minimum clear opening of 32 inches (815 mm) and must swing out, away from the stall enclosure. If a stall is less than 60 inches (1525 mm) deep, a 9-inch (225-mm) toe clearance is required under partitions. In planning toilet rooms, a 5-foot diameter (1525-mm) clear space should be allowed for.

Grab bars must also be provided as shown in Figure 11.10, mounted from 33 inches (838 mm) to 36 inches (915 mm) above the finished floor. Grab bars should be a minimum of 42 inches (1065 mm) long at a side wall and 36 inches (915 mm) at a rear wall. They should have a diameter of 1.5 inches (38 mm) and be located not more than 1.5 inches (38 mm) from the wall. In many toilets there is a lateral space to the side of the water closet, which only allows provision of a side horizontal rail. Toilet-paper dispensers are to be located below the grab bar, a minimum of 19 inches (483) above the finished floor.

In the absence of toilet stalls, the centerline of the toilet must still be 18 inches (455 mm) from a wall with back and side grab bars. A clear space should be provided in front of and to the side of open water closets.

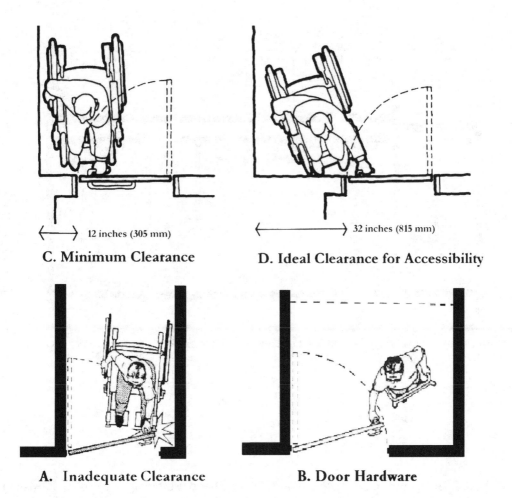

C. Minimum Clearance | 12 inches (305 mm)

D. Ideal Clearance for Accessibility | 32 inches (815 mm)

A. Inadequate Clearance

B. Door Hardware

Figure 11.8   Inadequate clearances can hamper accessibility (source: Designing for the Disabled: The New Paradigm by Selwyn Goldsmith).

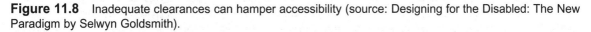

Where urinals are installed, stall or wall-mounted urinals must be used with an elongated rim no more than 17 inches (430 mm) above the floor. A clear floor space of 30 inches (762 mm) by 48 inches (1220 mm) must be provided in front of the urinal. This space may adjoin or overlap an accessible route.

Public lavatories must allow wheelchair users to move under the sink and easily use the basin and water controls. Any exposed piping below the lavatory must be insulated or otherwise protected. ADAAG makes a distinction between lavatories, which are basins for hand washing, and sinks, which are other

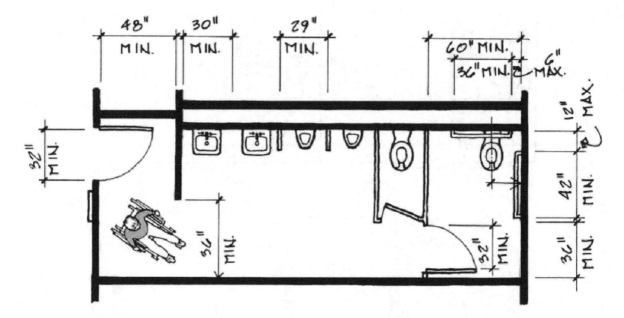

Figure 11.9   An example of a typical end toilet stall (source: Means ADA Compliance Pricing Guide).

types of basins. Faucets must be easy to operate with one hand without tight grasping or twisting of the wrist. Lever-operated, push-type, and automatically controlled mechanisms are acceptable.

Private residences are not typically subject to Title III of the ADA requirements; nevertheless, consultants should familiarize themselves with such requirements so as to be able to serve their clients better. In Figure 11.10 A we see a typical ADA-compliant residential bathroom in a senior living complex. Figure 11.10 B illustrates a prefabricated shower unit with strong grab bars in the shower, installed at different heights, along with a hand-held showerhead. These reflect some of the essentials of the accessible bathroom in the home.

Drinking water should be accessible with up-front spouts and controls that can be either hand- or floor-operated. Where only one drinking fountain is provided per floor, it should be accessible to people using wheelchairs, as well as persons who have difficulty bending or stooping. This can be resolved by the use of a "high-low" fountain, whereby one fountain is at a low level and accessible to those using wheelchairs and another is at the standard height for those who have difficulty bending.

Stairs and Ramps

Ramps should be installed as needed in areas of pedestrian-access level changes. They are required to provide a smooth transition between changes in elevation for both wheelchair-bound persons as well

Figure 11.10   A. A typical accessible bathroom in a senior living complex in Maryland. B. A prefabricated shower unit with strong grab bars (source: Charles A. Riley II).

as those whose mobility is otherwise restricted. In general, designers should use the least possible slope, but in no case should a ramp have a slope greater than 1:12 (1 inch in rise for every 12 inches in run). The maximum rise for any ramp is typically limited to 30 inches (762 mm), after which a level landing is required. A slope of up to 1:8 is permitted if the maximum rise does not exceed 3 inches (76 mm). In all cases a nonskid surface should be in place to enable traction in inclement weather.

A ramp's clear width must not be less than 36 inches, with landings that are at least as wide as the widest segment of the ramp leading to them. Landing lengths must not be less than 60 inches (1525 mm), and, if ramps change direction at a landing, the landing must be at least 60 inches square.

Handrails on both sides of ramps are to be incorporated if the ramps have a rise greater than 6 inches (152 mm) or a length exceeding 72 inches (1825 mm). The top of the handrail should be from 34 (864 mm) to 38 inches ((965 mm) above the ramp surface. Handrails must extend at least 12 inches (305 mm) beyond the top and bottom of the ramp segment and have a diameter or width of gripping surface from 1 1/2 (32 mm) to 1 1/2 inches (38 mm) for both ramps and stairs. Notice that the new ADAAG handrail guidelines are more flexible than the current guidelines (Figure 11.11).

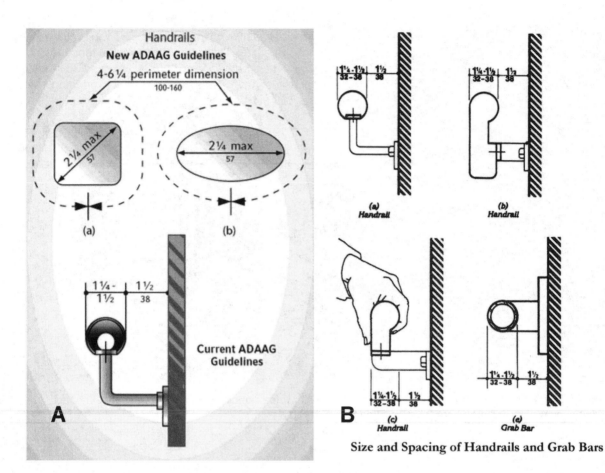

Figure 11.11 New and current handrail requirements for ramps and stairs (sources: Gerald J. Morgan; 28 CFR Ch.1, Pt. 36, App.A, Fig.39, 7-1-94 Edition).

The stairways should provide accessibility between building floors; when these stairs are not connected by an elevator, they must be designed according to certain standards specifying the configuration of treads, risers, nosings, and handrails. The maximum riser height is 7 inches (178 mm), and the treads must be a minimum of 11 inches (280 mm) as measured from riser to riser. Open risers are not permitted. The undersides of the nosings must not be abrupt and must conform to one of the styles shown in Figure 11.12. Stairway users are more likely to stumble or fall while going down stairs than when going up. Tread depth is pivotal in stair design. Typically when climbing a stairs, users place only part of their foot on the tread, whereas when descending the whole foot or most of the foot is placed on the tread.

Stairway handrails must be continuous on both sides of the stairs. The inside handrail on switchback or dogleg stairs must always be continuous as it changes direction. Other handrails must extend beyond the top and bottom riser. A handrail's top gripping surface must be between 34 (864 mm) and 38 inches (965 mm) above stair nosings. In addition, the handrail must have a diameter or width of gripping surface of between 1 1/4 inches (32 mm) and 1 1/2 inches (38 mm). There must also be a clear space between the handrail and the wall of at least 1 1/2 inches (38 mm) as shown in Figure 11.11.

Floor Surfaces and Tactile Pavings

Floor finishes in a facility must be firm and slip-resistant and should provide easy access throughout the building. If there is a change in level, the transition must meet the following requirements. If the change is less than 1/4 inch (6.4 mm), it may be vertical and without edge treatment. If the change in level is between 1/4 inch (6.4 mm) and 1/2 inch (12.7 mm), it must be beveled and its slope no greater than 1:2. Changes greater than 1/2 inch change the classification to a ramp, which must then meet the requirements outlined in the previous section. Bathroom floors should have a nonslip finish.

Door handles are also required to have a textured surface if they are part of a door that leads to an area that might prove dangerous to a blind person, including doors to loading platforms, boiler rooms, and stages.

Public Telephones

Telephones are one of the easiest building elements to make accessible. They should be positioned so that they can be reached by a person in a wheelchair. Accessible telephones may be designed for either front or side access. In either case a clear floor space of not less than 30 inches (762 mm) by 48 inches (1220 mm) is to be provided. Telephones should have pushbutton controls and telephone directories that are accessible by a person in a wheelchair.

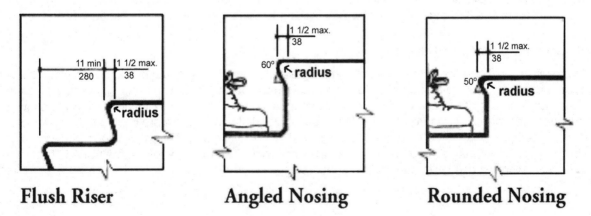

Figure 11.12   Tread and nosing requirements (source: CFR Ch.1, Pt. 36, App.A, Fig.18, 7-1-94 Edition).

Title III stipulates that in new construction at least one TTY is to be provided inside any building that has four or more public pay telephones (counting both interior and exterior phones). A TTY must also be provided whenever there is an interior public pay phone in a stadium, convention center, hotel with a convention center, covered shopping mall, or hospital emergency, recovery, or waiting room. Title III also stipulates that one accessible public phone must be provided for each floor of new construction, unless the floor has two or more banks of phones, in which case one accessible phone should be provided for each bank.

Protruding Objects

There are restrictions on objects and building elements that project into corridors and other walkways, because they present a hazard for visually impaired people. There are no restrictions on protruding objects where their lower edge is less than 27 inches (686 mm) above the floor, because these can be detected by a person using a cane. However, protruding objects cannot reduce the clear width required for an accessible route or maneuvering space, and a guardrail or other barrier must be provided for areas adjacent to accessible routes where the vertical clearance is reduced to less than 80 inches (2 m).

Signage and Alarms

Signage should give clear guidance for visually impaired people to emergency information and general circulation directions. Of importance in evaluating signage criteria is the ability to be viewed by people with low vision (20 percent of normal) from a distance of 30 feet (9.14 m). Signage is also required for elevators and handicapped-accessible entrances/exits, toilets, and other locations. For optimum clarity, adequate luminescence should be provided. Contrasting colors can also enhance legibility—70 percent or more contrast between letters and background is recommended.

The ANSI standards specify the width-to-height ratio of letters and how thick the individual letter strokes must be. They also require that characters, symbols, or pictographs on tactile signs be raised 1/32 inch (0.79 mm). If accessible facilities are identified, the international symbol of accessibility must be used (Figure 11.13). Braille characters must be Grade 2.

The ADA Accessibility Guidelines 4.1.3(14) state that, if emergency warning systems are provided, they shall include both audible alarms and visual alarms complying with 4.28. Audible alarms must produce a sound that exceeds the prevailing sound level in the room or space by at least 15 decibels. Visual alarms must be flashing lights that have a flashing frequency of about one cycle per second.

Elevators and Elevator Cars

All elevators must be accessible from the entry route and all public floors and must comply with the applicable codes for elevators and escalators. Elevators must be provided with handrails fixed 32 inches above the floor on all three sides of the cab. Minimum cab size should be 67 inches (1.7 m) to allow a wheelchair to maneuver (Figure 11.14). Both visual and audible hall signals are important to inform elevator passengers where an elevator is and in which direction it is going. This is particularly important at elevator banks comprising more than one car. Elevator controls should comply with ANSI A117.1 standards regarding visual, tactile, and audible controls.

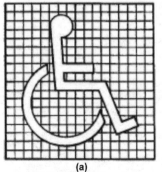

(a)
Proportions
International Symbol of Accessibility

(b)
Display Conditions
International Symbol of Accessibility

(c)
International TDD Symbol

(d)
International Symbol of Access for Hearing Loss

International Symbols

Figure 11.13 International symbols for accessibility (source: 28 CFR Ch.1, Pt. 36, App.A, Fig.43, 7-1-94 Edition).

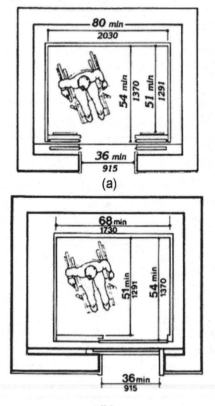

(a)

(b)

Minimum Dimensions of Elevator Cars

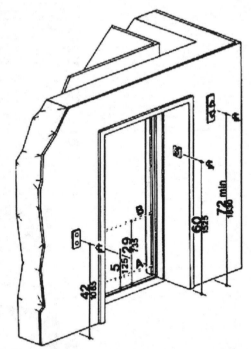

NOTE: The automatic door reopening device is activated if an object passes through either line A or line B. Line A and line B represent the vertical locations of the door reopening device not requiring contact.

(c)

Hoistway and Elevator Entrances

Figure 11.14   Minimum dimensions for elevator cabs (source: 28 CFR Ch.1, Pt. 36, App.A, Fig.22, 20, 7-1-94 Edition).

12

Construction Business Environment

12.1 GENERAL

Many skilled and experienced professionals are finding themselves on their own, seeking employment for the first time after being forced to abandon the safety of an organization that regularly delivered their paycheck each month. As a result, an increasing number of professionals are rethinking their employment strategy. Cash flow, health insurance, and retirement dominate this strategy. For seasoned veterans, there are also growing concerns over job satisfaction, location, and stress.

Let's say that you are considering the decision to incorporate. There are seductive attractions to being an independent contractor: being your own boss, having flexible hours, and seeing the family whenever you like. But before making a final determination to start a construction business, there are some things that need to be considered:

- Do you have the necessary qualifications to start a construction business? Having a few years of on-site experience in construction is extremely important.

- Do you possess any business training? Do you know anything about running a business?

- Starting a construction business from scratch is costly. You need to know how much it will cost to get the business up and running, and a sound business plan is crucial.

- Starting a construction business is not a 9 to 5 job. It will need someone who is willing to spend 12 to 15 hours a day to get the business off the ground.

- Do research to ensure that the city where you live is not inundated with this type of business. Discuss with others in the trade the potential for success of a construction business in your area. Try to determine if there is enough work to support your new company or if there is a niche that you can fill.

Make your dream of starting a construction company come true by doing adequate research and being prepared.

12.2 PREPARING A BUSINESS STRATEGY AND PLAN

Planning is the key to success in any business. The importance of planning cannot be overemphasized. It may take two or three weeks to complete a good plan. Most of that time is spent in research and reevaluating your ideas and assumptions. Taking an objective look at the business plan will identify areas of weakness and strength, pinpoint needs that might otherwise be overlooked, and spot opportunities early. It provides a road map to make sure you are going where you want to go to achieve your business objectives.

A business plan is an operating tool, which, when properly used, can help manage the business and work effectively toward its success. Moreover, lenders require one. A well-written business plan communicates ideas to others and provides the basis for a financial proposal. It will also determine the feasibility of a project and lay out the action necessary to complete it. A good plan can help convince a bank or potential investor that you are worthy of assistance in funding this new venture. The plan itself may not get you funding, since it is very difficult to find capital for startup businesses. As the business owner, you are expected to have sufficient startup capital from your savings or from a bank loan based on income other than the business.

Putting a successful business plan together is both an art and a science. While there are a number of ways to format a business plan, the following outline presents key elements to consider in developing your business plan:

Executive Summary

The executive summary provides an overview of the business plan and should not exceed two pages. Even though it will be at the beginning of the plan, it should be written after the rest of the plan is completed. It brings together the significant points of the project and should convey enthusiasm and professionalism. Typically, investors will not spend more than a few minutes to review a business plan to determine whether they should read it in detail or go on to another plan. It is therefore essential to prepare an appealing, convincing executive summary to capture the investor's attention and imagination and to make him or her more likely to read the remainder of the plan. It should be concise yet comprehensive and outline the fundamentals of the company, how it came into existence, and the people linked to it. If the business plan is part of a loan application, state clearly how much you want, precisely how it is going to be used, and how the money will increase the business's profits, thereby ensuring repayment. In essence, the executive summary is the most important part of the business plan, as it will dictate whether or not the remaining pages are read.

Company Description, Vision, and Mission

In any organizational venture, the first step is to develop a realistic vision for the business. This vision typically reflects an overall picture of the business as you see it in three or more years' time in terms of its likely physical form, size, activities, etc. Describe the company strengths and core competencies and the factors needed to make the company succeed. This should be followed by the company's stated mission. A mission statement should concisely reflect the direction of the company's business, its goals, and its expected achievements. Define both the key short-term and long-term goals and objectives and which factors are to be focused on in the short term and which in the long term. The company description should also indicate the structure of the company (sole proprietorship, partnership, corporation, etc.).

Management

Even if you are a one-person operation, a key ingredient of your potential business success is the strength of your management skills. When the business consists of more than one person, the business plan should identify the people who will be active in this business and include a short biography of principals and senior personnel as well as their backgrounds, positions, responsibilities, and strengths they bring to this new venture.

Market and Services Offered

Describe in depth the services offered, the market for your service, how you fit into that market, and your plans for achieving your share of the market. State the factors that give your firm a competitive advantage over others in this field. Examples may include level of quality or special skills or experience. Outline the pricing or fee structure of the services offered. Include any photos or sales brochures in an appendix.

The Successful Business Plan

A business plan is basically a written document that describes the business, its objectives, its strategies, its market, and its financial forecasts. Explain the type of company and services to be provided. If this is an existing business, give its history. If it is a new business, describe the product and/or service and note some of the qualifications to start this business. Also explain why this business is needed and what its chance for success is. Describe any unique features that will attract customers to this business.

An effective plan must outline the marketing strategy that is to be implemented. A business plan is thus a useful operating tool, which, if properly used, will help manage the business and facilitate its success. Although a business plan serves many functions from securing external funding to measuring success within the business, its main purpose is twofold: First, it helps ensure that you have researched and thought out the various aspects of running the business so you don't encounter any sudden unpleasant surprises. Second, lenders require it, and it can help convince banks or potential investors that your firm is worthy of receiving financial assistance. The concept here is to communicate ideas to others while providing the basis for a financial proposal.

Setting up a business is rarely easy; statistics show that over half of all new businesses fail within the first 10 years. The reason for such failure is often due to lack of planning and lack of funding. The best way to enhance the chances of success is to plan and follow through on that plan.

Financial Plan

This section of your business plan is critical. The credibility of your projections is essential to establishing the likelihood of success or failure for your business. Investors and lenders will use the information in this section to evaluate the financial prospects of your business. Be sure to check how many years of financial projections your lender requires—three years is the norm. State the business' financial requirements and how these funds will derive from project revenues, costs, and profits. Developing financial statements assists in understanding the cash flow of a business, its break-even point, and the sensitivity of the business to fluctuations in product costs and market factors. The financial plan would ideally consist of:

- a 12-month profit and loss projection
- three- to five-year profit and loss projection (optional)
- one-year month-by-month cash-flow projection
- projected balance sheet
- break-even point
- personal financial statement of owner

Together they constitute a reasonable portrayal of your company's financial potential. In all cases, it is strongly advised to consider seeking legal counsel to be sure the plan and business venture are legal and meet your requirements. In addition, the service of an experienced accountant is important. Unless you are prepared to construct spreadsheets and graphs explaining how you intend to use your money and your projections for the future, you might want to hire someone who knows all the financial ins and outs of a business.

Factors for Success

There are many factors that will impact the chances of a new company's success. They all require devoting considerable effort in personal networking with attorney groups and other potential clients in your area. Some of these factors include:

1. Extensive network: A single contact may yield a lucrative contract, but it takes a strong network to yield a continuing stream of work.
2. Excellent communication skills: Most executive-level professionals have excellent verbal skills, and this ability is one of the primary determinants for achieving success. Writing skills are an entirely different matter and can be a major challenge to those who largely depend on others to put pen to paper.
3. People skills: Fundamental to any successful business is relating effectively to others—clients, employees, suppliers, and consultants. Successful businesspeople invest in developing their communication and interpersonal skills.
4. Hard work: In private practice there is some flexibility in work hours, but this is no eight-to-five job, and hard, focused work and effort are needed to build the practice.
5. Self-direction: Some people have great difficulty in working on their own initiative and need a structured environment to perform. Independence can be freeing, but it can also be lonely; some people require daily, face-to-face interaction. This is especially true of individuals, who work out of their home office instead of renting space in a corporate office park.
6. Marketing skills: Some people are shy and introverted, but if you are not willing to engage in relentless self-promotion, you may not be able to bring in sufficient new business to succeed. Identify your target market. There should be specific target markets that will need your services and be willing to pay for it. Outline a marketing strategy with a competitive edge that draws customers to you and your company rather than the competition. Clients can come from a number

of different, unanticipated directions. You have to be flexible and be able to adapt to emerging situations.

7. Financial security: Owning your own business is one of the better ways to gain wealth, provided you know what is required. Starting a business is risky, but the chances of success are better if you understand the challenges you will meet and resolve them before you start. Likewise, it is essential that the new startup have a financial capability to survive the dry periods, which could easily last a year or more; otherwise it may be prudent to reconsider the decision to be an independent contractor.

In addition to the various bureaucratic and legal hurdles that an entrepreneur must overcome to incorporate and register a new firm, there are procedures as well as time and cost involved in launching a contracting firm. These need to be examined before attempting to launch such a venture.

12.3 STARTUP COSTS AND CAPITALIZATION

Start-up expenses can basically be segregated into: 1. investigatory, and 2. pre-opening costs. These costs can be incurred over a period of several weeks or several years, depending upon the type of industry under consideration and the length of the search process.

Start-up costs would normally include any amounts paid or incurred in connection with 1. investigating the creation or acquisition of an active trade or business, or 2. the actual creation of a new trade or business. However, distinguishing which costs should be classified start-up expenses can be a daunting task. Not only are the rules vague, but the various stages of the start-up process determine the tax classification of an expense.

In any case, there will be many startup expenses before you even begin operating your business. It would be futile to hope to establish, operate, and succeed in setting up a business without adequate funding. It is important to estimate these expenses accurately and then to plan on how to raise the required capital. Often, first-time business owners fail to consider or greatly misjudge the amount of money needed to get their small business off the ground, and they fail to include a contingency amount to meet unforeseen expenses. Consequently, they fail to secure sufficient financing to carry their business through the period before it reaches its breakeven status and starts to make money.

To avoid being "undercapitalized," you will need to do adequate cost planning during your pre-launch phase. Most experts recommend that startup funding be adequate to cover operating expenses for six months to a year. At the very least you will need several months to find customers and get established. But to determine how much in financing to seek, you will need to develop detailed cost projections. Experts suggest a two-part process. First, develop an estimate of your one-time startup costs. Second, put together a projection of your overhead and operating expenses for at least the first six months of operation. Performing these two exercises will help to ensure that you put into place the necessary financial cushion to start and stay in business.

Startup Cost Estimates

Estimating the amount needed to start a new business requires a careful analysis of several factors. The first step is to put together a list of realistic expenses of one-time costs for opening your doors, including furniture, fixtures, and equipment needed. The list would also include the cost—down payment or cash price or, if purchased on an installment plan, the amount of each monthly or periodic payment. Record them in the costs table below.

Down payment $_____
Amount of each payment $_____

The furniture, fixtures, and equipment required may include such things as desks, moveable partitions, storage shelves, file cabinets, tables, safe, special lighting, and signs.

TYPICAL STARTUP COSTS ITEMS TO BE PAID ONLY ONCE:
Furniture, fixtures, and equipment:
Interior decorating $_____
Installation of fixtures and equipment $_____
Starting inventory $_____
Deposits with public utilities $_____
Legal and other professional fees $_____
Licenses and permits $_____
Advertising and opening promotion $_____
Advance on lease $_____
Other miscellaneous cash requirements $_____

TOTAL ESTIMATED CASH NEEDED TO START = $

Estimated monthly expenses:
$_____Salary of owner-manager
$_____Other salaries and wages
$_____Payroll taxes and expense
$_____Rent or lease
$_____Advertising
$_____Office supplies
$_____Telephone
$_____Other utilities
$_____Insurance
$_____Property taxes
$_____Interest expense
$_____Repairs and maintenance
$_____Legal and accounting
$_____Miscellaneous

TOTAL ESTIMATED MONTHLY EXPENSES =
$_____ Multiply by 4 (4 months) or 6 (six months)
$_____ Add: total cash needed to start from above

TOTAL ESTIMATED CASH NEEDED
$_____

Once the approximate amount of cash required to start the business is determined, you can then estimate how much money is actually available or can be made available to put into the business and to think about where the rest of the money needed to start the business will be coming from.

Employees and Required Forms

If you intend to hire yourself or others as a full- or part-time employee of the company, you may have to register with the appropriate state agencies or obtain workers' compensation insurance or unemployment insurance (or both).

When hiring employees, personnel files are needed for each person. At a minimum an I-9 form, an IRS form W-4, and the state equivalent form for employee income-tax withholding need to be included. Independent subcontractors should sign IRS form W-9. You may also require a copy of the contractor's workers' compensation insurance. Some state laws require subcontractors to be included on the firm's policy.

Utilities

Utilities are necessary overhead expenses. Advance deposits are usually required when you sign up for power, gas, water, sewer, and phone services. Upon deciding to establish your own business, place a display ad in the telephone book's Yellow Pages (or at least a listing).

Expense Report

It saves time to have a standardized expense report form for employees so that they can request reimbursement for their business expenses. Even if the contractor has just started doing business, it is imperative to monitor expenditure, and a standard form is perhaps the best way to do so. The expense report needs to be neatly typed and organized, identifying each location, project name and number, and applicable dates. Original receipts and supporting documentation should all be attached to the standard form. Accounting should then process and record it in a timely manner before filing.

Office Equipment and Furniture

All businesses need some equipment and furniture, although no two businesses will have identical needs. Sometimes it may be better to preserve cash for inventories or working capital and purchase good used fixtures and equipment at less expense.

Phone and Internet Service

Get a phone number and domain name (for your Internet website) for your new business. The domain name should be simple and easy to remember. This is discussed in greater detail in the a later section of this chapter.

Suppliers

Identify key suppliers by their names and addresses, type and amount of inventory furnished, credit and delivery policies, history and reliability. It may be prudent to have more than one supplier for critical items. Suppliers are often reluctant to ship their products to new businesses. This is one reason why you should get to know your banker and request credit references acceptable to most firms. You may have to pay your suppliers C.O.D. during the early phase of startup, so take this fact into account when preparing your financial planning and startup costs.

Bookkeeping and Accounting

Organize your accounting and record-keeping system and learn about the taxes the new company is responsible for paying. Company documents generally are required to be kept for 3 years, including a list of all owners and addresses, copies of all formation documents, financial statements, annual reports, and amendments or changes to the company. All tax and corporate Filings should be kept for at least three years.

There is probably no reason why you cannot do your own record keeping, at least in getting started. If your bookkeeping capabilities are limited, use a part-time accountant to set up the company's books on the basis of the above simple method. The accountant can keep the books for the first few months while you learn the general procedure. Let your accountant be your advisor. After a short time you will probably feel comfortable doing all the accounting without outside help.

Miscellaneous Issues to Consider

In the construction industry, equipment and machinery are often essential to your business. Without the proper equipment, it would be difficult to get jobs done efficiently and on time. The goal is typically to complete a project in accordance with plans and specifications, on time, within budget, and at the lowest possible cost. When starting a construction business, it is crucial to have the appropriate equipment to execute the types of construction projects anticipated. As in all businesses, it is important that the equipment pay for itself to earn a significant profit. In other words, the cost to own and operate the equipment should be less than what the contractor charges for its use.

You will also want to consider both business and personal living expenses when determining how much cash you will need. If you are leaving a salaried job to start your business, you should include in your expense projection an estimate of your and your family's living costs for the months it will take to build your business. Talk to family members about the minimum amount of money your household will need each month to function.

Once you add up startup costs to your six-month tally of recurring costs, the total may amaze you and spur you to reconsider or to look for ways to economize. It probably makes sense to review certain categories, such as equipment, office supplies, or advertising/promotions, with cost control in mind.

You've estimated your startup costs to the best of your ability, but chances are that you've never owned or operated a business before. It would probably be wise to discuss with an established or experienced business owner whether you've made the correct assumptions in projecting your costs.

The U.S. Department of Commerce Minority Business Development Agency (www.mbda.gov) has articles that discuss how much money will be needed to start a business, including helpful checklists and referrals to other resources of information. The U.S. Small Business Administration (www.sba.gov) was created specifically to assist and counsel small businesses.

12.4 BUSINESS FORMS, TAXES, LICENSES, PERMITS, AND INSURANCE

As a business owner you are obliged to understand and comply with government laws and regulations that apply to your business and are designed to protect you, your customer, and any employees. Having taken the decision to start a new business, you may now need to obtain a number of licenses and permits from federal, state, and local governments. Since licensing and permit requirements for small businesses can vary from one jurisdiction to another, it is important that you contact your state and local government to determine the specific obligations of your new business. For example, in California you would need to take the contracting license exam before you can become licensed.

Name and Legal Structure

At this point, a decision must be made on the proposed name and legal structure of the new business. There are advantages and disadvantages of each type of business structure. The structure chosen will depend on the needs of your business and will have a fundamental effect on how you do business. You basically have four choices when selecting a legal structure: sole proprietorship, partnership, limited-liability company (LLC), and corporation or S-corporation.

An individual proprietorship is a business owned and operated by one person. It is the simplest form of business organization and the form of entity frequently used by small businesses at start-up. If you need additional capital or expertise, a partnership may be the best entity. You can always incorporate later if practical. The expense to incorporate a small business is nominal, but unless incorporating increases your chances of success or better protects your investment, there is not much benefit in forming a corporation. Even though there is limited liability as to your personal assets when obtaining outside financing or in the event funds are misused, you can still be personally liable.

Before deciding consult with an attorney and check with your Secretary of State—almost all states now are online—look for the corporations division to find the form for either Articles of Incorporation (to form your own corporation) or Articles of Organization (to form a Limited Liability Company). Many entrepreneurs today prefer forming a LLC to incorporation, partly because it requires less paperwork to maintain. While waiting for the Secretary of State to send the charter, you should contact each of the city, county, and state tax departments.

A limited-liability company is a new business entity that every entrepreneur should understand. It combines the best aspects of incorporation with the tax advantages of partnership without the red tape of either. This combination of benefits has never existed before in such a simple and effective way. Anyone starting a new business must separate his or her personal assets from his or her business ventures. A limited-liability company contains personal protection in its purest form.

Licenses and Permits

These refer to the various federal, state, and local licenses and permits you may need to acquire prior to opening for business.

A basic business-operation license is a license granting the company the authority to do business in that city or county and should be issued from the city in which your business will operate or from the local county (if the business will be operated outside of any city's limits). Most cities or counties require you to obtain a business license, even if you operate a home-based business.

A federal employer-identification number (EIN), also called a tax-identification number, is required for almost all types of businesses. Note: your business may also need to acquire a similar tax-identification number from your state's department of revenue or taxation. When you get your corporate charter (if a corporation), you must apply for a federal Employer Identification Number (EIN). To do so, first go to the IRS website, download Form SS-4, and fill it out. Then call toll-free (866) 816-2065 to get your EIN in 15 minutes or less.

Once you get the EIN number, download Form 2553 (S-election, if you want to avoid double taxation on your company earnings) and fill it out. Be sure to sign as a shareholder in the middle of the page and as an officer at the bottom. If you choose to form a LLC, you will also need to decide how you want to be taxed (as a sole proprietorship, partnership, S-corporation, or C-corporation) and make that election on IRS Form 8832.

A fictitious business name permit (also called "dba" or "doing business as" permit) is required for almost all types of businesses. To make the public better aware of just what your firm offers, it is generally good practice to choose a business name that describes your product or service. Apply for a fictitious business name with your state or county office when you plan on going into business under a name other than your own. The bank will require a certificate or resolution pertaining to your fictitious name at the time you apply for a bank account.

Persons entering into certain kinds of business will have to obtain an occupational license through the state or local licensing agency. These businesses include building contractors, real-estate brokers, those in the engineering profession, electricians, plumbers, insurance agents, and many others. Often they have to pass state examinations before they can get these permits and conduct business. Contact your state government offices to get a complete list of occupations that require licensing.

If you are planning to run your business from home, you should carefully investigate zoning ordinances. Residential neighborhoods tend to have strict zoning regulations preventing business use of the home. Even so, it's possible to get a variance or conditional-use permit; and in many areas attitudes toward home-based businesses are becoming more supportive, making it easier to obtain a variance.

Federal regulations control many kinds of interstate activities with license and permit requirements. In most cases, you won't have to worry about this. However, a few types of businesses do require federal licensing, including investment advisory services.

Tax Strategies

Tax planning is a year-round event if you want to minimize your business tax bill. Whether it's surviving an audit, capitalizing on business deductions, or finding tax-friendly ways to run your business, a good accountant can help reduce tax obligations and make paying taxes less anxiety-provoking.

Once you have the federal EIN number, you will need to obtain ID numbers for state sales- and income-tax withholding and for the state unemployment tax. Complete the forms for the tax registrations that you obtained; add the federal EIN number and you can now mail in these state forms. Having the federal EIN number in hand also allows you to open a corporate checking account.

If you are a sole proprietorship or partnership, you will have to file and pay federal estimated tax reports each quarter based on estimated annual income. Partnerships file an annual information return, and each partner's share of profits is included in their individual personal income-tax return. Corporations must also file for estimated taxes. Maximize what you can deduct and discover what you can write off by knowing what constitute legitimate business expenses.

Insurance

Insurance is an important prerequisite of any business. The premiums are usually expensive, especially business liability, but you cannot operate with peace of mind without full coverage. There are many types of insurance for businesses, but they are usually packaged as "general business insurance" or "business owner's policy." If you plan to offer your employees health insurance, talk to your agent about the up-front fee. Record the premium payment you will need to make before opening your business.

Bank Account

With the federal EIN number in hand, you can now open a corporate checking account. Get to know the manager of your bank. He will be one of your best references. Ask his or her advice on financial matters. The more he or she advises you, the better he or she will come to know you. Develop a line of credit so it will be there when you need it. The banks can't exist without making loans, so don't hesitate to apply. Get set up with American Express, Visa, or any other credit cards you will accept for payment and use for expenses. It will cost you from 3 to 5 percent, but it is necessary for attracting customers and entertaining.

To establish your bank account, you will need a Federal ID number or Social Security number along with your certificate of assumed (fictitious) business name. If you are incorporated, the bank will want a copy of the minutes and a corporate resolution authorizing the account. Contact the bank prior to opening the account to see what their specific requirements are for opening a business checking account.

12.5 CREATING A PROFESSIONAL IMAGE

Owning your own business is one of the better ways to gain wealth and personal satisfaction, provided you know what is required. Starting a business is risky, but your chances for making good increase if you understand the challenges you will face and work out as many of them as possible before commencement.

Develop a Business Identity

It is imperative to create a good corporate image and business identity that reflect confidence and efficiency. One of the first things to organize is a professionally designed logo, business card, letterhead, and promotional material for the business. A professionally created logo and letterhead can go a long way to giving clients the desired image.

Advertising and Promotion

Many new businesses start operating with a grand opening announcement, giving press releases to local papers and relevant business publications. It may also be prudent to print out a few hundred circulars to distribute to potential clients. Alternatively, grand-opening circulars can be placed in the newspaper to be distributed to subscribers. In any case, the dollar cost of any planned advertising and marketing initiative to announce the launch of the new business should be recorded and should include the cost of flyers, sales letters, phone calls, signs, brochures, and other promotional items.

Mr. John Doe   April 24, 2008
Real Estate Department,
First Union Bank,
1070 Chain Ridge Road,
McLain, VA 22101

(Tel)   (703) 760 0000
(Fax)   (703) 760 0001
Email   jdoe@abc-contractors.com

Re: Building Construction Services

Dear Mr. Doe:

I am taking this opportunity to apprise you of our contracting and project management
services that we offer for lending institutions.

ABC Contractors has recently been formed to provide general contracting and sub-
contracting services. Although ABC Contractors is a newly formed company, its
principals have over twenty five years of experience in construction and project
management. We are licensed in the states of Virginia, Maryland and Washington
D.C. For further information and an overview of our services, please visit our website
at: www.abc-construction.com.

We would be delighted to give you a free estimate for any proposed project including
renovation. We are able to travel anywhere within the United States to provide services to meet
your requirements. My direct telephone no. is: 703 852 4391.

Sincerely,

Sam Kubba, AIA, R.A., Ph.D.
Principal
ABC CONTRACTORS

SAK/bs
enclosures

cc: General Files

ABC/PROMO/LETTERS/First Union/ promo.doc

Figure 12.1   An example of a typical promotional letter that can be sent to clients to inform them that your
company is ready for business. Promotional material should accompany the letter.

Marketing Yourself

Develop professionally designed brochures and other marketing materials. As you are essentially selling a special service, be sure to know how to market it. A marketing plan and strategy are necessary for services that target your ideal customer. Now that you've set up the company for success, you need to get the word out. Figure 12.1 shows a typical marketing letter to let customers and potential clients know you are ready for business.

Time Management

A secretary/office manager could help make the operational aspect of the business as efficient as possible to allow you to concentrate on managing and growing the business. This will free you from having to process orders, pay bills, pay employees, pay taxes, maintain your permits, etc. The more organized you are, the more efficient you are.

12.6 SELLING YOURSELF

Professional performance means much more than doing your job effectively. The way you conduct yourself in a business environment not only reflects your position within a company but impacts your chances for career growth.

Many people are unsure of the proper protocol for various situations in a business setting.

Most potential clients can decide within minutes whether to trust you or your employees. Trust is pivotal in establishing relationships that can ultimately lead to business partnerships.

Awkward introductions, weak handshakes, poor communication, ineffective meetings, and lack of consideration can negatively affect your career and business relationships. When all else is equal, good manners can be your greatest strength.

Correspondence

Correspondence is a tricky subject since there are numerous ways to communicate today. The decision to communicate with a client via telephone, email, or face-to-face depends largely on one's personality. Introverts tend to prefer email because it is efficient and avoids direct contact; extroverts on the other hand prefer direct communication.

Confidence

If you are confident, you will appear as an accomplished professional to others. Confidence is attained by being organized and prepared and having the necessary knowledge to execute the job successfully. Being confident doesn't mean never asking questions or succumbing to challenges. Paradoxically, confident people know the importance of questions. Remember to be cool, calm, and collected, and, above all else, think before you speak. Whether you are entering a new work environment or attempting to reinvent yourself professionally, know that you can do it. The corporate world can be rough, but with the aforementioned skills you are already on a path for success.

Meetings

Meetings are an area where success is largely dependent upon good organization and adequate preparation. You will often meet people who are busy and usually inaccessible to you. Compile questions you want to ask or topics you want to cover and know in advance what you hope to accomplish.

Business meetings are one arena in which poor etiquette can have negative effects. By improving your business-meeting etiquette you automatically improve your chances of success. Comfort, trust, attentiveness, and clear communication are examples of the positive results of demonstrating good etiquette.

Informal meetings are generally more relaxed affairs and may not necessarily take place in the office or meeting room. Even so, a sense of professionalism and good business etiquette are still required. The business etiquette of formal meetings such as departmental meetings, management meetings, board meetings, negotiations, and the like can be perplexing. Such meetings usually have a set format. A summary of business etiquette guidelines for formal meeting include:

- Go prepared to the meeting as your contribution may be fundamental to the proceedings. If you are using statistics, reports or other information, hand out copies prior to the meeting with ample time to be studied.
- Mobile phones should be switched off during the meeting.
- Arrive on time and appropriately dressed as this reflects professionalism.
- Where an established seating pattern exists, accept it. If you are unsure, ask.
- It is courteous to begin introductions or opening remarks with a brief recognition of the chair and other participants.
- When discussions are under way it is good business etiquette to allow more senior participants to contribute first.
- Interrupting a speaker reflects bad manners - even if you strongly disagree. Note what has been said and return to it later with the chair's permission.
- When speaking, be brief and to the point and ensure that what you say is relevant.
- It is a breach of business etiquette to divulge information to others about a meeting. What has been discussed should be considered as confidential.

12.7 IDENTIFY AND TRACK SOURCES FOR LEADS

There are many methods to identify and track potential sources and project leads, depending on whether your business is a one-person organization or a well-organized firm with several employees. These methods include:

1. Send out flyers, brochures, emails, etc., to potential clients as an excellent start point.
2. Scan the Internet. Typically many of today's contractors and subcontractors have websites, and some of these firms have client lists (to build up potential customer confidence) on their websites. These lists can be researched to see which if any names are worth following up on.
3. Visit the various neighborhood commercial-real-estate agents to see which commercial properties are on the market. A list of all these possible leads should be made and followed up with letters and brochures offering the company's services.

4. Since many of the clients will be property owners, developers/investors, and lenders (e.g., banks, lending institutions), it would be prudent to make a list from the Yellow Pages, internet, and research in the public library.

12.8 BIDS, CONTRACTS, AND PAYMENTS

Construction jobs are usually awarded on the basis of a bid or by negotiation. The contractor will essentially estimate the cost to execute the project and then add a certain percentage for profit and contingencies. But in the final analysis, your rates should be set by the logic of most businesses—i.e., it should reflect what the market will bear.

In any case, ensure that you read the contract thoroughly. Many contractors and subcontractors often sign prepared contracts without adequately reading them or having an attorney review them. Many contractors are uncomfortable in dealing with the paperwork associated with the job. But, unfortunately, contracting is about contracts, and contracts include paperwork! As much as 50 percent of all profits made or lost on construction projects can be a result of managing the contract properly.

Assembling all the required documentation, change-order requests, notices, and information required by your contract may seem overwhelming. But once you get in the habit of following the contract, it will become a normal part of your construction methodology. In order to receive all you deserve while building a project for your customer, you must be timely in your requests. Missing the notice time requirements may result in a loss of your right to collect for items out of your scope of work or control. It would be wise to take a little extra time to be complete in your description of the event. Issues in question should be supported by photographs when necessary.

By not documenting conflicts or changes in a timely and complete manner, contractors inadvertently shift more responsibility onto their own shoulders, which may cost them their right to collect.

It is customary to visit the site beforehand (at no cost to the client) to obtain the information needed for preparing and submitting an estimate. This in turn helps to avoid surprises down the road, since the client already knows what to expect up front. When additional work is required, a separate invoice should be presented (Figure 12.2). Do not hesitate to get rates from other builders and compare them with your own, bearing in mind the supply and demand in your area.

12.9 THE INTERNET, THE WEBSITE, AND FORMING AN ENTITY

The Internet has created enormous opportunities to reach previously unimaginable numbers of people. In addition to making communication possible to more people than through print advertising and other media, it also makes it available to people who might not otherwise have access. A website for a small business or any size company, has become a high priority; not only is it a great marketing tool, but it allows you to develop your services and launch multiple marketing campaigns in a short period of time. Consider your website a platform to feature your services to your customers and the world.

But using the Internet is more than just creating a website and waiting for potential clients to find it. The website should be only one part of an overall Internet marketing strategy. And even before setting up a website, it is imperative to have email service in place.

BILLING FORM
ADDITIONAL WORK PERFORMED

EMPLOYEE NAME:

DATE:

PROJECT NO.

JOB NAME:

FOR PERIOD UP TO AND INCLUDING: _____

DESCRIPTION OF WORK PERFORMED, INCLUDING WHO WAS INVOLVED,
DATES, ETC.

HOURS: _____

EXPENSES: _____

APPROVED: SIGNATURE:

Figure 12.2 Typical form for additional work done.

Email Service

Email service is a high priority for communication and for sending promotional material. A growing number of clients consider email availability vital and find it burdensome and inefficient to communicate by posted mail. An email address should be professional and simple. Once a website is established, the email should reflect the domain name of the site.

The Website

The public increasingly expects a professional to have a website, just as he or she is expected to have a business card. But first it is important to understand why a website is necessary and what can be achieved from it. At a minimum, a website can market your services to a global audience. Additionally, it is an excellent vehicle to sell the company's services over the Internet. Remember to consider what information you want prospects to gather from visiting your website. A well designed website can typically be used to:

- Attract enquiries from potential customers
- Provide better service to customers
- Provide more information about the firm and its services
- Obtain feedback from customers on the company's services
- Recruit staff
- Improve efficiency

Using your website to provide a user with better access to your company can reap great benefits. Clients are happier and receive resolutions to problems quickly, and you can devote more of your time to other critical issues. The list of services you can offer via your website is enormous.

For example, it can include project facts and figures, including projects recently executed, company experience, etc. It can also include clarification regarding your firm's structure—who does what. The site can be used to draw attention to upcoming events or time-sensitive information.

Plan Your Approach

The Internet and your website are just another link to your target audience. You have already been implementing all the necessary steps you need to in your day-to-day business. You now need to translate this to the new medium. Think about the image you want to convey to site visitors, and make sure everything on your site contributes something toward that image. Developing and maintaining a website is no small accomplishment.

But whether you are creating a concept for your website for the very first time or trying to update a current site, make sure you look at your site from the user's point of view. For example, what will site visitors want to know when they log on? If you're working with an existing site, ask clients and prospects what they think of the information offered there. What else would they like to see featured on the site? Know your visitors (or future visitors) and what they want and need. If you're starting from scratch, a quick survey may help determine the answers to these questions. Check out other builder/contracting websites to discover additional services that your peers are providing to users.

For each of the objectives you have set for the company website and yourself, you need to decide:

- Who is the target audience?
- What are you trying to get them to do or obtain from your site?
- What do you need to have on your site to attract prospects to the site in the first place and to come back again? Website content and design are both vital for market success.
- What services do you need to provide on- or off-line to back up your "promise" to your visitors?
- How are you going to promote your website and contents to your target audience?

Setting Up the Website

To set up an acceptable commercial website (as opposed to a personal home page) you need the following:

- A domain name, such as www.mysite.com
- Web space—a home for your website's files, provided by a hosting company
- The website itself—a collection of pages and images, linked together to make a complete site

First, a web address or domain name is needed. There are scores of companies such as www.register.com and www.dotster.com that let you check that the name desired is available and, if so, register the domain name online. Domain registration is not expensive, and often the domain name will be registered with the same hosting company that provides the web space. Most companies will "park" your name until your website is ready.

Web space is space on a computer owned by a hosting company. It is set up so that anyone who types your domain name into their browser will be connected to your site. There are numerous hosting companies, but some research is needed on the Internet to find one that best serves your needs.

Anyone can put a website together, and it seems that just about everyone does, with greater or lesser success. If you can use a word processor, you can create a website. However, creating a good website is an entirely different matter and requires a lot of thinking and details.

Website Components and Details

The website should articulate the services offered. This may be outlined in a mission or vision statement. When you are starting out on the Internet, you are starting with something close to a clean sheet. The majority of Internet users probably won't actually know who you are. You can project any kind of image that you wish, and, moreover, you can emphasize any particular aspect of the organization that you wish to.

Sit down with some colleagues and some paper and brainstorm until you come up with a series of points that match your organization and what you do or wish to do. Next, you need to match these points to the kind of image you want to portray on your site. Consequently, when it comes time to actually put the site together, you can ignore the flashy graphics and animations and concentrate on guiding the user quickly and effectively to the information that they need. Use your site to display your particular abilities to their best advantage; in other words, play to your strengths.

Corporate image comes into play here as well. You will probably want to make sure that the image on the web pages matches that of the image displayed in other formats and media. However, don't forget that you are starting from scratch, so you have a free hand. It might actually make sense to have some subtle changes on the website; if your corporate color is blue, think about changing to another shade or be radical and go for red! Decide on a theme and stick to it. Use the same logo on your pages, in the same position, in the same size.

Today, it is easier than ever to create a website; you can even do it overnight. With a wealth of tools and options available on the web today, you don't need to be a graphic designer or web developer to capture the basics. Templates are a great way to get started in developing your own website, and you can choose from a variety of simple but attractive designs. There's always room to upgrade, and it's easier than ever to find a unique template so you're not simply imitating another site. No coding is involved—you don't even need to learn HTML.

It is illegal to copy another company's web page, but tracking down a free template is simple: just download it from the internet, and you'll have the framework ready to go. Once you've found the right match, take a look at your options for upgrading and customizing it. Not everyone needs a customized website; if you're looking for standard functions and presentation, you'll find plenty of attractive options on the Internet. After installing the templates using your favorite website builder, you can immediately publish to the web. Figure 12.3 shows an example of a template design that has been modified and that you can find on the Internet. These websites offer free web templates as well as design services. The button links may be on the top or on the left side of the page. Both easily take you to where you want to go (e.g., to the mission statement, jobs in progress, services offered, etc.). It is largely a matter of preference as to how the page is composed and designed. However, for a professional-looking website it is strongly recommended to use a professional for its design.

Who do you want to visit your site? This depends very much on the answers that you have already come up with. Existing customers and clients, potential customers, people interested in your subject area who may never have heard of you, organizations, individuals, groups, and so on. If you are a building firm, your groups might look a little like this:

Figure 12.3   Example of a sample home-page template, which can be freely downloaded from the Internet from sites such as *www.freewebsitetemplates.com* or *www.steves-templates.com*. This template has been modified to suit a contracting business (based on a *www.freewebsitetemplates.com* template).

- Property owners and facility managers
- Lenders
- Investors
- Architects and engineers

Create your list and then prioritize it. For example, if you want to promote an image of a contracting firm that specializes in electrical work, property owners may well take higher priority than casual users. This should be emphasized in the structure of the site's pages and the weighting that you give to it on the home page.

To let users know you are to be trusted, remember what was mentioned earlier—you cannot take it on trust that your viewer knows who you are. The message must be hammered home at every opportunity. Use your company logo, university crest, and so on to reinforce this. If you have won an award, make sure everyone who visits knows this! Your viewers need to be reassured that your firm is to be trusted and that its information is reliable. This must be emphasized on every single page that you publish—there is no telling which page a user will go to first, so be consistent in the positioning of your logo and place it on every page.

You've got less than 15 seconds to make an impact before your visitors leave. You have to work very hard in this small window of opportunity. Make it quite clear what your message is. If you don't know, your viewers won't. Design is vital. Give them enough guidance to let them work their own way(s) through your site. There has to be a reason for people to come and look at your website and a reason for them to keep coming back. You need to offer something of value such as a detailed list of your services with current prices. However, if you can think of some valuable free service (such as free estimates) that you can offer to bring people to your website, they are likely to browse around once they are there.

A successful website is not necessarily an attractive one or one full of the latest web technology. It is not even usually dependent upon how many people visit it. Rather, it is how many come back time and time again and how much business it generates.

Your site is up and you are waiting for the inquiries to come in, but they don't, because no one knows your site is out there. The site must be publicized. Potential customers need to know the website exists. All letterheads, brochures, cards, and advertising should mention the website address and email address.

Submit your site to all the search engines. Generally speaking, it takes at least a few months for a website to become recognized and start to generate responses. That's how long it usually takes the big search engines, especially Google, to index a new site. Also make sure that any expert directories you are listed in are linked to the site. Of course, the popularity of your site and the speed at which it becomes popular really depend on what is being offered and how it is promoted.

Appendix 1
Acronyms/Abbreviations

TERM	ACRONYM	Above Finished Grade	AF
Area	A	Silver	AG
Ampere	A, AMP	Aggregate	AGG, AGGR
Anchor Bolt	AB	Air-Handling Unit	AH, AHU
Aggregate Base Course	ABC	American Institute of Architects	AIA
Above	ABV	Aluminum	AL, ALUM
Air Conditioning	A/C, AC	Alternate	ALT
Alternate Current, Acoustical	AC	Amount	AMT
Access	ACC	Anchor, Anchorage	ANC
Air-cooled Condensing Unit	ACCU	Anodized	ANOD
Access Floor	ACFL	Access Panel	AP
Asbestos Containing Material	ACM	Approved	APPD
Acoustical Plaster	ACPL	Approximate	APPROX,APX
Acrylic Plastic	ACR	Apartment	APT
Acoustical Ceiling Tile	ACT	Architect (ural)	ARCH
Actual Age	ACT AGE	Asbestos	ASB
Area Drain	AD	Above Suspended Ceiling	ASC
ADA Architectural Guidelines	ADAAG	American Society of Mechanical	
Automatic Door Opener	ADO	Engineers	ASME
American Disabilities Act	ADA	Asphalt	ASPH
Addition, Addendum	ADD	Asphalt Concrete	ASPH CONC
Adhesive	ADH	American Society for Testing	
Adjacent, Adjoining	ADJ	and Materials	ASTM
Adjustable	ADJT	Ampere Trip	AT
Administration	ADMIN	Asphalt Tile	AT
Architect/Engineer	A/E	Attention	ATTN
Ampere Frame	AF	Automatic	AUTO
Above Finished Floor Level	AFF	Auxiliary	AUX

Air Vent	AV	Bedroom, Bearing	BR
Average	AVG	Bearing	BRG
American Wire Gauge	AWG	Brick	BRK
		Bracket	BRKT
Bathroom	BA	Broom Closet	BRM, B CL
Balcony	BALC	Bronze	BRZ
Basement	BASM	Both Sides	BS
Batten	BATT	Basement	BSMT
Baseboard	BB	British Thermal Unit	BTU
Bulletin Board	BBD	Built-up Roofing	BUR
Bare Copper	BC	Busway	BUS
Building Code	BC	Both Ways	BW
Board	BD		
Board Foot (Feet)	BD FT, BF	Celsius	C
Below	BEL	Channel	C
Between	BET	Compressed Air	CA
Beveled	BEV, BVL	Cabinet	CAB
Both Faces	BF	Cadmium	CAD
Bottom of Footing	BF	Caulking	CALK
Below Finish Floor	BFF	Cantilever	CANTIL
Boiler Feed Pump	BFP	Cavity	CAV
Bituminous	BIT, BITUM	Catch Basin	CB
Bed Joint	BJT	Circuit Breaker	CB, C/B
Base Line	BL	Corner Bead	CB
Building Line	BL	Center to Center, Cubic Centimeter	C-C, CC
Building	BLDG	Cast Concrete	C CONC
Black	BLK	Ceiling Diffuser	CD
Block	BLK	Construction	CD
Blocking	BLKG	Contract Documents	CD
Ballast	BLST	Chilled Drinking water	CDW
Built-In	BLT-IN	Cement	CEM
Beam	BM	Cement Finish	CEM FIN
Benchmark	BM	Cement Plaster	CEM PLAS
Bending Moment	BM	Ceramic	CER
Board Measure	BM, B/M	Certify	CERT
Building Officials and Code		Counterflashng	CFL, CFLG
Administrators	BOCA	Cubic Feet Per Minute	CFM
Bottom Of Duct	BOD	Cubic Feet Per Second	CFS
Bottom Of Pipe	BOP	Chemical Feed Unit	CFU
Bottom	BOT, BTM	Curb and Gutter	C&G
Building Paper	BP	Corner Guard	CG
Back Plaster (ed)	BP	Chiller	CH
Base Plate, Bearing Plate	BPL	Chamfer	CHAM

Chalkboard	CHBD	Continue, Continuous	CONT
Check	CHK	Contract (or)	CONTR
Chilled Water Supply	CHS	Corrugated	CORR,CORRUG
Ceiling Height	CHT	Carpet and Pad	C&P
Cast Iron	CI	Concrete Pipe	CP
Curb Inlet	CI	Control Panel	CP
Cinder Block	CIN BLK	Copper	CPR
Cast Iron Pipe	CIP	Cycles Per Second	CPS
Cast-In-Place	CIP	Carpet (ed)	CPT
Concrete-In-Place	CIP	Chromium (Plated)	CR
Circle	CIR	Cross Grain	CRG
Circuit (Electrical)	CIR, CKT	Course (s)	CRS
Circuit Breaker	CIR BKR	Cast Stone	CS
Circumference	CIRC	Control Switch	CS
Ceiling Joist	CJ	Countersink	CS
Control Joint	CJ, CJT	Casing	CSG
Calk (ing), Caulk (ing)	CK	Construction Standards Institute	CSI
Cladding	CLDG	Counter Sunk	CSK
Ceiling	CLG, CLNG	Casement	CSMT
Ceiling Diffuser	CLG DIFF	Cast Stone	CST
Ceiling Grille	CLG GRL	Casework	CSWK
Ceiling Height	CLG HT	Ceramic Tile	CT
Contract Limit Line	CLL	Coated	CTD
Clear	CLR	Ceramic Tile Floor	CTF
Closure	CLS	Center	CTR, CNTR
Centimeter (s)	CM	Contour	CTR
Ceramic Mosaic Tile	CMT	Counter	CTR
Concrete Masonry Unit	CMU	Control	CTRL
Conduit	CND	Copper	CU
Counter	CNTR	Cubic	CU
Carbon Monoxide	CO	Cubic Foot (Feet)	CU FT (ft3), CFT
Certificate of Occupancy	CO	Cubic Inch(es)	CU IN (in3)
Cleanout (Plumbing)	CO	Cubic Yard(s)	CU YD (yd3), CYD
Column	COL	Check Valve	CV, CKV
Combination	COMB	Casement Window	CW
Composition	COMP	Clockwise	CW
Compress (ed), (ion), (ible)	COMP	Cold Water	CW
Compartment	COMPT	Cold Water Riser	CWR
Composition, (Composite)	COMPO	Connection	CX
Condenser	COND	Cross Bracing	XBRA
Concrete	CONC	Dryer	D
Construction	CONST	Drain	D

Double Acting	DA	Door Frame	DR FR
Door Air Curtain	DAC	Disconnect Switch	DS
Decibel	DB, dB	Downspout	DS
Double	DBL	Design	DSGN
Double Glaze	DBL GLZ	Drain Tile	DT
Direct Current	DC	Dovetail Anchor	DTA
Degree	o , DEG	Detail	DTL
Demolish, Demolition	DEM, DEMO	Dovetail	DVTL
Depressed	DEP	Dishwasher	DW, D/W
Department	DEPT	Drinking Water	DW
Detail	DET	Drywall	DW
Diesel Fuel	DF	Dumbwaiter	DW
Drinking Fountain	DF	Drawing	DWG
Double Hung (Door, Window)	DH	Dowel	DWL
Domestic Hot Water	DHW	Drawer	DWR
Diameter	DIA, DIAM	Duplex Outlet	DX OUT
Diagonal	DIAG	East	E
Diagram	DIAG	Modulus of Elasticity	E
Diameter	DIAM	Each	EA
Diffuser	DIF, DIFF	Expansion Bolt	EB
Dimension	DIM	Edge of Curb	EC
Dimmer	DIM	Empty Conduit	EC
Ductile iron pipe	DIP	Energy Efficiency Ratio	EER
Directive	DIR	Each Face	EF
District	DIS	Exhaust Fan	EF
Divide, Division	DIV	Effective Age	EFF AGE
Double Joist	DJ	Exterior Finish System	EFS
Decking	DK	Equipment Ground	EG
Dead Load	DL	Electric Heater	EH
Damp Proofing	DMPF	Exterior Insulation & Finish System	EIFS
Demountable	DMT	Easement Line	EL
Down	DN	Elbow	EL
Ditto (Same As)	DO	Elevation	EL, ELEV
Dozen	DOZ	Elbow	ELB
Deep or Depth	DP	Electric (al)	ELEC
Dampproofing	DP	Elevator	ELEV
Damper	DPR	Emergency	EMER
Dispenser	DPR	Energy Management System	EMS
Demountable Partitions	DPTN	Enamel	ENAM
Dining Room	DR	Enclose (ure)	ENC, ENCL
Door	DR	Entrance	ENT, ENTR
Drainboard	DRB	Edge of Slab	EOS

Electric Panel (Panel Board)	EP	Fan Coil Unit	FCU
Environmental Protection Agency	EPA	Floor Drain	FD
Equal	EQ	Fire Extinguisher	FE
Equipment	EQP, EQUIP	Fire Extinguisher Cabinet	FEC
Escalator	ESC	Factory Finish	FF
Easement	ESMT	Face to Face	F/F
Estimate	EST	Furniture (fixtures) & Equipment	FF&E
Et Cetera, and so forth	ETC	Finished Floor Elevation	FFE
Expected Useful Life	EUL	Federal Fair Housing Act	FFHA
Evacuate	EVAC	Finished Floor Line	FFL
Evaporate	EVAP	Fuel Gas	FG
Each Way	EW	Fiberglass	FGL
Electric Water Cooler	EWC	Fair Housing Act	FHA
Electric Water Heater	EWH	Fire Hose Cabinet	FHC
Example	EX	Flathead Machine Screw	FHMS
Excavate, Excavation	EXCAV, EXC	Fire Hose Station	FHS
Existing	EXG, EXIST	Flathead Wood Screw	FHWS
Exhaust	EXH	Figure	FIG
Exhibit	EXH	Finish(ed)	FIN
Expanded Metal Plate	EXMP	Finish Floor	FIN FLR
Expansion	EXP	Finish Grade	FIN GRD
Exposed	EXP	Fixture	FIX
Expansion Joint	EXP JT, EJ	Flush Joint	FJT
Exterior, External	EXT	Flashing	FL, FLASH, FLG
Extinguisher	EXT	Flammable	FLAM
Exterior Grade	EXT GR	Floor (ing)	FL, FLR
Exit Light	EXT LT	Floor Cleanout	FLCO
Extension	EXTN	Folding	FLDG
Fahrenheit	F	Flooring	FLG
Fire Alarm	FA	Flange	FLG
Fabric	FAB	Fluorescent	FLUOR, FLUR
Fabricate	FAB	Flexible	FLX
Fascia	FAS	Fence	FN
Fasten, Fastener	FAS	Free-On-Board, Factory-On-Board	FOB
Fire Alarm Station	FAS	Face of Concrete	FOC
Forced Air Unit	FAU	Face of Curb	FOC
Fiberboard	FBD	Face of Finish	FOF
Foot Board Measure	FBM	Freedom of Information Letter	FOIL
Furnished By Others	FBO	Face of Masonry	FOM
Fire Brick	FBRK	Face of Slab	FOS
Foot Candle	FC	Flush on Slab	FOS
Facing Brick	FC BRK	Face of Stud	FOS

Foundation	FOUND, FDN	Ground Fault Circuit Interrupter	GFCI
Face of Wall	FOW	Glass-Fiber-Reinforced Concrete	GFRC
Fireplace	FP	Galvanized Iron	GI
Freezing Point	FP	Girder	GIRD
Fireproof	FP	Gasket	GKT
Fireplace	FPL	Glass, Glazing	GL
Floor Plate	FPL	Ground Level	GL
Feet Per Minute	FPM	Glass Block	GLB, GL BLK
Frame (d), (ing)	FR	Glass Fiber	GLF
Fresh Air	FRA	Glue Laminated	GLU-LAM
Fire-resisting Coating	FRC	Glazing	GLZ
Frequency	FREQ	Ground	GND, G
Forged	FRG	Galvanized Pipe	GP
Fire-retardant	FRT	Gypsum Plaster Ceiling	GPC
Federal Specification	FS	Gypsum Dry Wall	GPDW
Full Scale, Full Size	FS	Gypsum Lath	GPL
Flame Safeguard Control Panel	FSCP	Gypsum Plaster	GPPL
Fused Safety Switch	FSS	Gypsum Tile	GPT
Foot or Feet	FT	Grille	GR
Fire Treated	FT	Grade	GRD, GD, G
Footing	FTG	Ground Floor	GR FL, GF
Foot Pound(s)	FT LB	Granite	GRN
Furred (ing)	FUR	Grout	GRT
Furnishing	FURN	Grating	GRTG
Furnace	FURN	Galvanized Steel Sheet	GSS
Future	FUT	Glazed Structural Tile	GST
Flush Valve	FV	Glazed Structural Unit	GSU
Forward	FWD	Grout	GT
Frostproof Wall Hydrant	FWH	Ground Terminal Box	GTB
Fixed Glass	FX GL, FX	Gutter	GUT
Gas	G	Gravel	GVL
Gram	G	Gas Fired Water Heater	GWH
Gage, Gauge	GA	Gypsum	GYP
Gallon	GAL	Gypsum Board	GYP BD
Galvanize(d)	GALV, GV	Hazardous Materials	HAZMAT
Garage	GAR	Hose Bibb	HB
Grab Bar	GB	Hardboard	HBD
General Contractor	GC	Heating Coil	HC
Glazed Concrete Masonry Units	GCMU	Hollow Core	HC
Ground (Earth/Electric)	GD	Handicap	HCP
Generator	GEN	Heavy Duty	HD
Ground Face	GF	Hardboard	HDB

Header	HDR	Insulate (d), (ion)	INSUL
Hardware	HDW	Interior	INT
Hardwood	HDWD	Intermediate	INTM
Heat Exchanger	HE	Invert	INV
Hemlock	HEM	Iron Pipe	IP
Handhole	HH	Iron Pipe Size	IPS
Head Joint	HJT	International Standards Organization	ISO
Hook (s)	HK	Indirect Vent	IV
Hollow Metal	HM		
Humidity	HMD	Joist	J, JSTS
Horizontal	HOR, HORIZ	Junction Box	JB
Horsepower	HP	Janitor's Closet	JC
Heat-recovery Unit	HRU	Junction	JCT
Height	HT	Junction Box	J-Box
Heating	HTG	Joint Filler	JF
Heater	HTR	Jamb	JMB
Heating/Ventilating/Air Conditioning	HVAC	Joist	JST
Hot and Cold Water	H&CW	Joint	JT
Hot Water	HW	Kiln Dried	KD
Hardwood	HWD	Knocked Down	KD
Hot Water Heater	HWH	Kilogram	KG
Hot Water Tank	HWT	Kitchen	KIT
Hexagonal	HX	Kiloliter	KL
Hydrant	HYD	Kilometer	KM
Hertz	HZ	Knockout	KO
		Kick Plate	KPL
Iron	I	Kilovolt-Ampere	KVA
Indoor Air Quality	IAQ	Kilowatt	KW
I Beam	IB	Kilowatt-Hour	KWH
Inside Diameter	ID		
Interior Design	ID	Left	L
Inside Face	IF	Length	L, LGTH
Interlock	ILK	Line	L
Illuminate	ILLUM	Laboratory	LAB
Inch(es)	IN	Ladder	LAD
Incinerator	INCIN	Laminate (d)	LAM
Include (d), (ing)	INCL	Laminated Glass	LAM GL
Inch Pound(s)	IN LB	Laundry	LAU
Incandescent	INCAN	Lavatory	LAV
Inflammable	INFL	Lag Bolt	LB
Information	INFO	Pound	LB.
Insulating Concrete	INSC	Label	LBL
Insulating Fill	INSF	Labor	LBR

Lumber	LBR	Mechanical Engineer	ME
Light Control	LC	Mechanic (al)	MECH
Load-Bearing	LD BRG	Mechanical Room	MECH RM
Landing	LDG	Medium	MED
Leader	LDR	Membrane	MEMB
Level	LEV	Metal	MET
Linear Foot (Feet)	LF	Metal Flashing	METF
Loose Fill Insulation	LF INS	Metal Roof	METR
Left Hand	LH	Mezzanine	MEZZ
Lineal	LIN	Metal Floor Decking	MFD
Linen Closet	LIN, LCL	Manufacture (er)	MFR
Linoleum	LINO	Milligram	MG
Liquid	LIQ	Management	MGT
Live Load	LL	Manhole	MH
Limestone	LMS	Malleable Iron	MI
Location	LOC	Minimum	MIN
Lightproof	LP	Mirror	MIR, MIRR
Low Pressure	LP	Miscellaneous	MISC
Low Point	LPT	Mixture	MIX
Living Room	LR	Mark (Identifier)	MK
Light	LT, LITE	Millwork	MKWK
Lath	LTH	Molding (Moulding)	MLDG
Lintel	LTL	Millimeter	MM
Lumber	LUM, LBR	Membrane	MMB
Louver	LV, LVR	Main	MN
Level	LVL	Masonry Opening	MO
Lightweight	LW	Model	MOD
Lightweight concrete	LWC	Modify	MOD
		Modular	MOD
Meter	M	Modified bitumen	MOD BIT
Magnesium	MAGN	Movable	MOV
Maintenance	MAIN	Moisture Resistant	MR
Manual	MAN	Marble	MRB
Machine Bolt	MB	Metal Roof Deck (ing)	MRD
Medicine Cabinet	MC	Metal Threshold	METHR
Manufacturer	MFR	Mount (ed), (ing)	MT, MTD
Masonry	MAS	Metal Furring	MTFR
Material	MATL	Material (s)	MTL
Maximum	MAX	Metal	MTL
Master Bedroom	MBR, MBED	Mullion	MULL
Member	MBR	Millwork	MWK
Manhole Cover	MC		
Main Distribution Panel	MDP	North	N

Not Applicable	NA, N/A	Opposite	OPP
National Association of		Opposite Surface	OPS
Home Builders	NAHB	Optimum	OPT
Natural	NAT	Original	ORIG
Natural Grade	NAT GR	Ounce	OZ
National Bureau of Standards	NBS	Pound	LB
Normally Closed	NC	Parallel	PAR
National Electric Code	NEC	Panic Bar	PB
Negative	NEG	Pushbutton	PB
National Fire Protection Association	NFPA	Particleboard	PBD
Non-Fused Safety Switch	NFSS	Piece	PC
Nickel	NI	Pull Chain	PC
Not In Contract	NIC	Property Condition Assessment	PCA
Nonmetallic	NMT	Precast Concrete	PCC
Normally Open	NO	Pounds Per Cubic Foot	PCF
Number	NO	Cement Plaster (Portland)	PCPL
Net Operating Income	NOI	Property Condition Report	PCR
Nominal	NOM	Porcelain Enamel	PE
Nominal Pipe Size	NPS	Pedestal	PED
Noise Reduction	NR	Perimeter	PER
Noise Reduction Coefficient	NRC	Perforate (d)	PERF
Not to Scale	NTS	Perimeter	PERI
Orange	O	Perpendicular	PERP
Oxygen	O	Prefabricated (d)	PFB
Outside Air	OA	Preferred	PFD
Overall (Measure)	OA	Pounds Per Lineal Foot	PFL
Outside Air Intake	OAI	Prefinished	PFN
Obscure	OBS	Plate Glass	PG
On Center	OC, O/C	Pressure Gauge	PG
Outside Diameter	OD	Parking	PK
Open-end Drain	OED	Plate	PL, PLT
Open-end Duct	OED	Property Line	PL
Overflow	OF	Plastic Laminate	PLAM
Outside Face	OF	Plastic	PLAS
Overhang (Eave Line)	O/H	Plaster	PLAS, PLAST, PLS
Overhead	OH, OVHD	Platform	PLAT
Overhead Door	OH Door	Plumbing	PLBG
Ovalhead Machine Screw	OHMS	Plywood	PLYWD, PLY, PWD
Ovalhead Wood Screw	OHWS	Protected Membrane	
Open Web Joist	OJ	Roofing System	PMRS
Opaque	OP	Panel	PNL
Opening	OPG	Paint (ed)	PNT

Power Operated Damper	POD	Road	RD
Polyethylene	POLY	Roof Drain	RD
Polypropylene (Plastic)	PP	Round	RD, RND
Pair	PR	Ridge	RDG
Prefabricate (d)	PREFAB	Rigid Insulation	RDG INS
Preliminary	PRELIM	Redwood	RDWD
Preformed	PRF	Reinforce (d)	RE
Property	PROP	Reinforced Steel Bar	REBAR
Pressure-reducing Valve	PRV	Recessed	REC
Prestressed Concrete	PSC	Receptacle	RECEPT, RECP
Pounds Per Square Foot	PSF	Reference	REF
Pounds Per Square Inch	PSI	Refrigerator	REF, REFR
Painted	PTD	Register	REG
Point	PT	Reinforcement	REINF
Pressure-Treated	PT	Remove	REM
Post-tensioned Concrete	PTC	Requirement, Required	REQ
Paper Towel Dispenser	PTD	Retain(ing)	RET
Partition	PTN	Return	RET
Paper Towel Receptor	PTR	Revision	REV
Pave (d), (ing)	PV	Roof	RF
Polyvinylchloride	PVC	Roofing	RFG
Pavement	PVMT	Roof Hatch	RFH
Power	PWR	Reflect (ed), (ive), (or)	RFL
		Request for Proposal	RFP
Quadrant, Quadrangle	QUAD	Rough	RGH
Quarry Tile	QT	Right Hand	RH
Quart	QT	Rail (ing)	RL
Quality	QTY	Room	RM
Quantity	QTY	Rough Opening	RO
		Right of Way	ROW
Radius	R, RAD	Revolutions Per Minute	RPM
Right	R	Remaining Useful Life	RUL
Riser	R	Rooftop Package Unit	RTU
Return Air	RA	Relief Valve	RV
Riser	R, RIS	Reverse (side)	RVS
Radiator	RAD	Rivet	RVT
Return Air Duct	RAD	Rainwater Conductor	RWC
Rubber Base	RB		
Rubble Stone	RBL	South	S
Rabbet, Rebate	RBT	Switch	S, SW
Rubber Tile	RBT	Safety	SAF
Reinforced Concrete	RC	Sanitary	SAN
Reflected Ceiling Plan	RCP	Suspended Acoustical Tile Ceiling	SATC
Reinforced Concrete Pipe	RCP		

Southern Building Code	SBC	Shutoff Valve	SOV
Self-Closing	SC	Soil Pipe (Plumbing)	SP
Solid Core	SC	Soundproof	SP
Schedule	SCH, SCHED	Sump Pit	SP
School	SCH	Swimming Pool	SP
Screw	SCR	Spacer	SPC
Screen	SCRN, SCN	Specification(s)	SPEC(S)
Screened	SCRND	Sprinkler	SPKLR
Shop Drawings	SD	Special	SPL
Smoke Detector	SD	Square	SQ
Storm Drain	SD	Square Foot (Feet)	SQ FT (ft^2)
Supply Duct	SD	Square Inch(es)	SQ IN (in^2)
Siding	SDG	Square Yard(s)	SQ YD (yd^2)
Second	SEC	Safety Switch	SS
Section	SEC, SECT	Sanitary Sewer	SS
Select	SEL	Service Sink	SSK
Separate	SEP	Stainless Steel	SST
Service (Utility)	SERV	Stairs	ST
Sewer	SEW	Steel	ST
Special Floodway Hazard Area	SFHA	Street	ST
Square Foot	SF	Station	STA
Square Foot Gross	SFG	Sound Transmission Coefficient	STC
Safety Glass	SFGL	Standard	STD
Softwood	SFTWD	Stirrup (Rebar)	STIR
Sheet Glass	SG	Stock	STK
Sliding Glass Door	SGD	Steel	STL
Shelf, Shelving	SH	Steam	STM
Single Hung	SH	Stone	STN
Shingles	SH	Storage	STO
Shower	SH, SHR	Stirrup (Rebar)	STR, STIR
Shore (d), (ing)	SHO	Structural	STR, STRUCT
Sheet	SHT	Story	STY
Sheathing	SHTH, SHTG	Substitute	SUB
Shower	SHWR	Supply	SUP
Similar	SIM	Surface	SUR
Skylight	SKL	Survey	SURV
Sleeve	SL	Surface Two Sides	S2S
Sliding	SL	Surface Four Sides	S4S
Sealant	SLNT	Suspended	SUS
Sealant	SNT	Suspended Ceiling	SUSP CLG
Slab-on-Grade	SOG	Symbol	SYM
SOIL And Waste	S&W	Symmetry (ical)	SYM

Synthetic	SYN	Terrazzo	TZ
System	SYS	U.S. Department of Housing &	
Ton	T	Urban Development	HUD
Tread	T, TR	Urinal	U, UR
Tangent	TAN	Uniform Building Code	UBC
Tar and Gravel	T&G	Undercut	UC
Tongue and Groove	T&G	Under Floor	UF
Top of Beam	TB	Underground feeder (Electrical)	UF
Towel Bar	TB	Unit Heater	UH
Terra Cotta	TC	Underwriter's Laboratories	UL
Time-delay Fuse	TDF	Unexcavated	UNEX, U-EXC
Technical	TECH	Unfinished	UNF, UNFIN
Telephone	TEL	Uniform	UNIF
Temperature	TEMP	Unless Noted Otherwise	UNO
Thermometer	TH	Utility	UTIL
Thick(ness)	THK	Valve	V
Threshold	THOLD, THR	Vent Line	V
Thermostat	THRM, TSTAT	Volt	V
Through	THRU	Variable Air Volume	VAR, VAV
Tackboard	TKBD	Varnish	VAR
Tackstrip	TKS	Vinyl Asbestos Tile	VAT
True North	TN	Vapor Barrier	VB
Top of Beam	TOB	Vinyl Base	VB
Top of Curb	TOC	Vitreous Clay Tile	VCT
Top of Footing	TOF	Velocity	VEL
Toilet	TOIL	Ventilation, Ventilator	VENT
Tolerance	TOL	Vertical	VERT
Top of Parapet	TOP	Vinyl Fabric	VF
Topography	TOPO	Vertical Grain	VG
Toilet Paper Dispenser	TPD	Verify In Field	VIF
Toilet Partition	TPTN	Vinyl	VIN
Transition	TR	Vitreous	VIT
Transom	TR, TRANS	V-joint (ed)	VJ
Transformer	TRANS	Veneer	VNR
Track, Truck	TRK	Volume	VOL
Tensile Strength	TS	Vent Pipe	VP
Top of Slab	TS, TSL	Vapor Retarder	VR
Top of Steel	TST	Voltage Regulator	VR
Tubing	TUB	Vermiculite	VRM
Television	TV	Vent Stack	VS
Top of Wall	TW	Vinyl Tile	VT
Typical	TYP	Vent Through Roof	VTR

Vinyl Wall Covering	VWC	Welded Wire Fabric	WWF
Water	H_2O	Welded Wire Mesh	WWM
Watt	W	Extra Heavy	XH
West	W		
Width, Wide	W	Yard	Y, YD
Wire	W	Yellow	Y
Wardrobe	WARD	Yard Drain	YD
Wet Bulb	WB	Yellow Pine	YP
Wood Base	WB	Year	YR
Water Closet (Toilet)	WC	Zinc	Z, ZN
Wall Cabinets	W CAB		
Wood	WD		
Wood Door	WD		
Wood Door and Frame	WDF		
Wood Paneling	WDP		
Window	WDW, WIN		
Wired Glass	WG		
Wall Hung	WH		
Water Heater, Domestic Hot Water Heater	WH, DHWH		
Weep Hole	WH		
Wheel Bumper	WHB		
Wrought Iron	WI		
Washing Machine	WM		
Wire Mesh	WM		
Without	W/O		
Waterproof, Weatherproof	WP		
Water Pump	WP		
White Pine	WP		
Waterproof Membrane	WPM		
Working Point	WP		
Wash Room	WR		
Water Repellent	WR		
Waste Stack	WS		
Water Stop	WS		
Weather Strip	WS		
Wainscot	WSCT		
Weight	WT		
Window Unit	WU		
Wall Vent	WV		
Water Valve	WV		
Wall to Wall	W/W, WTW		

Appendix 2
Glossary

ABS   (Acrylonitrile Butadiene Styrene) Plastic pipe used in plumbing construction.

Abut   Joining end to end.

Accessible   Describes a site, building, facility, or portion thereof that complies with these guidelines.

Accessible Route   A continuous unobstructed path connecting all accessible elements and spaces of a building or facility. Interior accessible routes may include corridors, floors, ramps, elevators, lifts, and clear floor space at fixtures. Exterior accessible routes may include parking access aisles, curb ramps, crosswalks at vehicular ways, walks, ramps, and lifts.

AC Current   Electrical current that reverses direction at regular intervals (cycles).

Addendum   A written or graphic instruction issued by the architect prior to the execution of the contract which modifies or interprets the bidding documents by additions, deletions, clarifications or corrections. An addendum becomes part of the contract documents when the contract is executed.

Adhesive   A bonding material used to bond two materials together.

Adobe   A heavy clay soil used in many southwestern states to make sun-dried bricks.

Aggregate   Fine, lightweight, coarse, or heavyweight grades of sand, vermiculite, perlite, or gravel added to cement for concrete or plaster.

Air Handling Unit   A mechanical unit used for air conditioning or movement of air as in direct supply or exhaust of air within a structure.

Aligned Section   A section view in which some internal features are revolved into or out of the plane of the view.

Alligatoring   A pattern of rough cracking on a coated surface, similar in appearance to alligator skin.

American Bond   Brickwork pattern consisting of five courses of stretchers followed by one bonding course of headers.

Analog   The processing of data by continuously variable values.

Anchor Bolt   Metal rods, varying in diameter, to join and secure one material to another.

Angle   A figure formed by two lines or planes extending from, or diverging at, the same point.

ANSI   The American National Standards Institute. ANSI is an umbrella organization that administers and coordinates the national voluntary consensus standards system. http://www.ansi.org.

Application Block A part of a drawing of a subassembly showing the reference number for the drawing of the assembly or adjacent subassembly.

Approved Equal Material Equipment or method approved by the architect for use in the work as being acceptable as an equivalent in essential attributes to the material, equipment or method specified in the contract documents.

Arc A portion of the circumference of a circle.

Architect's Scale The scale used when dimensions or measurements are to be expressed in feet and inches.

Asphalt Shingles They are shingles made of asphalt or tar-impregnated paper with a mineral material embedded; very fire resistant.

Assumed Liability It is liability which arises from an agreement between people, as opposed to liability which arises from common or statutory law. See also Contractual Liability.

ASTM International Formerly the American Society for Testing Materials. They develop and publish testing standards for materials and specifications used by industry. http://www.astm.org.

Auxiliary View An additional plane of an object, drawn as if viewed from a different location. It is used to show features not visible in the normal projections.

Axial Load A weight that is distributed symmetrically to a supporting member, such as a column.

Axis The center line running lengthwise through a screw.

Axonometric Projection A set of three or more views in which the object appears to be rotated at an angle, so that more than one side is seen.

Backfill Any deleterious material (sand, gravel, etc.) used to fill an excavation.

Balloon Framing A system in wood framing in which the studs are continuous without an intermediate plate for the support of second-floor joists.

Baluster A vertical member that supports handrails or guardrails.

Balustrades A horizontal rail held up by a series of balusters.

Banister That part of the staircase which fits on top of the balusters.

Bar Chart A calendar that graphically illustrates a projected time allotment to achieve a specific function.

Base A trim or molding piece found at the interior intersection of the floor and the wall.

Beam A weight-supporting horizontal member.

Base Building The core (common areas) and shell of the building and its systems that typically are not subject to improvements to suit tenant requirements.

Base Flashing Consists of flashing that covers the edges of a membrane.

Batten A narrow strip of wood used to cover a joint.

Batt Insulation An insulating material formed into sheets or rolls with a foil or paper backing to be installed between framing members.

Bearing Wall A wall which supports any vertical loads in addition to its own weight.

Benchmark A point of known elevation from which the surveyors can establish all their grades.

Bend Allowance An additional amount of metal used in a bend in metal fabrication.

Bill Of Material A list of standard parts or raw materials needed to fabricate an item.

Bisect To divide into two equal parts.

Blistering The condition that paint presents when air or moisture is trapped underneath and makes bubbles that break into flaky particles and ragged edges.

Blocking The use of internal members to provide rigidity in floor and wall systems. Also used for fire draft stops.

Block Diagram   A diagram in which the major components of a piece of equipment or a system are represented by squares, rectangles, or other geometric figures, and the normal order of progression of a signal or current flow is represented by lines.

Blueprints   Documents containing all the instructions necessary to manufacture a part. The key sections of a blueprint are the drawing, dimensions, and notes. Although blueprints used to be blue, modem reproduction techniques now permit printing of black-on-white as well as colors.

Body Plans   An end view of a ship's hull, composed of superimposed frame lines.

Board foot   A unit of lumber of measure equaling 144 cubic inches; the base unit (B.F.) is 1 inch thick and 12 inches square or 1 x 12 x 12 = 144 cubic inches.

Bond   In masonry, the interlocking system of brick or block to be installed.

Border Lines   Dark lines defining the inside edge of the margin on a drawing.

Boundary Survey   A mathematically closed diagram of the complete peripheral boundary of a site, reflecting dimensions, compass bearings and angles. It should bear a licensed land surveyor's signed certification, and may include a metes and bounds or other written description.

Break Lines   They are lines used to define the boundary of an imaginary broken-out section or to shorten dimensions that are excessively long. There are two types: long, thin ruled line with freehand zigzag and the short, thick wavy freehand line.

Breezeway   A covered walkway with open sides between two different parts of a structure.

Brick Pavers   A term used to describe special brick to be used on the floor surface.

Buck   A frame found around doors.

Building Codes   Rules and regulations adopted by the governmental authority having jurisdiction over the commercial real estate, which govern the design, construction, alteration and repair of such commercial real estate. In some jurisdictions trade or industry standards may have been incorporated into, and made a part of, such building codes by the governmental authority. Building codes are interpreted to include structural, HVAC, plumbing, electrical, life-safety, and vertical transportation codes.

Building Envelope   The enclosure of the building that protects the building's interior from outside elements, namely the exterior walls, roof and soffit areas.

Building Inspector   A representative of a governmental authority employed to inspect construction for compliance codes, regulations and ordinances.

Building Line   An imaginary line determined by zoning departments to specify on which area of a lot a structure may be built (also known as a setback).

Building Paper   Also called tar paper, roofing paper, and a number of other terms; paper having a black coating of asphalt for use in weatherproofing.

Building Permit   A permit issued by appropriate governmental authority allowing construction of a project in accordance with approved drawing and specifications.

Building Systems   Interacting or independent components or assemblies, which form single integrated units, that comprise a building and its site work, such as, pavement and flatwork, structural frame, roofing, exterior walls, plumbing, HVAC, electrical, etc.

Build-Out   The interior construction and customization of a space (including services, space and stuff) to meet the tenant's requirements; either new construction or renovation (also referred to as fit-out or fit-up).

Buttock Line   The outline of a vertical, longitudinal section of a ship's hull.

Cabinet Drawing   A type of oblique drawing in which the angled receding lines are drawn to one-half scale.

Caisson   A below-grade concrete column for the support of beams or columns.

Cantilever   A horizontal structural condition where a member extends beyond a support, such as a roof overhang.

Capillary   The action by which the surface of a liquid, where it is in contact with a solid, is elevated or depressed.

Casement   A type of window hinged to swing outward.

Casting   A metal object made by pouring melted metal into a mold

Catch Basin   A complete drain box made in various depths and sizes; water drains into a pit, then from it through a pipe connected to the box.

Caulk   Any type of material used to seal walls, windows, and doors to keep out the weather.

Cavalier Drawing   A form of oblique drawing in which the receding sides are drawn full scale, but at 45° to the orthographic front view.

Cavity Wall   A masonry wall formed with an air space between each exterior face.

Cement Plaster   A plaster that is comprised of cement rather than gypsum.

Center Lines   Such lines that indicate the center of a circle, arc, or any symmetrical object; consist of alternate long and short dashes evenly spaced.

Certificate For Payment   A statement from the architect to the owner confirming the amount of money due the contractor for work accomplished or materials and equipment suitably stored, or both.

Certificate of Insurance   A document issued by an authorized representative of an insurance company stating the types, amounts and effective dates of insurance in force for a designated insured.

Certificate of Occupancy   Document issued by governmental authority certifying that all or a designated portion of a building complies with the provisions of applicable statutes and regulations, and permitting occupancy for its designated use.

Certificate of Substantial Completion   A certificate prepared by the architect on the basis of an inspection stating that the work or a designated portion thereof is substantially complete, which established the date of substantial completion; states the responsibilities of the owner and the contractor for security, maintenance, heat, utilities, damage to the work, and insurance; and taxes the time within which the contractor shall complete the items listed therein.

Cesspool   An underground catch basin for the collection and dispersal of sewage.

Change Order   A written and signed document between the owner and the contractor authorizing a change in the work or an adjustment in the contract sum or time. The contract sum and time may be changed only by change order. A change order may be in the form of additional compensation or time; or less compensation or time known as a "Deduction."

Checklist   List of items used to check drawings.

Circle   A plane closed figure having every point on its circumference (perimeter) equidistant from its center.

Circuit Breaker   A safety device that opens and closes an electrical circuit.

Circumference   The length of a line that forms a circle.

Clear Floor Space   The minimum unobstructed floor or ground space required to accommodate a single, stationary wheelchair and occupant.

Cleat   Any strip of material attached to the surface of another material to strengthen, support, or secure a third material.

Clerestory   A window or group of windows that are placed above the normal window height, often between two roof levels.

Clevis   An open-throated fitting for the end of a rod or shaft, having the ends drilled for a bolt or a pin. It provides a hinging effect for flexibility in one plane.

Column   A vertical weight-supporting member.

Column Pad   An area of concrete in the foundation for the support of weight distributed into a column.

Common Use   Refers to those interior and exterior rooms, spaces, or elements that are made available for the use of a restricted group of people (for example, occupants of a homeless shelter, the occupants of an office building, or the guests of such occupants).

Component A fully functional portion of a building system, piece of equipment, or building element.

Computer-Aided Drafting (CAD)   A method by which engineering drawings may be developed on a computer.

Computer-Aided Manufacturing (CAM)   A method by which a computer uses a design to guide a machine that produces parts.

Concrete Block   A rectangular concrete form with cells in them.

Condensation   The process by which moisture in the air becomes water or ice on a surface (such as a window) whose temperature is colder than the air's temperature.

Cone   A solid figure that tapers uniformly from a circular base to a point.

Construction Documents   A term used to represent all drawings, specifications, addenda, and other pertinent construction information associated with the construction of a specific project.

Construction Lines   Lightly drawn lines used in the preliminary layout of a drawing.

Contingency Allowance   A sum included in the project budget designated to cover unpredictable or unforeseen items of work, or changes in the work subsequently required by the owner. See Budget, Project.

Contour Line   A line that represents the change in level from a given datum point.

Contract   A legally enforceable promise or agreement between two or among several person. See also Agreement.

Convection   Transfer of heat through the movement of a liquid or gas.

Cornice   The projecting or overhanging structural section of a roof.

Cost Appraisal   Evaluation or estimate (preferably by a qualified professional appraiser) of the market or other value, cost, utility or other attribute of land or other facility.

Cost Estimate   A preliminary statement of approximate cost, determined by one of the following methods: 1. Area and volume method; cost per square foot or cubic foot of the building. 2. Unit cost method; cost of one unit multiplied by the number of units in the project; for example, in a hospital, the cost of one patient unit multiplied by the number of patient units in the project. 3. In-place unit method; cost in-place of a unit, such as doors, cubic yards of concrete, and squares of roofing.

Coving   The curving of the floor material against the wall to eliminate the open seam between floor and wall.

Crawl Space   The area under a floor that is not fully excavated; only excavated sufficiently to allow one to crawl under it to get at the electrical or plumbing devices.

Crest   The surface of the thread corresponding to the major diameter of an external thread and the minor diameter of an internal thread.

Cross-section   A slice through a portion of a building or member that depicts the various internal conditions of that area.

CSI   Construction Specifications Institute. Membership organization of design professionals, construction professionals, product manufacturers, and building owners. Develops and promotes industry communication standards and certification programs. http://www.csinet.org.

Cube   Rectangular solid figure in which all six faces are square.

Cul-de-Sac   A curved turnaround with the radius determined by the traffic load, located at the end of a street.

Curtain Wall An exterior wall that provides no structural support.

Cutting Plane Line   A line showing where a theoretical cut has been made to produce a section view. Cutting plane lines consist of two short dashes alternating with a longer dash.

Cylinder   A solid figure with two equal circular bases.

Dado   A rectangular groove cut into a board across the grain.

Dangerous or Adverse Conditions   These are essentially conditions which may pose a threat or possible injury to the field observer, and which may require the use of special protective clothing, safety equipment, access equipment, or any other precautionary measures.

Date of Agreement   The date stated in the Agreement. If no date is stated, it could be the date on which the Agreement is actually signed, if this is recorded, or it may be the date established by the award.

Date of Commencement of the Work   The date established in a notice to the contractor to proceed or, in the absence of such notice, the date of the owner contractor agreement or such other date and may be established therein.

Date of Substantial Completion   The date certified by the architect when the work or a designated portion thereof is sufficiently complete, in accordance with the contract documents, so the owner can occupy the work or designated portion thereof for use for which it is intended.

Datum Point   Reference point.

Dead Load   The weight of a structure and all its fixed components.

Defective Work   The work not conforming with the contract requirements.

Deferred Maintenance   Physical deficiencies that cannot be remedied with routine maintenance, normal operating maintenance, etc., excluding de minimis conditions that generally do not present a material physical deficiency to the subject property.

Depth   The distance from the root of a thread to the crest, measured perpendicularly to the axis.

Design-Build Construction   When an owner contract with a prime or main contractor to provide both design and construction services for the entire construction project. Use of the design-build project delivery system has grown from 5 percent of U.S. construction in 1985 to 33 percent in 1999, and is projected to surpass low-bid construction in 2005. If a design-build contract is extended further to include the selection, procurement, and installation of all furnishings, furniture, and equipment, it is called a "turnkey" contract.

Details   An enlarged drawing to show a structural aspect, an aesthetic consideration, a solution to an environmental condition, or to express the relationship among materials or building components.

Digital The processing of data by numerical or discrete units.

Dimension Line A thin unbroken line (except in the case of structural drafting) with each end terminating with an arrowhead; used to define the dimensions of an object. Dimensions are placed above the line, except in structural drawing where the line is broken and the dimension placed in the break.

Dimensions   Numerical values used to indicate size and distance.

Direct Costs (Hard Costs)   The aggregate costs of all labor, materials, equipment and fixtures necessary for the completion of construction of the Improvements.

Direct Costs Loan; Indirect Costs Loan   That portion of the loan amount applicable and equal to the sum of the loan budget amounts for direct costs and indirect costs, respectively, shown on the borrower's project cost statement.

Dormer   A structure that projects from a sloping roof to form another roofed area. This new area is typically used to provide a surface to install a window.

Downspouts   Pipes connected to the gutter to conduct rainwater to the ground or sewer.

Drawing Number   An identifying number assigned to a drawing or a series of drawings.

Drawings   The collection of lines illustrating the shape and features of a part. The original graphic design from which a blueprint may be made; also called plans.

Drywall   An interior wall covering installed in large sheets made from gypsum board.

Duct   Usually sheet metal forms used for the distribution of cool or warm air throughout a structure.

Due Diligence   The process of conducting a walkthrough survey and appropriate inquiries into the physical condition of a commercial real estate's improvements, usually in connection with a commercial real estate transaction. The degree and type of such survey or other inquiry may vary for different properties and different purposes.

Dwelling Unit   A single unit which provides a kitchen or food preparation area, in addition to rooms and spaces for living, bathing, sleeping, and the like. Dwelling units include a single family home or a townhouse used as a transient group home; an apartment building used as a shelter; guestrooms in a hotel that provide sleeping accommodations and food preparation areas; and other similar facilities used on a transient basis. For purposes of these guidelines, use of the term "Dwelling Unit" does not imply the unit is used as a residence.

Easement   The right or privilege to have access to or through another piece of property such as a utility easement.

Eave   That portion of the roof that extends beyond the outside wall.

Egress   A continuous and unobstructed way of exit travel from any point in a building or facility to a public way. A means of egress comprises vertical and horizontal travel and may include intervening room spaces, doorways, hallways, corridors, passageways, balconies, ramps, stairs, enclosures, lobbies, horizontal exits, courts and yards. An accessible means of egress is one that complies with these guidelines and does not include stairs, steps, or escalators. Areas of rescue assistance or evacuation elevators may be included as part of accessible means of egress.

Elastomeric   A material having characteristics which allow it to expand or contract its shape and still return to its original dimensions without losing stability.

Electromechanical Drawing   A special type of drawing combining electrical symbols and mechanical drawing to show the position of equipment that combines electrical and mechanical features.

Element   An architectural or mechanical component of a building, facility, space, or site, e.g., telephone, curb ramp, door, drinking fountain, seating, or water closet.

Elementary Wiring Diagram   (1) A shipboard wiring diagram showing how each individual conductor is connected within the various connection boxes of an electrical circuit system. (2) A schematic diagram; the term elementary wiring diagram is sometimes used interchangeably with schematic diagram, especially a simplified schematic diagram.

Elevations   A four-view drawing of a structure showing front, sides, and rear.

Engineer's Scale   The scale used whenever dimensions are in feet and decimal parts of a foot, or when the scale ratio is a multiple of 10.

Epicenter   The point of the earth's surface directly above the focus or hypocenter of an earthquake.

Expansion Joint   A joint often installed in concrete construction to reduce cracking and to provide workable areas.

Expected Useful Life (EUL)   The average amount of time in years that an item, component or system is estimated to function when installed new and assuming routine maintenance is practiced.

Exploded View   A pictorial view of a device in a state of disassembly, showing the appearance and interrelationship of parts.

Extension Line   A line used to visually connect the ends of a dimension line to the relevant feature on the part. Extension lines are solid and are drawn perpendicular to the dimension line.

External Thread   A thread on the outside of a member. Example: a thread of a bolt.

Façade   The exterior covering of a structure.

Face of Stud (F.O.S.)   Outside surface of the stud. Term used most often in dimensioning or as a point of reference.

Fascia   A horizontal member located at the edge of a roof overhang.

Facility   All or any portion of buildings, structures, site improvements, complexes, equipment, roads, walks, passageways, parking lots, or other real or personal property located on a site.

Falsework   Temporary supports of timber or steel sometimes required in the erection of difficult or important structures.

Felt   A tar-impregnated paper used for water protection under roofing and siding materials. Sometimes used under concrete slabs for moisture resistance.

Fillet   A concave internal corner in a metal component, usually a casting.

Final Completion   Term denoting that the work has been completed in accordance with the terms and conditions of the contract documents.

Final Inspection   Final review of the project by the architect to determine final completion, prior to issuance of the final certificate for payment.

Final Payment   Is the payment made by the owner to the contractor, upon issuance by the architect of the final certificate for payment, of the entire unpaid balance of the contract sum as adjusted by change orders.

Finish Grade   The soil elevation in its final state upon completion of construction.

Finish Marks   These are marks used to indicate the degree of smoothness of finish to be achieved on surfaces to be machined.

Fire Barrier   A continuous membrane such as a wall, ceiling, or floor assembly that is designed and constructed to specified fire-resistant rating to hinder the spread of fire and smoke. This resistant rating is based on a time factor. Only fire-rated doors may be used in these barriers.

Fire Compartment of Fire Zone   An enclosed space in a building that is separated from all other parts of the building by the construction of fire separations having fire resistance ratings.

Fire Department Records   These are records maintained by or in the possession of the local fire department in the area in which the subject property is located.

Fire Door   A door used between different types of construction that has been rated as being able to withstand fire for a certain amount of time.

Fire Draft Stop   A member of varying materials placed in walls, floors, and ceilings to prevent a rapid spread of fire.

Fire Resistance Rating   Sometimes called fire rating, fire resistance classification, or hourly rating. A term defined in building codes, usually based on fire endurance required. Fire resistance ratings are assigned by building codes for various types of construction and occupancies, and are usually given in half hour increments.

Fire-stop   Blocking placed between studs or other structural members to resist the spread of fire.

Flashing   A thin, impervious sheet of material placed in construction to prevent water penetration or to direct the flow of water. Flashing is used especially at roof hips and valleys, roof penetrations, joints between a roof and a vertical wall, and in masonry walls to direct the flow of water and moisture.

Floor Joist   Structural member for the support of floor loads.

Floor Plan   A horizontal section taken at approximately eye level.

Flue Liner   A terra-cotta pipe used to provide a smooth flue surface so that unburned materials will not cling to the flue.

Flush   Even, level, or aligned.

Footings   Weight-bearing concrete construction elements poured in place in the earth to support a structure.

Forging   The process of shaping heated metal by hammering or other impact.

Format   The general makeup or style of a drawing.

Foundation Plan   A drawing that graphically illustrates the location of various foundation members and conditions that are required for the support of a specific structure.

Frame Lines   The outline of transverse plane sections of a hull.

Framing Connectors   Metal devices, varying in size and shape for the purpose of joining wood framing members together.

French Curve   An instrument used to draw smooth irregular curves.

Frieze   A decoration or ornament shaped to form a band around a structure.

Frost Line   The depth at which frost penetrates the soil.

Full Section   A sectional view that passes entirely through the object.

Furred Ceiling   The construction for a separate surface beyond the main ceiling or wall that provides a desired air space and modifies interior dimensional conditions.

Furring   Wood strips attached to structural members that are used to provide a level surface for finishing materials when different-sized structural members are used.

Gable   A type of roof with two sloping surfaces that intersect at the ridge of the structure.

Galvanized   Steel products that have had zinc applied to the exterior surface to provide protection from rusting.

Gambrel Roof   A barn-shaped roof.

Gauge   The thickness of metal or glass sheet material.

General Conditions   (When used by contractors) Construction project activities and their associated costs that are not usually assignable to a specific material installation or subcontract. Example: temporary electrical power. (When used by everyone else) The contract document (often a standard form) that spells out the relationships between the parties to the contract. Example: the AIA Document A201.

General Contract   Any contract (together with all riders, addenda and other instruments referred to therein as "contractor or any other person, which requires the general contractor or such other person to provide, or supervise or manage the procurement of, substantially all labor and material needed for completion of the Improvements.

Girder   A horizontal structural beam for the support of secondary members such as floor joists.

Glue-Laminated Beams   A beam comprised of a series of wood members glued together.

Grade   The designation of the quality of a manufactured piece of wood.

Grading   The moving of soil to effect the elevation of land at a construction site.

Gravel Stop   A metal strip used to retain gravel at the edge of built-up roofs.

Ground Floor   Any occupiable floor less than one story above or below grade with direct access to grade. A building or facility always has at least one ground floor and may have more than one ground floor as where a split level entrance has been provided or where a building is built into a hillside.

Grout   A mixture of cement, sand, and water used to fill joints in masonry and tile construction.

Guardrail   A horizontal protective railing used around stairwells, balconies, and changes of floor elevation greater than 30 inches.

Gusset   A plate added to the side of intersecting structural members to help form a secure connection and to reduce stress.

Gutter   A metal or wooden trough set below the eaves to catch and conduct water from rain and melting snow to a downspout.

Half Section   A combination of an orthographic projection and a section view to show two halves of a symmetrical object.

Hanger   A metal support bracket used to attach two structural members.

Hatching   The lines that are drawn on the internal surface of sectional views. Used to define the kind or type of material of which the sectioned surface consists.

Head   The top of a window or door frame.

Header   A horizontal structural member spinning over openings, such a doors and windows, for the support of weight above the openings.

Header Course   In masonry, a horizontal masonry course of brick laid perpendicular to the wall face; used to tie a double wythe brick wall together.

Helix   The curve formed on any cylinder by a straight line in a plane that is wrapped around the cylinder with a forward progression.

Hidden Lines   A line used to define a part feature that is not visible in a specific view. Hidden lines consist of a series of short dashes.

Hip Roof   A roof shape with four sloping sides.

Hose Bibb   A faucet used to connect a hose.

Indemnification   A contractual obligation by which one person or entity agrees to secure another against loss or damage from specified liabilities. See also Contractual Liability.

Indirect Costs (Soft Costs)   Certain costs (other than direct costs) of completion of the improvements, including but not limited to, architects', engineers' and Lender's attorneys' fees, ground rents, interest on and recording taxes and title charges in respect of building loan mortgages, real estate taxes, water and sewer rents, survey costs, loan commitment fees, insurance and bond premiums and such other non-construction costs as are part of the "cost of improvements," as such quoted term as defined in the applicable state's lien law.

Indirect Cost Statement   A statement by the borrower in a form approved by the lender of indirect costs incurred and to be incurred.

Inscribed Figure   A figure that is completely enclosed by another figure.

Inspection   Examination of work completed or in progress to determine its conformance with the requirements of the contract documents. The architect ordinarily makes only two inspections of work, one to determine substantial completion and the other to determine final completion. These inspections should be distinguished from the more general observations made by the architect on visits to the site during the progress of the work. The term is also used to mean examination of the work by a public official, owner's representative, or others.

Insulation   Any material capable of resisting thermal, sound, or electrical transmission.

Interconnection Diagram   A diagram showing the cabling between electronic units, as well as how the terminals are connected

Interior Elevation   A straight-on view of the surface of the interior walls of a structure.

Internal Thread   A thread on the inside of a member. Example: the thread inside a nut.

Isometric Drawing   A form of a pictorial drawing in which the main lines are equal in dimension. Normally drawn using 30 degree, 90 degree angle.

Isometric Projection   A set of three or more views of an object that appears rotated, giving the appearance of viewing the object from one corner. All lines are shown in their true length, but not all right angles are shown as such.

Isometric Wiring Diagram   A diagram showing the outline of a ship, an aircraft, or other structure, and the location of equipment such as panels and connection boxes and cable runs.

Jamb   The side portion of a door, window, or any opening.

Joist   A horizontal beam used to support a ceiling.

Joist Hanger   A metal connector for the end support of floor and ceiling joist.

Key   A small wedge or rectangular piece of metal inserted in a slot or groove between a shaft and a hub to prevent slippage.

Key Plan   A plan, reduced in scale used for orientation purposes.

Keyseat   A slot or groove into which the key fits.

Keyway   A slot or groove within a cylindrical tube or pipe into which a key fitted into a key seat will slide.

King Stud   A full-length stud placed at the end of a header.

Knee Wall   A wall of less than full height.

Lattice   A grille made by criss-crossing strips of material.

Lead   The distance a screw thread advances one turn, measured parallel to the axis. On a single-thread screw the lead and the pitch are identical; on a double-thread screw the lead is twice the pitch; on a triple-thread screw the lead is three times the pitch.

Leader Line   A thin line with an arrow head that is often positioned at an angle and is used to tie a dimension, note or reference to a feature, especially when there are space limitations.

Ledger   Structural framing member used to support ceiling and roof joists at the perimeter walls.

LEED   Leadership in Energy and Environmental Design. A sustainable design building certification system promulgated by the United States Green Building Council. Also an accrediting program for professionals (LEED APs) who have mastered the certification system. http://www.usgbc.org.

Legend   A description of any special or unusual marks, symbols, or line connections used in the drawing.

Lien   A monetary claim on a property.

Limit of Liability   The maximum amount which an insurance company agrees to pay in case of loss.

Lintel   A load-bearing structural member supported at its ends. Usually located over a door or window.

Live Load   A temporary and changing load superimposed on structural components by the use and occupancy of the building, not including the wind load, earthquake load, or dead load.

Load-bearing Wall   A support wall that holds floor or roof loads in addition to its own weight.

Logic Diagram   A type of schematic diagram using special symbols to show components that perform a logic or information processing function.

Major Diameter   The largest diameter of an internal or external thread.

Manifold   A fitting that has several inlets or outlets to carry liquids or gases.

Mansard   A four-sided, steep-sloped roof.

Masonry Opening   The actual distance between masonry units where an opening occurs. It does not include the wood or steel framing around the opening.

Masonry Veneer   A layer of masonry units that are bonded to a frame or masonry wall.

Master Specification   A resource specification section containing options for selection, usually created by a design professional firm, which once edited for a specific project becomes a contract specification.

Master Format   Industry standard for organizing specifications and other construction information, published by CSI and Construction Specifications Canada. Formerly a 5-digit number system with 16 divisions. Now a 6- or 8-digit numbering system with 49 divisions.

Masterspec®   Subscription master guide specification library published by ARCOM and owned by the American Institute of Architects. http://www.specguy.com/www.masterspec.com

Mastic   An adhesive used to hold tiles in place; also refers to adhesives used to glue many types of materials in the building process.

Mechanical Drawing   Applies to scale drawings of mechanical objects.

Mechanics' Lien   A lien on real property created by statute in all states in favor of person supplying labor or materials for a building or structure for the value of labor or materials supplied by them. In some jurisdictions a mechanic's lien also exists for the value of professional services. Clear title to the property cannot be obtained until the claim for the labor, materials or professional services is settled.

Mesh   A metal reinforcing material placed in concrete slabs and masonry walls to help resist cracking.

Metal Wall Ties   Corrugated metal strips used to bond brick veneer to its support wall.

Mezzanine or Mezzanine Floor   That portion of a story which is an intermediate floor level placed within the story and having occupiable space above and below its floor.

Modules   A system based on a single unit of measure.

Modulus of Elasticity (E)   The degree of stiffness of a beam.

Moisture Barrier   Typically a plastic material used to restrict moisture vapor from penetrating into a structure.

Moldings   Trim mounted around windows, floors, and doors as well as closets.

Monolithic One-Pour System   A method of poured-in-place concrete construction that is without joints.

Mortar   The mixture of cement, sand, lime, and water that provides a bond for the joining of masonry units.

Multifamily Dwelling   Any building consisting of more than two dwelling units.

Nailer   A wood member bolted to concrete or steel members to provide a nailing surface for attaching other wood members.

Negligence   Failure to exercise due care under the circumstances. Legal liability for the consequences of an act or omission frequently depends upon whether or not there has been negligence. See also Due Care.

Net Size   The actual size of an object.

Newel   The end post of a stair railing.

Nominal Size   The call-out-size. May not be the actual size of the item.

Nonbearing Wall   A wall that supports no loads other than it own weight. Some building codes consider walls that support only ceiling loads as nonbearing.

Nonconforming Work   Work that does not fulfill the requirements of the contract documents.

Nonferrous Metal   They are metals such as copper or brass, that contain no iron.

Nosing   The rounded front edge of a tread that extends past the riser.

Notes   An additional instruction or general comment added to a blueprint. Notes contain information about the material, finish, tooling, tolerances,

Oblique Drawing   A type of pictorial drawing in which one view is an orthographic projection and the views of the sides have receding lines at an angle.

Oblique Projection   A view produced when the projectors are at an angle to the plane the object illustrated. Vertical lines in the view may not have the same scale as horizontal lines.

Occupiable   A room or enclosed space designed for human occupancy in which individuals congregate for amusement, educational or similar purposes, or in which occupants are engaged at labor, and which is equipped with means of egress, light, and ventilation.

Offset Section   A section view of two or more planes in an object to show features that do not lie in the same plane.

On Center   A measurement taken from the center of one member to the center of another member.

Opinions of Probable Costs   Determination of probable costs, a preliminary budget, for a suggested remedy.

Orthographic Projection   A view produced when projectors are perpendicular to the plane of the object. It gives the effect of looking straight at one side.

Outlet   An electrical receptacle that allows for current to be drawn from the system.

Pad   Term used for isolated concrete piers.

Parapet   A portion of wall extending above the roof level.

Parging   A thin coat of plaster used to smooth a masonry surface.

Parquet Flooring   Wood flooring laid to form patterns.

Partial Occupancy   The occupancy by the owner of a portion of a projection prior to final completion. See Final Completion.

Partial Section   A sectional view consisting of less than a half section. Used to show the internal structure of a small portion of an object. Also known as a broken section.

Partition   An interior wall.

Party Wall   A wall dividing two adjoining spaces such as apartments or offices

Patent Defect   A defect in materials, equipment of completed work which reasonably careful observation could have discovered; distinguished from a latent defect, which could not be discovered by reasonable observation.

Penny (d)   The unit of measure of the nails used by carpenters.

Performance Specifications   The written material containing the minimum acceptable standards and actions, as may be necessary to complete a project.

Perpendicular   Vertical lines extending through the outlines of the hull ends and the designer's waterline.

Perspective   The visual impression that, as parallel lines project to a greater distance, the lines move closer together.

Phantom View   A view showing the alternate position of a movable object, using a broken line convention.

Phase   An impulse of alternating current. The number of phases depends on the generator windings. Most large generators produce a three-phase current that must be carried on at least three wires.

Physical Deficiency   Conspicuous defects or significant deferred maintenance of a subject property's material systems, components, or equipment as observed during the field observer's walk-through survey. Included within this definition are material life-safety/building code violations and, material systems, components, or equipment that are approaching, have reached, or have exceeded their typical EUL or whose RUL should not be relied upon in view of actual or effective age, abuse, excessive wear and tear, exposure to the elements, lack of proper or routine maintenance, etc. This definition specifically excludes deficiencies that may be remedied with routine maintenance, miscellaneous minor repairs, normal operating maintenance, etc., and excludes de minimis conditions that generally do not constitute a material physical deficiency of the subject property.

Pictorial Wiring Diagram   A diagram showing actual pictorial sketches of the various parts of a piece of equipment and the electrical connections between the parts.

Pier   A vertical support for a building or structure, usually designed to hold substantial loads.

Pilaster   A rectangular pier integral with a wall for the purpose of strengthening the wall and supporting axial loads.

Pile   A steel or wooden pole driven into the ground sufficiently to support the weight of a wall and building.

Pillar   A pole or reinforced wall section used to support the floor and consequently the building.

Pitch   The distance from a point on a screw thread to a corresponding point on the next thread, measured parallel to the axis.

Planes   Any two dimensional surface.

Planking   A term for wood members having a minimum rectangular section of 1 1/2 inch to 3 1/2 inch in thickness. Used for floor and roof systems.

Plans   All final drawings, plans and specifications prepared by the borrower, borrower's architects, the general contractor or major subcontractors, and approved by lender and the construction consultant, which describe and show the labor, materials, equipment, fixtures and furnishings necessary for the construction of the Improvements, including all amendments and modifications thereof made by approved change orders (and also showing minimum grade of finishes and furnishings for all areas of the Improvements to be leased or sold in ready-for-occupancy conditions).

Pictorial Drawing   A drawing that gives the real appearance of the object, showing general location, function, and appearance of parts and assemblies.

Plan View   A view of an object or area as it would appear from directly above.

Plaster   A mixture of cement, water and sand.

Plat   A map or plan view of a lot showing principal features, boundaries, and location of structures.

Plenum   An air space (above the ceiling) for transporting air from the HVAC system.

Plumb   True vertical.

Polarity   The direction of magnetism or direction of flow of current.

Polyvinyl Chloride (PVC)   A plastic material commonly used for pipe and plumbing fixtures.

Portico   A roof supported by columns instead of walls.

Post   A vertical wood structural member generally 4 x 4 (100 mm) or larger.

Post-and-beam Construction   A type of wood frame construction using timber for the structural support.

Precast   A concrete component which has been cast in a location other than the one in which it will be used.

Primer   The first coat of paint or glue when more than one coat will be applied.

Progress Payment   Partial payment made during progress of the work on account of work completed and or materials suitably stored.

Progress Schedule   A diagram, graph or other pictorial or written schedule showing proposed and actual of starting and completion of the various elements of the work.

Project Cost   Total cost of the project including construction cost, professional compensation, land costs, furnishings and equipment, financing and other charges.

Projection   A technique for showing one or more sides of an object to give the impression of a drawing of a solid object.

Project Manual   The volume(s) prepared by the architect for a project which may include the bidding requirements, sample forms and conditions of the contract and the specifications.

Projector   The theoretical extended line of sight used to create a perspective or view of an object.

Purlin   A horizontal roof member that is laid perpendicular to rafters to help limit deflections.

Quarry Tile   An unglazed, machine made tile.

Quick Set   A fast-curing cement plaster.

Rabbet   A rectangular groove cut on the edge of a board.

Rafter   A sloping or horizontal beam used to support a roof.

Radius   A straight line from the center of a circle or sphere to its circumference or surface.

Rake Joint   A recessed mortar joint.

Ramp   A walking surface which has a running slope greater than 1:20.

Readily Accessible   Describes areas of the subject property that are promptly made available for observation by the field observer at the time of the walk-through survey and do not require the removal of materials or personal property, such as furniture, and that are safely accessible in the opinion of the field observer.

Record Drawings   Construction drawing revised to show significant changes made during the construction process, usually based on marked-up prints, drawing and other data furnished by the contracts to the architect. Preferable to as built drawings.

Reference Designation   A combination of letters and numbers to identify parts on electrical and electronic drawings. The letters designate the type of part, and the numbers designate the specific part. Example: reference designator R-12 indicates the 12th resistor in a circuit.

Reference Numbers   They consist of numbers used on a drawing to refer the reader to another drawing for more detail or other information.

Reference Plane   The normal plane that all information is referenced

Register   An opening in a duct for the supply of heated or cooled air.

Relative Humidity   The amount of water vapor in the atmosphere compared to the maximum possible amount at the same temperature.

Release of Lien   Instrument executed by a person or entity supplying labor, materials or professional services on a Project which releases that person's or entity's mechanic' lien against the Project property.

Remaining Useful Life (RUL)   A subjective estimate based upon observations, or average estimates of similar items, components, or systems, or a combination thereof, of the number of remaining years that an item, component, or system is estimated to be able to function in accordance with its intended purpose before warranting replacement. Such period of time is affected by the initial quality of an item, component, or system, the quality of the initial installation, the quality and amount of preventive maintenance exercised, climatic conditions, extent of use, etc.

Removed Section   A drawing of an object's internal cross section located near the basic drawing of the object.

Requisition   A statement prepared by the Borrower in a form approved by the Lender setting forth the amount of the Loan advance requested in each instance and including, if requested by the Lender:

Retainage   A sum withheld from progress payments to the contractor in accordance with the terms of the Owner-Contractor Agreement.

Retaining Wall   A masonry wall supported at the top and bottom designed to resist soil loads.

R-Factor   A unit of thermal resistance applied to the insulating value of a specific building material.

Revision Block   This block is located in the upper right corner of a print. It provides a space to record any changes made to the original print.

Revolved Section   A drawing of an object's internal cross section superimposed on the basic drawing of the object.

Rheostat   An electrical control device used to regulate the current reaching a light fixture. A dimmer switch.

Ridge   The highest point on a sloped roof.

Ridge Board   A horizontal member that rafters are aligned against to resist their downward force.

Riser   The vertical member of stairs between the treads.

Roof Drain   A receptacle for removal of roof water.

Roof Pitch   The ratio of total span to total rise expressed as a fraction.

Root   The surface of the thread corresponding to the minor diameter of an external thread and the major diameter of an internal thread

Rotation   A view in which the object is apparently rotated or turned to reveal a different plane or aspect, all shown within the view.

Rough In   To prepare a room for plumbing or electrical additions by running wires or piping for a future fixture.

Rough Opening   A large opening made in a wall frame or roof frame to allow the insertion of a door or window.

Round   The rounded outside corner of a metal object.

R-Value   The unit that measures the effectiveness of insulation; the higher the number, the better the insulation qualities.

Sanitary Sewer   A conduit or pipe carrying sanitary sewage.

Scale   The relation between the measurement used on a drawing and the measurement of the object it represents. A measuring device, such as a ruler, having special graduations.

Schedule of Values   A statement furnished by the Contractor to the Architect reflecting the portions of the Contract sum allocated to the various of the Work and used as the basis for reviewing the Contractor's Applications for Payment.

Schematic Diagram   A diagram using graphic symbols to show how a circuit functions electrically.

Scratch Coat   The first coat of stucco, which is scratched to provide a good bond surface for the second coat.

Scupper   An opening in a parapet wall attached to a downspout for water drainage from the roof.

Scuttle   Attic or roof access with cover or door.

Section   A view showing internal features as if the viewed object has been cut or sectioned

Section Lines   Thin, diagonal lines used to indicate the surface of an imaginary cut in an object.

Seismicity   The worldwide or local distribution of earthquakes in space and time; a general term for the number of earthquakes in a unit of time, or for relative earthquake activity.

Septic Tank   A tank in which sewage is decomposed by bacteria and dispersed by drain tiles.

Shake   A hand-split wooden roof shingle, usually western cedar.

Sheet Steel   Flat steel weighing less than 5 pounds per square foot.

Shear Distribution   The distribution of lateral forces along the height or width of a building.

Shear Wall   A wall construction designed to withstand shear pressure caused by wind or earthquake.

Shoring   Temporary support made of metal or wood used to support other components.

Short-Term Costs   Opinions of probable costs to remedy physical deficiencies, such as deferred maintenance, that may not warrant immediate attention, but require repairs or replacements that should be undertaken on a priority basis in addition to routine preventive maintenance. Such opinions of probable costs may include costs for testing, exploratory probing, and further analysis should this be deemed warranted by the consultant. The performance of such additional services are beyond this guide. Generally, the time frame for such repairs is within one to two years.

Sill   A horizontal structural member supported by its ends.

Single-Line Diagram   A diagram using single lines and graphic symbols to simplify a complex circuit or system.

Single Prime Contract   The most common form of construction contracting. In this process, the bidding documents are prepared by the architect/engineer for the owner and made available to a number of qualified bidders. The winning contractor then enters into a series of subcontract agreements to complete the work. Increasingly, owners are opting for a "design-build contract" under which a single entity provides design and construction services.

Site   A parcel of land bounded by a property line or a designated portion of a public right-of-way.

Site Improvement   Landscaping, paving for pedestrian and vehicular ways, outdoor lighting, recreational facilities, and the like, added to a site.

Slab-on-grade   The foundation construction for a structure with no basement or crawl space.

Sole Plate   A horizontal structural member used as a base for studs or columns.

Spandrel Beam   The beam in an exterior wall of a structure.

Special Conditions   A section of the conditions of the contract, other than general conditions and Supplementary Conditions, which may be prepared to describe conditions unique to a particular project.

Specifications   A part of the contract documents contained in the project manual consisting of written requirements for material, equipment, construction systems, standards and workmanship. Under the Uniform Construction Index, the specifications comprise sixteen Divisions.

Split-Level   A house that has two levels, one about a half a level above or below the other.

Stack   A vertical plumbing pipe.

Standpipe   A vertical pipe generally used for the storage and distribution of water for the extinguishing purposes.

Statute of Limitations   A stature specifying the period of time with which legal action must be brought for alleged damage or injury, or other legal relief. The lengths of the periods vary from state to state and depend upon the types of legal action. Ordinarily, the period commences with the occurrence of the damage or injury, or discovery of the act resulting in the alleged damage or injury. In construction industry cases, many jurisdictions define the period as commencing with completion of Work or services performed in connection therewith.

Steel Plate Flat   Steel weighing more than 5 pounds per square foot.

Storm Sewer   A sewer used for conveying rain water, surface water condensate, cooling water, or similar liquid wastes exclusive of sewage.

Stretch-out Line   The base or reference line used in making a development.

Stringer   The inclined support member of a stair that supports the risers and treads.

Stucco   A type of plaster made from Portland cement, sand, water, and a coloring agent that is applied to exterior walls.

Structural Frame   The components or building system that supports the building's nonvariable forces or weights (dead loads) and variable forces or weights (live loads).

Stud   A light vertical structure member, usually of wood or light structural steel, used as part of a wall and for supporting moderate loads.

Subcontract   Agreement between a prime contractor and a subcontractor for a portion of the work at the site.

Subcontractor   A person or entity who has a direct or indirect contract with a subcontractor to perform any of the work at the site.

Substitution   A material, product or item of equipment offered in lieu of that specified.

Superintendent   Contractor's representative at the site who is responsible for continuous field supervision, coordination, completion of the work and, unless another person is designated in writing by the contractor to the owner and the architect, for the prevention of accidents.

Supervision   Direction of the work by contractor's personnel. Supervision is neither a duty nor a responsibility of the architect as part of professional services.

Surety Bond   A legal instrument under which one party agrees to answer to another party for the debt, default or failure to perform of a third party.

Survey   Observations made by the field observer during a walk-through survey to obtain information concerning the subject property's readily accessible and easily visible components or systems.

Symbol   Stylized graphical representation of commonly used component parts shown in a drawing.

System (a process)   A combination of interacting or interdependent components assembled to carry out one or more functions.

Tangent   A straight line from one point to another, which passes over the edge of a curve.

T-Beam   A beam which has a T-shaped cross section.

Tee   A fitting, either cast or wrought, that has one side outlet at right angles to the run.

Temper   To harden steel by heating and sudden cooling by immersion in oil, water, or other coolant.

Template   A piece of thin material used as a true-scale guide or as a model for reproducing various shapes.

Tensile Strength   The maximum stretching of a piece of metal (rebar, etc.) before breaking; calculated in kps.

Termite Shield   Sheet metal placed in or on a foundation wall to prevent intrusion.

Terrazzo   A mixture of concrete, crushed stone, calcium shells, and/or glass, polished to a tile-like finish.

Thermal Resistance (R)   A unit used to measure a material's resistance to heat transfer. The formula for thermal resistance is: $R = $ Thickness (in inches)/k.

Thermostat An automatic device controlling the operation of HVAC equipment.

Three-Phase Power   A combination of three alternating currents in a circuit with their voltages displaced at 120 degrees or one-third of a cycle.

Threshold   The beveled member directly under a door.

Tie   A soft metal wire that is twisted around a rebar or rod and chair to hold in place until concrete is poured.

Timely Access   Entry provided to the consultant at the time of the site visit.

Timely Completion The completion of the work or designated portion thereof on or before the date required.

Title Insurer The issuer (s), approved by interim lender and permanent lenders, of the title insurance policy or policies insuring the mortgage.

Title Block A blocked area in the lower right corner of the print. Provides information to identify the drawing, its subject matter, origins, scale, and other data. Title blocks are unique to each manufacturer.

Tolerance The amount that a manufactured part may vary from its specified size.

Tongue and Groove A joint where the edge of one member fits into a groove in the next member.

Top Chord The topmost member of a truss.

Topographic Survey The configuration of a surface including its relief and the locations of its natural and man-made features, usually recorded on a drawing showing surface variations by means of contour lines indicating height above or below a fixed datum.

Top Plate A horizontal member at the top of an outer building wall; used to support a rafter.

Toxicity A reflection of a material's ability to release poisonous particulate.

Tracing Paper High-grade, white, transparent paper that takes pencil well; used when reproductions are to be made of drawings. Also known as tracing vellum.

Transient Lodging A building, facility, or portion thereof, excluding inpatient medical care facilities and residential facilities, that contains sleeping accommodations. Transient lodging may include, but is not limited to, resorts, group homes, hotels, motels, and dormitories.

Tread The part of a stair on which people step.

Trap A fitting designed to provide a liquid seal which will prevent the back passage of air without significantly affecting the flow of waste water through it.

Triangulation A technique for making developments of complex sheet metal forms using geometrical constructions to translate dimensions from the drawing to the pattern.

Trimmer A piece of lumber, usually a 2 x 4, that is shorter than the stud or rafter but is used to fill in where the longer piece would have been normally spaced except for the window or door opening or some other opening in the roof or floor or wall.

Truss A prefabricated sloped roof system incorporating a top chord, bottom chord, and bracing.

Turbulence Any deviation from parallel flow in a pipe due to rough inner wall surfaces, obstructions, etc.

UL Underwriters Laboratories, Inc. A private testing and labeling organization that develops test standards for product compliance. UL standards appear throughout specifications, often in roofing requirements, and always in equipment utilizing or delivering electrical power. http://www.ul.com.

Union Ell An ell with a male or female union at one end.

Unfaced Insulation Insulation that does not have a facing or plastic membrane over one side of it.

Union Joint A pipe coupling, usually threaded, which permits disconnection without disturbing other sections.

Union Tee A tee with a male or female union at one end of the run.

Utility Plan A floor plan of a structure showing locations of heating, electrical, plumbing and other service system components.

Vacuum Any pressure less than that exerted by the atmosphere.

Valley The area of a roof where two sections come together and form a depression.

Vapor Barrier The same as a moisture barrier.

Vehicular Way A route intended for vehicular traffic, such as a street, driveway, or parking lot.

Veneer   A thin layer or sheet of wood.

Veneered Wall   A single-thickness (one-wythe) masonry unit wall with a backup wall of frame or other masonry; tied but not bonded to the backup wall.

Ventilation   The exchange of air, or the movement of air through a building; may be done naturally through doors and windows or mechanically by motor-driven fans.

Vent   Usually a hole in the eaves or soffit to allow the circulation of air over an insulated ceiling; usually covered with a piece of metal or screen.

Vent Stack   A system of pipes used for air circulation and prevent water from being suctioned from the traps in the waste disposal system.

Vertical Pipe   Any pipe or fitting installed in a vertical position or which makes an angle of not more than 45 degrees with the vertical.

View   A drawing of a side or plane of an object as seen from one point.

Volt (E) or (V)   The unit of measurement of electrical pressure (force).

Voltage Drop   Voltage reduction due to resistance.

Waiver of Lien   An instrument by which a person or organization who has or may have a right of mechanic's lien against the property of another relinquishes such right.

Walk   An exterior pathway with a prepared surface intended for pedestrian use, including general pedestrian areas such as plazas and courts.

Warranty   Legally enforceable assurance of quality or performance of a product or work, or of the duration of satisfactory performance. Warranty guarantee and guaranty are substantially identical in meaning; nevertheless, confusion frequently arises from supposed distinctions attributed to guarantee (or guaranty) being exclusively indicative of duration of satisfactory performance or of a legally enforceable assurance furnished by a manufacturer or other third party. The Uniform Commercial Code provisions on Sales (effective in all states except Louisiana) use warranty but recognize the continuation of the use of guarantee and guaranty.

Waste Pipe   Discharge pipe from any fixture, appliance, or appurtenance in connection with a plumbing system which does not contain fecal matter.

Water-Cement Ratio   The ratio between the weight of water to cement.

Water Hammer   The noise and vibration which develops in a piping system when a column of non-compressible liquid flowing through a pipe line at a given pressure and velocity is abruptly stopped.

Water Main   The water supply pipe for public or community use.

Waterproofing   Materials used to protect below-and on-grade construction from moisture penetration.

Weep Holes   Small holes in a wall to permit water to exit from behind.

Welded Wire Fabric (WWF)   A reinforcement used for horizontal concrete strengthening.

Wind Lift (Wind Load)   The force exerted by the wind against a structure.

Wiring (Connection) Diagram   A diagram showing the individual connections within a unit and the physical arrangement of the components.

Working Drawings   A set of drawings which provide the necessary details and dimensions to construct the object. May include specifications.

Wye (Y)   A fitting that has one side outlet at any angle other than 90 degrees.

Wythe   A continuous masonry wall width.

X Brace   Cross brace for joist construction.

Zinc   Non-corrosive metal used for galvanizing other metals.

Zone Numbers   They consist of numbers and letters on the border of a drawing to provide reference points to aid in indicating or locating specific points on the drawing.

Zoning   The legal restriction that deems that parts of cities be for particular uses, such as residential, commercial, industrial and so forth.

Zoning Permit   A permit issued by appropriate governmental authority authorizing land to be used for a specific purpose.

Appendix 3
Common Conversion Factors

AREA

FROM	TO	MULTIPLY BY
acres (ac)	hectares (ha)	0.4047
acres (ac)	square meters (m^2)	4047
hectares (ha)	square meters (m^2)	10000
hectares (ha)	acres (ac)	2.471
perches (p)	square meters (m^2)	25.2929
roods (rd)	square meters (m^2)	1011.7
squares	square meters (m^2)	9.2903
square centimeters (cm^2)	square inches (in^2)	0.155
square feet (ft^2)	square meters (m^2)	0.0929
square inches (in^2)	square meters (m^2)	0.00064516
square inches (in^2)	square millimeters (mm^2)	645.16
square inches (in^2)	square centimeters (cm^2)	6.4516
square meters (m^2)	square yards (yd^2)	1.196
square meters (m^2)	square feet (ft^2)	10.76
square meters (m^2)	square inches (in^2)	1550
square meters (m^2)	acres (ac)	0.000247
square meters (m^2)	roods (rd)	0.000988
square meters (m^2)	perches (p)	0.0395
square meters (m^2)	hectares (ha)	0.0001
square meters (m^2)	squares	0.1076
square millimeters (mm^2)	square inches (in^2)	0.00155
square yards (yd^2)	square meters (m^2)	0.8361

LENGTH

FROM	TO	MULTIPLY BY
centimeters (cm)	inches (in)	0.3937
chains (ch)	meters (m)	20.117
feet (ft)	meters (m)	0.3048
inches (in)	meters (m)	0.0254
inches (in)	centimeters (cm)	2.54
inches (in)	millimeters (mm)	25.4
international nautical miles (n mile)	meters (m)	1852
light years (l.y.)	meters (m)	9.4605×10^{15}
links (lk)	meters (m)	0.2012
meters (m)	miles	0.00062137
meters (m)	chains (ch)	0.0497
meters (m)	yards (yd)	1.0936
meters (m)	links (lk)	4.9709
meters (m)	feet (ft)	3.2808
meters (m)	inches (in)	39.37
meters (m)	international nautical miles (n mile)	0.000539
meters (m)	light years (l.y.)	1.057×10^{-16}
meters (m)	perches (p)	0.1988
miles	meters (m)	1609.3
millimeters (mm)	inches (in)	0.03937
perches (p)	meters (m)	5.0292
yards (yd)	meters (m)	0.9144

MASS

FROM	TO	MULTIPLY BY
drams (dr)	kilograms (kg)	0.0017718
grains (gr)	kilograms (kg)	0.000064798
hundredweights (cwt)	kilograms (kg)	50.8023
kilograms (kg)	tons	0.000984
kilograms (kg)	tonnes (t)	0.001
kilograms (kg)	short tons (sh tn)	0.0011023
kilograms (kg)	hundredweights (cwt)	0.197
kilograms (kg)	quarters (qr)	0.078736
kilograms (kg)	stones (st)	0.1575
kilograms (kg)	pounds (lb)	2.2046
kilograms (kg)	ounces (oz)	35.274
kilograms (kg)	drams (dr)	564.3
kilograms (kg)	grains (gr)	15432
kilograms (kg)	slugs	0.0685
kilograms (kg)	troy ounces (oz tr)	32.1507
kilograms (kg)	pennyweight (dwt)	643.01
ounces (oz)	kilograms (kg)	0.02835
pennyweight (dwt)	kilograms (kg)	0.0015552
pounds (lb)	kilograms (kg)	0.4536
quarters (qr)	kilograms (kg)	12.7006
short tons (sh tn)	kilograms (kg)	907.18
short tons (sh tn)	tonnes (t)	0.9072
slugs	kilograms (kg)	14.5939
stones (st)	kilograms (kg)	6.3502
tons	kilograms (kg)	1016
tons	tonnes (t)	1.016
tonnes (t) (metric ton)	kilograms (kg)	1000
tonnes (t) (metric ton)	tons	0.984
tonnes (t) (metric ton)	short tons (sh tn)	1.1023
troy ounces (oz tr)	kilograms (kg)	0.0311

VOLUME

FROM	TO	MULTIPLY BY
cubic centimeters (cm^3)	cubic inches (in^3)	0.06102
cubic feet (ft^3)	cubic meters (m^3)	0.0283
cubic feet (ft^3)	liters (L)	28.317
cubic inches (in^3)	cubic millimeters (mm^3)	16387
cubic inches (in^3)	cubic centimeters (cm^3)	16.387
cubic inches (in^3)	cubic meters (m^3)	0.000016387
cubic meters (m^3)	cubic yards (yd^3)	1.3079
cubic meters (m^3)	cubic feet (ft^3)	35.315
cubic meters (m^3)	cubic inches (in^3)	61024
cubic meters (m^3)	U.S. gallons (U.S. gal)	264.172
cubic meters (m^3)	gallons (gal)	219.9
cubic meters (m^3)	liters (L)	1000
cubic meters (m^3)	quarts (qt)	879.88
cubic meters (m^3)	pints (pt)	1759
cubic meters (m^3)	gills	7039
cubic meters (m^3)	fluid ounces (fl oz)	35195
cubic meters (m^3)	fluid drachms (fl dr)	281560.6
cubic meters (m^3)	minims (min)	16893604
cubic millimeters (mm^3)	cubic inches (in^3)	0.00006102
cubic yards (yd^3)	cubic meters (m^3)	0.7646
fluid ounces (fl oz)	cubic meters (m^3)	0.000028413
fluid ounces (fl oz)	milliliters (mL)	28.4131
fluid drachms (fl dr)	cubic meters (m^3)	0.0000035516
gallons (gal)	cubic meters (m^3)	0.0045461
gallons (gal)	liters (L)	4.54609
gills	cubic meters (m^3)	0.00014206
liters (L)	cubic meters (m^3)	0.001
liters (L)	cubic feet (ft^3)	0.0353
liters (L)	gallons (gal)	0.22
liters (L)	US gallons (US gal)	0.2642
liters (L)	minims (min)	16894
milliliters (mL)	fluid ounces (fl oz)	0.035195
minims (min)	cubic meters (m^3)	0.000000059194
minims (min)	liters (L)	0.000059194
pints (pt)	cubic meters (m^3)	0.00056826
quarts (qt)	cubic meters (m^3)	0.0011365
U.S. gallons (U.S. gal)	cubic meters (m^3)	0.0037854
U.S. gallons (U.S. gal)	liters (L)	3.7854

TEMPERATURE

FROM	TO	MULTIPLY BY
Celsius (degrees C)	Fahrenheit (degrees F)	$9/5 \times C + 32$
Celsius (degrees C)	Kelvin (K)	$C + 273.15$
Fahrenheit (degrees F)	Celsius (degrees C)	$5/9 \times (F - 32)$
Fahrenheit (degrees F)	Kelvin (K)	$5/9 \times (F + 459.67)$
Kelvin (K)	Celsius (degrees C)	$K - 273.15$
Kelvin (K)	Fahrenheit (degrees F)	$9/5 \times K - 459.67$

PRESSURE

FROM	TO	MULTIPLY BY
millibars (mb)	pascals (Pa)	100
millibars (mb)	standard atmospheres (atm)	0.00098692
millimeters of mercury (mm Hg)	pascals (Pa)	133.32
pascals (Pa)	millimeters of mercury (mm Hg)	0.0075
pascals (Pa)	millibars (mb)	0.01
pascals (Pa)	standard atmospheres (atm)	0.0000098692
standard atmospheres (atm)	pascals (Pa)	101325
standard atmospheres (atm)	millibars (mb)	1013
pounds-force per square inch (lbf/in^2)	kilopascals (kPa)	6.8947573
kilopascals (kPa)	pounds-force per square inch (lbf/in^2)	0.14503774

POWER

FROM	TO	MULTIPLY BY
horsepower (hp)	Watts (W)	745.7
Watts (W)	horsepower (hp)	0.001341

VELOCITY AND SPEED

FROM	TO	MULTIPLY BY
feet per minute (ft/min)	meters per second (m/s)	0.00508
kilometers per hour (km/h)	miles per hour (mile/h)	0.6214
knots, international	meters per second (m/s)	0.51444
meters per second (m/s)	knots, international	1.9438
meters per second (m/s)	miles per hour (mile/h)	2.2369
meters per second (m/s)	feet per minute (ft/min)	196.8
miles per hour (mile/h)	meters per second (m/s)	0.447
miles per hour (mile/h)	kilometers per hour (km/h)	1.6093

ENERGY

FROM	TO	MULTIPLY BY
British thermal units (Btu)	joules (J)	1055.1
calories (cal)	joules (J)	4.1868
joules (J)	kilocalories (kcal)	0.00023884
joules (J)	calories (cal)	0.23884
joules (J)	British thermal units (Btu)	0.0009478
joules (J)	kilowatt hours (kW.h)	0.00000027778
kilocalories (kcal)	joules (J)	4186.8
kilowatt hours (kW.h)	joules (J)	3600000

AREA EQUIVALENTS

1 acre = 43,560 ft squared = 4840 yd^2 = 0.4047 hectares = 160 rods squared = 4047 m^2 = 0.0016 sq. mile

1 acre-inch = 102.8 m^3 = 27,154 gal = 3630 ft^3

1 hectare (ha) = 10,000 m^3 = 100 are = 2.471 acres = 107,639 ft squared

1 cubic foot (ft^3) = 1728 in^3 = 0.037 yd^3 = 0.02832 m^3 = 28,320 cm^3

1 square foot (ft^2) = 144 in^2 = 929.03 cm^2 = 0.09290 m^2

1 square yard (yd^2) = 9 ft^2 = 0.836 m^2

1 cubic yard (yd^3) = 27 ft^3 = 0.765 m^3

UNITS	SQ. IN.	SQ. FT.	SQ YD.	SQ. CM	SQ. M
Sq. In.	1	0.006944	0.0007716	6.452	0.000645
Sq. Ft.	144	1	0.1111	929	0.0929
Sq Yd.	1296	9	1	8361	0.8361
Sq. cm	0.155	0.001076	0.0001196	1	0.0001
Sq. m	1550	10.76	1.196	10,000	1

MISCELLANEOUS CONVERSIONS

FROM	TO	MULTIPLY BY
acres	sq. feet	43,560
acres	sq. kilometer	0.00405
acres	sq. meter	4047
acres	sq. yards	4840
acre-feet	sq. feet	325,851
acre-feet	cu. feet	43560
acre-feet	m (cubed)	1233.5
bar	lb/sq. in	14.5
bar	g/cm (cubed)	1019.7
bar	inches Hg at 0	29.53
centimeters (cm)	m squared	0.03281
centimeters	feet	0.3937
centimeters	inches	0.1094
centimeters	yards	0.01
centimeters	meters	10
cm/sec	millimeters (ml)	1.9685
cm/sec	ft/min	0.0223694
cm (cubed)	mph	0.0610237
cubic feet	inch (cubed)	0.0283
cubic feet	cu. meter	7.4805
cubic feet	gallons	1728
cubic feet	cubic inches	0.037
feet (ft)	fl oz	30.48
feet	centimeters	0.3048
feet per minute	meters	0.01136
foot candle	psi	10.764
gallons (gal)	lux	3.785
gal	liters	3785
gal	millimeters	128
gal/acre	ounces (liquid)	9.354
gal/acre	liters/hectare	2.938
gal/1000 ft squared	ounces/1000 ft squared	4.0746
gal/minute	(liquid)	$2.228 \times 10 (-3)$
grams (g)	l/100 m squared	0.002205
gram	cubic feet/second	0.035274
grams per liter	oz	1000
grams per liter	lbs/a	10
grams/sq.meter	ppm	0.00020481
g/cm (cubed)	percent	0.036127
g/cm (cubed)	lb/sq. ft	62.428
hectares (ha)	lb/in (cubed)	2.471
inches	lb/ft (cubed)	2.540
inches	acres	0.0254

FROM	TO	MULTIPLY
inches	centimeters	25.40
in squared	meters	6.4516
kilograms (kg)	pounds	2.2046
kg/hectare	pounds/acre	0.892
kg/ha	lb/1000 ft sq.	0.02048
kg/l	lb/gal	8.3454
kilometers (km)	centimeters	100,000
kilometers	feet	3281
kilometers	meters	1000
kilometers	miles	0.6214
kilometers	yards	1094
km/h	mph	0.62137
km/h	ft/min	54.6807
kilopascals (kPa)	pounds/sq. in (psi)	0.145
liters (l)	gallons	0.2642
liters	ounces	33.814
liters	pints	2.113
liters	quarts	1.057
l/100 m squared	gal/1000 ft squared	0.2454
liters/hectare	gal/acre	0.107
meters (m)	feet	3.281
meters	inches	39.37
meters	yards	1.094
meters	centimeters	100
meters	kilometers	0.001
meters	millimeters	1000
meters/sec	mph	2.2369
m squared	ft squared	10.764
m (cubed)	ft (cubed)	35.3147
m (cubed)	yd (cubed)	1.30795
miles (statute)	centimeters	160,900
miles	feet	5280
miles	kilometers	1.609
miles	yards	1760
miles/hour (mph)	feet/second	1.467
miles/hour	feet/minute	88
miles/hour	kilometers/hour	1.61
miles/hour	meters/second	0.447
milliliters (ml)	ounces (fluid)	0.0338
milliliters	gallons	0.0002642
millimeters (mm)	inches	0.03937
1 mm Hg @ 0 C	kpa	0.13332
ounces (fluid)	liters	0.02957
ounces (fluid)	milliliters	29.573
ounces (weight)	grams	28.35

Appendix 4

Test Questions and Answers

CHAPTER 1 (BLUEPRINT STANDARDS)

Questions

1. Are blueprints today generally blue?
2. Diazo prints can also be produced with sepia lines. (True or False)
3. Use of photocopy machines is increasing because ordinary paper that does not require coating is more common. (True or False)
4. What is the main American drafting-standard-setting organization?
5. Which drafting-standard-setting organization is generally used in Europe for sheet sizes?

Answers

1. No
2. True
3. True
4. American National Standards Institute (ANSI)
5. International Organization for Standardization (ISO)

CHAPTER 2 (BLUEPRINTS AND CONSTRUCTION DRAWINGS: A UNIVERSAL LANGUAGE)

Questions

1. One of the first things to do upon receipt of a set of contract documents is to verify that the set is complete and that the documents are the most current. True or False?

2. What do designers use a t-square for?
3. Lettering is used on construction drawings as a means to provide written information. True or False?
4. What are the four basic classifications that the majority of today's various alphabets and type-faces fit into?
5. What does CADD stand for?
6. What is one of the main advantages of using CADD systems?
7. Based on the function they serve, what are the five main categories of blueprints commonly used in construction?
8. What are detail drawings?
9. Falsework drawings show temporary supports of timber or steel that are required sometimes in the erection of difficult or important structures. True or False?
10. Title blocks are only used on the cover sheet of construction documents. True or False?

Answers

1. True
2. Drawing/drafting
3. True
4. Roman, gothic, script and text.
5. CADD stands for computer-aided design and drafting.
6. It allows quick alterations to drawings.
7. Preliminary drawings, presentation drawings, working drawings, shop/assembly drawings, specialized, and miscellaneous drawings.
8. Detail drawings provide information about specific parts of the construction and are on a larger scale than general drawings.
9. True
10. False. It is standard practice to include a title block on each page of a set of blueprints or construction documents.

CHAPTER 3 (UNDERSTANDING LINE TYPES)

Questions

1. Border lines are roughly as bold as dimension lines. True or False?
2. What are the dimensions for an A1 sheet of tracing paper?
3. What are object lines?
4. What is the purpose of hidden lines in construction drawings?
5. Center lines consist of thin, broken lines of alternating long and short dashes. True or False?
6. Approximately how much space is usually allowed between the object and the beginning of the extension line?
7. What is a chain dimension?
8. Leader lines are fine lines terminating with an arrowhead or dot at one end to relate a note or callout to its feature. True or False?

9. What are break lines?
10. Contour lines define the boundaries of a property. True or False?

Answers

1. False
2. Dimensions for an A1 sheet is 841 mm x 594 mm (33.1 inches x 23.4 inches).
3. Object lines, also known as visible lines, are solid lines used mainly to define the shape and size of a structure or object.
4. They are used to represent hidden surfaces or intersections of an object. On floor plans they are also used to represent features that lie above the plane of the drawing, such as high wall cabinets in a kitchen.
5. True
6. Approximately 1/16 inch.
7. A chain dimension is when dimensions can be added together to come up with one overall dimension.
8. True
9. Break lines are fine straight lines with zig-zag-zig offsets to show a break or termination of a partial view or to omit portions of an object.
10. False. Contour lines consist of fine lines that are used mainly to delineate variations in a site's elevation.

CHAPTER 4 (UNDERSTANDING DIMENSIONS)

Questions

1. What does a dimension represent?
2. What is a radial dimension?
3. Some of the many advantages of the metric system include the elimination of fractions on drawings, simpler calculations, and international uniformity. True or False?
4. Where is the scale of a drawing usually noted?
5. What are the three commonly used drawing scales for reading construction drawings and for the development of plans?
6. Drawings produced in metric, such as floor plans, elevations, and sections, are normally drawn to a scale of 1:50 or 1:100, as opposed to the 1/4 inch = 1 inch, 0 inches or 1/8 inch = 1 foot, 0 inches scale used in the imperial system. True or False?
7. What are the two methods of dimensioning in common use?
8. How are angular dimensions used?
9. What are reference dimensions used for?
10. What does GDT stand for?

Answers

1. A dimension represents a numerical value expressed in an appropriate unit of measure.
2. Radial dimensions are used for measuring the radius of arcs, circles, and ellipses and displaying it with a leader line.

3. True
4. The scale is usually noted in the title block or just below the view when it differs in scale from that given in the title block.
5. They are the architectural scale, the engineering scale, and the metric scale.
6. True
7. Unidirectional and aligned
8. Angular dimensions are used for measuring and displaying inside and outside angles.
9. Reference dimensions are used only for information purposes.
10. It stands for geometric dimensioning and tolerancing (GDT).

CHAPTER 5 (TYPES OF VIEWS)

Questions
1. It is generally possible to read a blueprint just by looking at a single view. True or False?
2. What is an oblique projection?
3. To manually draw an isometric you will need a 30 / 60 degree set square. True or False?
4. How do circles appear in isometric?
5. What is an auxiliary view?
6. First-angle projection is as if the object were sitting on the paper and, from the "face" (front) view, it is rolled to the right to show the left side or rolled up to show its bottom. True or False?
7. What is axonometric drawing?
8. How are pictorial drawings used?
9. What is the main use of perspective drawings?
10. What are the three main types of perspective drawings?

Answers
1. False
2. Oblique projection is a simple type of graphical projection used for producing pictorial, two-dimensional images of three-dimensional objects.
3. True
4. They appear elliptical.
5. An auxiliary view is an angle at which one can view an object that is not one of the primary views for an Orthographic projection.
6. True
7. An axonometric drawing is one that is accurately scaled and depicts an object that has been rotated on its axes and is inclined from a regular parallel position to give it a three-dimensional appearance.
8. Pictorial drawings are used to depict a three-dimensional view of an object and to convey as much information as possible while helping the reader visualize the object drawn.
9. Perspective drawings are used mainly for illustrative purposes.
10. The main types of perspective drawings are: one-point perspective, two-point perspective and three-point perspective.

CHAPTER 6 (UNDERSTANDING CONSTRUCTION DRAWINGS)

Questions

1. Construction drawings are often referred to as working drawings. True or False?
2. What do architects generally use a time schedule for?
3. Site plans are drawn to locate a building on a site but typically does not include the surrounding area. True or False?
4. What is a floor plan?
5. An elevation is a drawing that represents a view of the finished structure as you would see it from the front, back, left, or right. True or False?
6. What are the most important sections used in construction drawings?
7. What do mechanical systems basically deal with?
8. Electrical systems involve electricity to provide power for lighting, outlets and equipment. True or False?
9. What two major components are plumbing systems involved in?
10. What does millwork refer to?

Answers

1. True
2. They generally use them for programming the different phases of a project.
3. False. It also typically includes the surrounding area.
4. A floor plan is a two dimensional view of a space, such as a room or building.
5. True
6. They are the wall sections.
7. They deal mainly with the heating and cooling of buildings or spaces.
8. True
9. They are involved in water supply and drainage.
10. Millwork refers to custom, shop-built, wood components for interior finish construction.

CHAPTER 7 (UNDERSTANDING INDUSTRIAL BLUEPRINTS)

Questions

1. What does tolerance mean in machine blueprints?
2. What are fillets?
3. A key is a small wedge or rectangular piece of metal inserted in a slot or groove between a shaft and a hub to prevent slippage. True or False?
4. The acceptable roughness of a surface depends upon its size. True or False?
5. What is an assembly drawing?
6. What are the two most widely used screw-thread series?
7. What is meant by Pitch diameter (PD) with relation to gears?
8. How many classifications of helical springs are there?

9. In machine drawings dimensions should not be duplicated. Reference dimensions can be used if necessary to duplicate. True or False?
10. There are no existing standards for creating technical drawings of mechanical parts and assemblies. True or False?

Answers

1. Tolerance is the total amount a dimension may vary.
2. Fillets are concave metal corner (inside) surfaces.
3. True
4. False. It depends on how the part will be used.
5. It consists of a drawing that conveys the completed shape of the product as well as its overall dimensions, and the various parts of the product assembled in their relative working positions.
6. They are (1) Unified or National Form Threads, which are called National Coarse, or NC, and (2) National Fine, or NF threads.
7. The diameter of the pitch circle (or line), which equals the number of teeth on the gear divided by the diametral pitch.
8. There are three classifications of helical springs: compression, extension, and torsion.
9. True
10. False

CHAPTER 8 (THE MEANING OF SYMBOLS)

Questions

1. What is one of the main functions of graphic symbols on blueprint drawings?
2. The American National Standards Institute (ANSI) specifies exactly how the "North Point" should be represented on a drawing. True or False?
3. Why do electrical drawings typically include a symbol legend or list?
4. Piping used in commercial and residential plumbing applications must not include plastic material. True or False?
5. HVAC drawings show the location of the heating, cooling, and air-conditioning units as well as piping and ducting diagrams. True or False?
6. How can one improve HVAC drafting productivity and efficiency.
7. A symbol should be scaled to determine its size. True or False?

Answers

1. It is to reference other drawings within the set.
2. False
3. Because electrical symbols are not standardized throughout the industry.
4. False
5. True
6. Using custom HVAC CADD programs can increase drafting productivity and efficiency.
7. False

CHAPTER 9 (UNDERSTANDING SCHEDULES)

Questions

1. What is the primary purpose of incorporating schedules into a set of construction documents?
2. Schedules are basically tables that list information about specific items (e.g., doors and room finishes). True or False?
3. VectorWorks Architect is a CAD software program that can be used for generating schedules. True or False?
4. What are door schedules used for?
5. What information does a finish schedule provide?
6. What are some of the different types of HVAC schedules used in blueprints?
7. While reviewing a grille/diffuser schedule you see the abbreviation (CFM). What does this stand for?
8. Who typically prepares the various types of HVAC schedules?
9. Material schedules represent the exact amount of materials needed for a project. True or False?
10. What are the two basic types of notes used in blueprints?

Answers

1. It is to provide clarity, location, sizing, materials, and other information on the designation of the various elements and components of a project.
2. True
3. True
4. They are typically used to indicate amongst other things, the door's model, quantity, size, thickness, type, material, function, frame material, fire rating, and remarks.
5. It provides information on the finishes for the walls, floors, ceilings, baseboards, doors, window trim, etc.
6. HVAC schedules may include air-handling units, fan-coil units, chiller schedules, etc.
7. It stands for the volume of air in cubic feet per minute that each grille/diffuser will handle.
8. The mechanical consultant.
9. False
10. General notes and key notes.

CHAPTER 10 (INTERPRETING SPECIFICATIONS)

Questions

1. What does CSI stand for?
2. In 2004 how many divisions was the MasterFormat™ expanded to?
3. There are two broad categories of specifications, _____ or open.
4. When the specification does not allow for any substitution of materials, what is it know as?
5. Specifications are legal documents. True or False?
6. Drawings typically convey information that can be most readily and effectively expressed graphically by means of drawings and diagrams. True or False?

7. Coordination of the specifications with the construction drawings is not necessary, as they do not complement each other.
8. What does division 09 of the MasterFormat 2004 Edition relate to?
9. Building Systems Design's (BSD) SpecLink System is an automated software system for designing buildings. True or False?
10. A designer who fails to reject defective work by a contractor or supplier is not considered to be professionally negligent or in breach of contract. True or False?

Answers

1. Construction Specifications Institute (CSI)
2. MasterFormat™ in 2004 was expanded to 50 divisions.
3. Closed
4. It is known as a base bid proprietary specification.
5. True
6. True
7. False
8. Finishes
9. False. It is an electronic specification system that uses CSI master-guide format in preparing construction specifications.
10. False

CHAPTER 11 (BUILDING CODES AND BARRIER-FREE DESIGN)

Questions

1. Building codes are international laws. True or False?
2. What does the ICC stand for?
3. How often will the International Building Code (IBC) be updated?
4. What does an occupancy classification refer to?
5. It is estimated that roughly 75 percent of all codes deal with fire and life-safety issues. True or False?
6. When a building code states that a 1-hour-rated partition assembly is required between an exit corridor and an adjoining tenant space, what must a designer do?
7. Exiting is comprised of three main categories. What are they?
8. In what years did the International Code Council (ICC) published the first International Plumbing Code (IPC) and International Mechanical Code?
9. How many titles does the Americans with Disabilities Act (ADA) consists of?
10. What is the maximum slope of an accessible route before it is classified as a ramp?
11. What is the minimum clear internal turning space that toilet rooms and toilet stalls must have to meet ADA requirements?
12. A handrail's top gripping surface must be between 30 and 32 inches above the stair nosings. True or False?
13. Telephones are one of the more difficult building elements to make accessible. True or False?

14. All elevators must be accessible from the entry route and all public floors. True or False?

Answers

1. False. Building codes are essentially local laws.
2. The International Code Council
3. Every three years.
4. It refers to the type of use a building or interior space is put to, such as a residence, an office, a school, or a restaurant.
5. True
6. The designer must select and detail a design that incorporates the requirements for 1-hour construction.
7. They are: exit access, exit, and exit discharge.
8. The International Plumbing Code in 1997 and the International Mechanical Code in 1998.
9. The Americans with Disabilities Act (ADA) consists of five titles, they are: Title I. Employment, Title II. Public Services, Title III. Public Accommodations, Title IV. Telecommunications, Title V. Miscellaneous.
10. The slope of an accessible route must not exceed 1:20; otherwise it is classified as a ramp and must meet other requirements.
11. A minimum of 60-inch (1525-mm) clear internal turning space is required.
12. False. It must be between 34 and 38 inches above the stair nosings.
13. False
14. True

CHAPTER 12 (CONSTRUCTION BUSINESS ENVIRONMENT)

Questions

1. Starting a new construction business is basically a 9 to 5 job. True or False?
2. What is the key to success in starting a new business?
3. What is the use of an executive summary?
4. Why do first-time business owners often fail?
5. All tax and corporate Filings should be kept for at least three years. True or False?
6. What does "SBA" stand for?
7. What is an "EIN" number?
8. Insurance is not needed for a new business. True or False?
9. Sending out flyers, brochures, emails, etc., to potential clients is one way to identify and track possible sources and potential project leads. True or False?
10. Construction jobs are usually awarded on the basis of a bid or by negotiation. True or False?

Answers

1. False
2. Planning
3. It important because it provides an overview of the business plan.

4. Mainly because they greatly misjudge the amount of money needed to get their small business off the ground, and fail to include a contingency amount to meet unforeseen expenses.
5. True
6. Small Business Administration
7. It is a federal employer-identification number, also called a tax-identification number.
8. False
9. True
10. True

Index